CSR, Sustainability, Ethics & Governance

More information about this series at http://www.springer.com/series/11565

Samuel O. Idowu • Mara Del Baldo
Editors

Integrated Reporting

Antecedents and Perspectives for Organizations and Stakeholders

Springer

Editors
Samuel O. Idowu
Guildhall Faculty of Business and Law
London Metropolitan University
London, UK

Mara Del Baldo
Department of Economics, Society and Politics
University of Urbino Carlo Bo
Urbino, Italy

ISSN 2196-7075 ISSN 2196-7083 (electronic)
CSR, Sustainability, Ethics & Governance
ISBN 978-3-030-01718-7 ISBN 978-3-030-01719-4 (eBook)
https://doi.org/10.1007/978-3-030-01719-4

Library of Congress Control Number: 2018965941

This Springer imprint is published by the registered company Springer Nature Switzerland AG
The registered company address is: Gewerbestrasse 11, 6330 Cham, Switzerland

Foreword

Integrated Reporting

Accounting, even for the educated public in general, has a reputation for being a dry and boring form of clerical work. Many business students view it as a necessary evil. On the contrary, accounting is the most challenging and underappreciated of all business dimensions and disciplines. The overarching classical concern has always been whether accounting reports or rules for them should be interpreted strictly and uniformly (mostly to avoid fraud or misrepresentation) or more loosely to accommodate the complexities of contemporary commerce. Do we need to revise everything every 10 years or can we reach some semblance of definitive finality?

Integrated reporting adds an additional dimension. Integrated reporting attempts to clarify how every form of value-relevant information enables companies to make decisions for the long term. This will benefit all stakeholders, but most especially those engaged in financial capital allocation decisions.

In a remarkable collection put together by Samuel O Idowu and Mara Del Baldo, all of these issues are addressed in a clear and comprehensive way. What are all the different elements, material and non-material, that contribute to sustained long-term growth? Are these reports going to adhere to uniform rules or will they be principles based? Can we have international standards or will the standards have to be adjusted country by country? Will the same standards apply to large and small firms? What would such a report look like?

The entire world is now influenced by dynamic, global, market economies that continue to evolve in unpredictable ways. Can we introduce some semblance of order and transparency into our understanding of and interaction with that world? Read this collection of insightful essays to find out.

Global Corporate Governance Institute,
New Orleans, LA, USA

Nicholas Capaldi

Preface

Although integrated reporting (IR) is a relatively new area of policy and practice in corporate reporting, it has rapidly gained considerable prominence since the formation of the International Integrated Reporting Committee (IIRC) in 2010. Interestingly, both the public policy and organisational practices in this area have developed rapidly during the last decade. IR is a new corporate reporting approach that reflects "integrated thinking" and represents "how an organisation creates value" (IIRC 2013, p. 4). It follows then that an integrated report aims to bring together detailed financial information, operational data and sustainability information in the quest by companies to focus only on those matters "that have a material bearing on the ability to create value in the short, medium and long term" (IIRC 2013, p. 26).

As a rapidly developing accounting regulatory arena which involves corporate CEOs, CFOs, directors, regulators, professional accounting bodies and consultants as well as auditors, it could become the biggest development in external reporting. It is designed to provide both financial and non-financial information in one comprehensive report. It provides forward-looking information, which focuses on strategic issues in a concise, integrated, connected and understandable format. Nevertheless, it could also fail to live up to the required expectations of both its many proponents and users (De Villiers et al. 2014; Cheng et al. 2014 ; Brown and Dillard 2014).

As an emerging phenomenon, IR has also attracted a great deal of academic attention, and a body of literature has been increasing since it came on the reporting scene globally. Studies to date have focused on the outcomes and benefits of integrated reporting, in particular when compared with sustainability reporting (Adams and Simnett 2011; Krzus 2011; Watson 2011; Hampton 2012; Roberts 2011). Nevertheless, integrated reporting has been criticised for its main focus on the interests of financial capital providers—shareholders and lenders to the detriment of the information demands and needs of other key stakeholders (Cheng et al. 2014; Frìs-Aceituno et al. 2014; Brown and Dillard 2014)

The need for a better and deeper understanding of those issues and benefits derivable from the diffusion of integrated reporting and integrated report and its implementation in practice bring to the table some important questions for both academics and practitioners who are active on the interface between business and society. Therefore, a critical examination and analysis of the implementation of integrated reporting and adoption of integrated report are needed to critically assess and answer a number of questions.

This volume provides answers to many of these pertinent questions about integrated reporting; we hope that our readers find the book to be a good companion regardless of what they do and how they do these things in integrated reporting.

London, UK
Urbino, Italy
Summer 2018

Samuel O. Idowu
Mara Del Baldo

Acknowledgements

Like many of these books we have been fortunate to edit, a number of the activities involved in putting them together always pose many challenges to us; this book is not an exception to that assertion. However, with a number of reliable and hard-working scholars behind us we knew that those challenges were surmountable. We therefore like to express our gratitude to thirty scholars who supported us in realising the goal of publishing the book. We are eternally grateful to you all.

We would also like to thank the President and Chief Executive Officer of the Global Corporate Governance Institute—Professor Nicholas Capaldi for writing an excellent foreword to the book.

We would like to thank our publishing team at Springer headed by the Executive Editor, Christian Rauscher, Barbara Bethke and other members of the publishing team who have supported this project and all our other projects.

We are grateful to our respective families—Samuel's—Olufunmilola Sarah, Josiah Opeyemi and Hannah Ayomide and Mara's—Giovanni, Diletta, Benedetta and Fiammetta.

Finally, we apologise for any errors or omissions that may appear anywhere in this book; please be assured that no harm was intended to anybody. Causing harm or discomfort to others is simply not the spirit of corporate social responsibility.

Contents

Editors and Contributors

About the Editors

Mara Del Baldo is associate professor of financial accounting and economics of sustainability and accountability at the University of Urbino (Italy), Department of Economics, Society and Politics, School of Economics. Her main research interests include, among others, corporate social responsibility; entrepreneurs and SMEs' business ethics; territorial responsibility; SMEs' strategies of qualitative development and networking strategies; social and environmental accounting; financial reporting; integrated reporting; and accounting and gender. She was visiting professor at the University of Vigo (Spain), Department of Finances and Accounting, the New Bulgarian University of Sofia (Bulgaria), the Juraj Dobrila University of Pula (Croatia), the Corvinus University of Budapest (Hungary) and the University of Craiova (Romania). She lectured on invitation in various Italian and foreign conferences and universities. Mara is a member of several Italian (including SIDREA, AIDEA and SISR) and international scientific associations: European Council for Small Business (ECSB), Centre for Social and Environmental Accounting Research (CSEAR, University of St Andrews), the SPES Institute (Spirituality in Economics and Society, Leuven, Belgium), the CRIMPI (Research Centre on Entrepreneurship and Small-medium Firms (Italy), the AFECA (Association des Formations Européennes à la Comptabilité et à l'Audit), the Italian Network on Business Reporting (NIBR) and the European Business Ethics Network (EBEN Italy). Moreover, she is a member of the academic advisory committee in the Global Corporate Governance Institute (USA) and Global network member of Corporate Governance Experts Global Repository (Virtuesinterpress). She serves as editorial board member, associate editor and reviewer of various international scientific journals, including *International Journal of Corporate Social Responsibility* (Springer Open Access Journal), *Piccola Impresa/Small Business*, *International Journal of Society Systems Sciences* (IJSSS), *Journal of Business Administration Research* (JBAR), *Journal of Modern Accounting and Auditing* (IJMAA), *International Journal of Business and*

Management (IJBM), *Sustainability*, *Review of Economics & Finance* (REF), *China-USA Business Review*, *Chinese Business Review*, *Journal of Management Development*, *Management Research Review* (MRR), *International Journal of Social Ecology and Sustainable Development* (IJSESD) and *Journal of Small Business Management* (JSBM). She serves as Editorial Board Member of the *Dictionary of Corporate Social Responsibility* (Springer, 2015). She published in several journals, such *as Accounting History*, *Meditari Accountancy Research*, *European Journal of International Management*, *International Journal of Social Ecology and Sustainable Development*, *International Journal of Society Systems Science*, *Journal of Management and Governance* and *International Journal of Corporate Social Responsibility*, among others. She authored several scientific publications, including articles, book chapters on Springer's Series *CSR, Sustainability, Ethics and Governance* and conference proceedings.

Samuel O. Idowu is a senior lecturer in accounting and corporate social responsibility at London Guildhall School of Business and Law, London Metropolitan University, UK. He researches in the fields of corporate social responsibility (CSR), corporate governance, business ethics and accounting and has published in both professional and academic journals since 1989. He is a freeman of the City of London and a Liveryman of the Worshipful Company of Chartered Secretaries and Administrators. Samuel is the Deputy CEO and First Vice President of the Global Corporate Governance Institute. He is the Editor-in-Chief of three Springer's reference books—the *Encyclopedia of Corporate Social Responsibility*, the *Dictionary of Corporate Social Responsibility* and the *Encyclopedia of Sustainable Management* (forthcoming); he is an Editor-in-Chief of the *International Journal of Corporate Social Responsibility* (IJCSR), the Editor-in-Chief of the *American Journal of Economics and Business Administration* (AJEBA) and an Associate Editor of the *International Journal of Responsible Management in Emerging Economies* (IJRMEE). He is also a Series Editor for Springer's books on CSR, Sustainability, Ethics and Governance. One of his edited books won the most Outstanding Business Reference book Award of the American Library Association (ALA) in 2016 and another was ranked 18th in the 2010 Top 40 Sustainability Books by *Cambridge University Sustainability Leadership Programme.* Samuel is a member of the Committee of the Corporate Governance Special Interest Group of the British Academy of Management (BAM). He is on the Editorial Boards of the *International Journal of Business Administration*, Canada, and *Amfiteatru Economic* journal, Romania. Samuel has delivered a number of keynote speeches at national and international conferences and workshops on CSR and has on two occasions 2008 and 2014 won Emerald's Highly Commended Literati Network Awards for Excellence. To date, Samuel has edited several books in the field of CSR, Sustainability and Governance and has written seven forewords to CSR books. Samuel has served as an external examiner to the following UK universities—Sunderland, Ulster, Anglia

Ruskin, Plymouth, Robert Gordon University, Aberdeen, Teesside University, Middlesbrough, Sheffield Hallam University and Leicester De Montfort University.

List of Contributors

Maria Aluchna She is associate professor at the Department of Management Theory, Warsaw School of Economics (SGH), Poland, and Academic Director of CEMS MIM at SGH. She studied marketing and management at the Warsaw School of Economics where she graduated in 1998 with an MA (Econ). She specialises in corporate governance (ownership structure, board, executive compensation, transition economies) as well as in strategic management and corporate social responsibility. She was awarded Deutscher Akademischer Austauschdienst (DAAD) scholarship for a research stay at Universität Passau and Polish-American Fulbright Commission scholarship for a research stay at Columbia University. She received Polish Science Foundation award for young researchers (2004, 2005). Since 1998 she has been working at the Department of Management Theory, at the Warsaw School of Economics obtaining PhD degree (2004) and completing habilitation procedure (2011). Currently, Maria Aluchna teaches "Corporate governance" (both in Polish and English for CEMS programme), "Transition in Central and Eastern Europe" (in English in cooperation with the University of Illinois, Springfield) and "Strategic management" (in English). She also serves as the faculty advisor for case competition of Warsaw School of Economics student teams and is the lecturer of the Summer University Warsaw. She is the member of the editorial team of *Journal of Knowledge Globalization* and *European Journal of Economics and Management* as well as of the Polish journals—*Przegląd Organizacji* [Organization Review] and *e-Mentor*. Maria Aluchna is the member of European Corporate Governance Institute (ECGI) and European Academy of Management (EURAM). She is also the Academic Director of the Community of European Management Schools (CEMS) programme at Warsaw School of Economics. She is the team member at the law firm Głuchowski, Siemiątkowski i Zwara.

Maria-Gabriella Baldarelli PhD, is an associate professor of accounting at the University of Bologna, Department of Management, and visiting professor at the following universities: State University of Tirana, Albania (2017); La Trobe, Melbourne, Australia (2015); teaching staff mobility at the New Bulgarian University of Sofia, Bulgaria, from 22nd to 27th November 2010; University of Sao Paulo, Brazil, from the end of May–1st June 2011; University of Diocese of Buia (UDEB), Cameroon—from 4th to 8th February 2012; University of Pula, Croatia, in May 2006; and University of Vlore (Albania) from 12th to 15th May 2009. She is a member of the Editorial Board International Review "Economic Research" (UDK 338; ISSN 1331-677X). Her research interests include financial statement in tour operator and travel agencies; corporate social responsibility; ethical, social and environmental accounting and accountability; sustainability in tourist enterprises; responsible and accessible tourism for blind people; economy of communion enterprises; and gender (pink) accounting. She is a member of the Board-Center of Advanced Studies on Tourism (CAST; http://www.turismo.unibo.it), as delegate of the Department of Management, University of Bologna. She has been a member of the SIDREA council from December 2009. Website: www.unibo.it/docenti/maria.baldarelli.

Silvia Ruiz Blanco is an associate professor at the Business School of the University of Vigo (Spain). She graduated in economics and business administration from the University of Santiago de Compostela. She worked as external auditor, and she became collaborator of the Department of Finance and Accounting at the University of Vigo. Her research interest is focused on corporate social responsibility reporting, and she has presented her research results in international conferences and journals.

Adriana Schiopoiu Burlea is a professor of human resources management at the University of Craiova, Romania. Professor Burlea's teaching and research areas are in human resources management, project management, social audit, corporate social responsibility and negotiation techniques at the University of Craiova, Faculty of Economics and Business Administration in Romania. She is visiting professor teaching and research in cross-cultural management at BEM Bordeaux Management School, 680, Cours de la Libération, Talence, France; visiting professor teaching and research in human resources management at the University of Corsica "Pasquale Paoli", IAE Corte, Corte Corsica, France; and visiting professor teaching and research in human resources management and intercultural management at Saint-Etienne School of Management (SESOM). Professor Adriana Schiopoiu Burlea is Romanian ambassador for ARGH (Association Francophone de Gestion des Ressources Humaines—http://www.agrh.fr/les-ambassadeurs/) and member of the Académie des Sciences du Management de Paris—http://www.asmparis.org/csdr/view/30. She is author and co-author of more than 50 books and chapters in the books and of more than 130 articles in prestigious journals. She is involved in managerial and administrative activities (as vice-president of Ethics Commission at the University of Craiova and as a member of Doctoral University School). She is Expert for the Eurostars Programme—helping SMEs to innovate and compete internationally—2017; Expert for the "Eugen Ionescu" programme, L'Agence Universitaire de la Francophonie—2018; Expert for European Commission for Human Resources Strategy for Researchers (HRS4R) projects—2016/2017/2018; Member of the GLOBAL VALUE expert crowd; Expert for International Projects European Union Horizon 2020: Erasmus+ "Capacity Building in the field of Youth" and "Large scale EVS events"—2014; Expert for International Projects European Union—FP7—PEOPLE—2008, 2012, 2013—registered in 2002—with the following application number EX2002B002264; Expert for Programme "Diversity, inequalities and social inclusion", Programme area "Research within Priority Sectors" of the EEA Financial Mechanism 2009–2014, Hellenic Republic, Ministry of

Culture, Education and Religious Affairs; and Expert for William Harvey Research Institute (WHRI)—ACADEMY (COFUND Marie Curie)—2014, 2015. She is member in many organising committees for national and international conferences. She stimulated and encouraged the cooperation between her university and the other universities around the world (e.g. cooperation with staff and teachers from other international universities in various scientific projects—for example, universities from France, the USA, the UK, Switzerland, Belgium, Italy, Portugal, Norway and Holland).

Mark Anthony Camilleri is a resident academic at the University of Malta's Department of Corporate Communication. He received his PhD from the University of Edinburgh in Scotland—where he was also nominated for his "Excellence in Teaching". Dr. Camilleri has published four books, including Springer's *Corporate Sustainability, Social Responsibility and Environmental Management: An Introduction to Theory and Practice with Case Studies*. He has also published his research in peer-reviewed journals and chapters. Currently, Dr. Camilleri is a member on the editorial board of Springer's *International Journal of Corporate Social Responsibility* and Inderscience's *International Journal of Responsible Management in Emerging Economies*. He is a frequent speaker and reviewer at the British Academy of Management and American Marketing Association's (AMA) Marketing & Public Policy annual gatherings. Mark is also a member of the academic advisory committee in the Global Corporate Governance Institute (USA).

Federica Casonato received a master's degree in international management at the University of Bologna in 2015. Her thesis "Integrated Reporting: the Evolution in an Australian Company" was written during her stay at Macquarie University, Sydney, as a visiting scholar. She has been working for 2 years at Ernst & Young Spa in Milan where she occupies a position of Senior Accountant, working, for example, on SEC regulation.

Miriam Corrado obtained master's degree in business administration in December 2016 at Roma Tre University, where she also received the bachelor's degree in October 2014. In 2015, she participated in the Exchange European Program (Erasmus+), attending for 6 months the Master in Finance and Banking at Liechtenstein University. In 2016, she won a scholarship for 3 months to conduct the master thesis research at Macquarie University in Sydney (Australia) investigating the role of assurance on integrated reporting. From March 2017, she has been selected for the Graduate Program Deals at PwC in the advisory line of service. She is currently a Phd student at the Università Politecnica delle Marche.

Antonio Corvino earned his PhD at the University of Bari. He is currently an associate professor in corporate strategy and business sustainability at the Department of Economics—University of Foggia (Italy)—where he currently teaches corporate strategy. In his academic research, he investigates corporate strategy as well as integrated and sustainability reporting. In 2015, he also received the Award of the National Scientific Qualification ("Abilitazione Scientifica Nazionale", ASN 2013) as full professor in business administration.

Antonio has the membership of the "European Institute for Advanced Studies in Management" (EIASM), the "Strategic Management Society" (SMS), the "European Academy of Management" (EURAM), the "British Accounting and Finance Association" (BAFA), the "Centre for Social and Environmental Accounting Research" (CSEAR), University of Saint Andrews (UK), the "International Academy of Business and Economics" (IABE), the "Italian Academy of Business Administration and Management" (AIDEA) and the "Italian Society of Professors in Business Administration" (SIDREA). He published several papers on academic peer-reviewed journals. Furthermore, he is a member of the Editorial Board of the *International Journal of Business Research* (IJBR) and other academic journals as well as the member of scientific committees and/or the referee of relevant conferences organised by distinguished scientific bodies, such as SMS, EURAM, SIDREA, IABE, GLOBE and so on.

Tiziana De Cristofaro was conferred a bachelor's degree in "Economia e Commercio" and a PhD degree in "Economia delle Aziende e degli Intermediari Finanziari" from the University "G. d'Annunzio" of Chieti-Pescara (Italy). Since 2001, she is assistant professor of "Economia Aziendale" at the University "G. d'Annunzio" of Chieti-Pescara (Italy). From 2002, she teaches modules of the "Economia Aziendale" Italian disciplinary sector. In particular, she taught modules related to cost accounting ("Analisi e Contabilità dei Costi"), intermediate accounting ("Ragioneria II"), introductory accounting ("Ragioneria") and business administration ("Economia Aziendale"). In the period 2003–2009, she also taught at the School of Specialisation for Secondary School Teaching of the University "G. d'Annunzio" of Chieti-Pescara. At present, she teaches "Economia Aziendale" and "Ragioneria" at the "Scuola delle Scienze Economiche, Aziendali, Giuridiche e Sociologiche", of the above-mentioned University. She has published monographs and articles on her main research fields of interest that are financial reporting and accounting education.

Paola Demartini is full professor of accounting at the University of Rome TRE. Her research interest includes corporate financial communication, managerial accounting and intellectual-based management. She has published several articles in refereed journals, and she is Head of the Corporate Governance Lab at Roma Tre University. She is also a member of the editorial boards and advisory boards of several national and international journals.

Federica Doni earned her PhD at the University of Pisa. She is currently an assistant professor in business administration at the Department of Business and Law, University of Milano-Bicocca, where she currently teaches financial accounting. Her main research interests include intangibles and intellectual capital, sustainability and integrated reporting and corporate governance. In 2015, she also received the Award of the National Scientific Qualification ("Abilitazione Scientifica

Nazionale", ASN 2013) as associate professor in business administration. Federica is a member of the "European Institute for Advanced Studies in Management" (EIASM), the "European Accounting Association" (EAA), the "Centre for Social and Environmental Accounting Research" (CSEAR), University of Saint Andrews (UK), "British Accounting and Finance Association" (BAFA) SIG on Corporate Governance and SIG on Accounting History, the "International Academy of Business and Economics" (IABE), the "Italian Academy of Business Administration and Management" (AIDEA), the "Italian Society of Professors in Business Administration" (SIDREA), "Italian Society of Accounting History" (SISR) and "European Roundtable for Sustainable Consumption and Production Society" (ERSCP). She is a member of the Scientific Committee "IPAZIA", i.e. a scientific observatory on gender culture and issues. She is involved in the "Principles for Responsible Investment's (PRI) Academic Network", PRI Association, London, UK, and in "<IR> Academic Network", IIRC, London, UK. She published several papers on academic peer-reviewed journals. Moreover, she is a member of the Editorial Board of the *International Journal of Business Research* (IJBR) and other academic journals as well as the member of scientific committees and/or the referee of relevant conferences organised by distinguished scientific bodies, such as EURAM, IFKAD, IABE and so on.

Federica Doni is the corresponding author and can be contacted at: federica.doni@unimib.it.

Ioana Dragu is a PhD Student at Babes-Bolyai University, Faculty of Economics and Business Administration, Accounting and Audit Department. She graduated in finance and banking (English Line of Study) from the Faculty of Economics and Business Administration and then attended a 2-year MBA in managerial accounting, audit and control. She started her research in corporate reporting and disclosure when she was still a student. Currently, she is writing her thesis on integrated reporting as the mixture between financial and non-financial information together, including in her topic sustainability disclosure practices. She published a series of papers on reporting practices and disclosure regarding financial and non-financial information. Her research interests incorporate sustainability insights and their integration into standard reporting practices and information disclosure.

John Dumay is an associate professor of accounting at Macquarie University, Sydney. He completed a PhD in intellectual capital in 2008. His research activities link closely to management, accounting and scholarly practice. He has published over 50 peer-reviewed academic journal articles, publishing in important journals such as *Accounting Auditing and Accountability Journal* and *The British Accounting Review*. His work is highly cited and has won several important awards.

Teodora Viorica Farcas is an assistant researcher at Babeş-Bolyai University, Faculty of Economics and Business Administration, in Cluj-Napoca (Romania) holding a PhD in accounting at the University of Nantes, France, and at Babeş-Bolyai University, Romania. Her research focus is on corporate reporting with a specific focus on integrated reporting and non-financial reporting. Also, she published papers in accounting history and accounting profession.

Federica Farneti is an associate professor in accounting at the University of Bologna, Italy. She has gained the national qualification as full professor until 2021. She holds a PhD in economics and management of public organisations. Her research interests include sustainability reporting, Directive 2014/95/UE, integrated reporting, sustainable management and disclosure in SMEs, local governments and in higher education.

Belen Fernandez-Feijoo is an associate professor in the Business School of the University of Vigo (Spain). She graduated from the University of Santiago de Compostela (Spain) and finished her PhD at the University of Vigo. She worked in the public sector and in private companies for 12 years. In 1986, she became collaborator of the Department of Finance and Accounting at the University of Vigo (Spain). Her main research area is corporate social responsibility reporting. She has presented her research results in international conferences and has published numerous book chapters and papers in journals of impact, such as *Journal of Cleaner Production*, *Corporate Social Responsibility and Environmental Management* and *Journal of Business Ethics*, among others.

Sumona Ghosh has been associated with St. Xavier's College Kolkata since 2002. Presently, she is the Head of Business Regulatory Framework in the Department of Commerce. After completing her post-graduation in commerce with rare distinction, Prof. Ghosh has been conferred with the Degree of Philosophy in Business Management by the University of Calcutta on 31st of July 2014. Her area of research was on corporate social responsibility (CSR). The title of her doctoral dissertation was *"Pattern of participation of Public and Private sector companies in Corporate Social Responsibility Activities"*. She has published in journals of national and international repute. Dr Ghosh has been highly acclaimed for her guest lectures on CSR in premier institutes of higher learning including the Indian Institute of Management (Calcutta and Shillong). She has taken sessions in Management Development Programmes conducted by premier institutes on CSR. She has presented papers on CSR at various national and international conferences. Dr Ghosh is also the recipient of the "Bharat Jyoti Award" for meritorious services, outstanding performance and remarkable role given by the India International Friendship Society for the year 2012 given by Dr Bhishma Narain Singh, Former Governor of Tamil Nadu and Assam, on 20th of December 2012. Her research interests lie in corporate social responsibility, sustainable development, integrated reporting and corporate ethics.

Zanellato Gianluca is a PhD candidate in accounting, at University Babes Bolyai, Faculty of Economics and Business Administration. He is writing his thesis on "The benefits brought by Integrated Reporting in the cases of the European State-Owned Enterprises". Zanellato is interested in corporate reporting, integrated reporting and accountability in state-owned enterprises. He believes integrated reporting will be the future of corporate reporting, for developing a more sustainable society and not least more sustainable businesses, and proving the benefits will lead to widespread integrated reporting across the business environment.

Carmela Gulluscio is associate professor of "Economia Aziendale" at the "Unitelma Sapienza" University of Rome since 2015. After completing her degree in business studies ("Economia Aziendale"), in 2005 she was awarded a PhD in "Economia Aziendale" from the University "Roma Tre". She teaches Economia Aziendale, public accounting and auditing. Her main research interests include accounting history, financial accounting, financial and non-financial reporting, corporate social responsibility and accounting education, on which she has published books and national and international journal articles and book chapters.

Diana-Lavinia Martin is a third-year PhD candidate in accounting at Babeş-Bolyai University, Faculty of Economics and Business Administration in Cluj-Napoca (Romania). Prior to her PhD studies she had a master's degree in audit, accounting convergence and harmonisation at the same university. Her research focuses on non-financial reporting in the context of corporate governance, and she is interested in financial accounting, integrated reporting and CSR. Diana also has a professional experience for more than 10 years in the finance and accounting field as CFO (chief financial officer).

Silvio Bianchi Martini earned his PhD at the University of Pisa. He is currently a full professor in business administration at the Department of Economics and Management—University of Pisa (Italy). Nowadays, he is also the Dean of the Department of Economics and Management at the University of Pisa (Italy) where, since November 2002, he teaches corporate strategy. He also taught competitive strategy at the Bocconi University (Italy). His main scientific interests are focused on the following fields: strategic management and corporate governance.

Silvio is a member of the "European Institute for Advanced Studies in Management" (EIASM), the "Strategic Management Society" (SMS), the "European Academy of Management" (EURAM), the "International Academy of Business and Economics" (IABE), the "Italian Academy of Business Administration and Management" (AIDEA), the "Italian Society of Professors in Business Administration" (SIDREA) and the "Italian Society of Accounting History" (SISR). He published several papers on academic peer-reviewed journals. Additionally, he is a member of the Editorial Board of the *International Journal of Business Research* (IJBR) and other academic journals as well as the member of scientific committees and/or the referee of relevant conferences organised by distinguished scientific bodies, such as SMS, EURAM, SIDREA, IABE and so on.

Ninel Nesheva-Kiosseva is a State Associate Professor of Economics and State PhD. Ninel Nesheva-Kiosseva currently works like a professor at the Department of

Business Administration, New Bulgarian University, Sofia. Ninel does research in business administration, financial economics and accounting.

Her research interests at present are in the fields of environmental, social and sustainability accounting and reporting, integrated reporting, pricing of water and methods of regulatory pricing of water and energy.

She has publications of articles and books at valuated scientific journals and publishers. She teaches different courses on accounting, economic development and valuation pricing, CSR and social entrepreneurship. Ninel was a visiting professor in the University of Bologna, University of Urbino, University of Aveiro and others.

Her most recent paper is "Water related price risks: Implications for company competitiveness" and most recent books: *Introduction in Environmental and Social Accounting* and *Environmental Accounting and Reporting Theory and Practice* with Baldarelli, Maria-Gabriella, Del Baldo, Mara.

Tudor Oprisor is a research assistant at Babes-Bolyai University, Faculty of Economics and Business Administration, Accounting and Audit Department. In 2017, he successfully defended his doctoral thesis on implementation perspectives for integrated reporting in public sector entities. He is also interested in the regulating activities conducted by the governing bodies and professional organisations regarding the integrated reporting initiative in both the private and public sector, as well as the implementation perspectives in public higher education. His research interests also relate to human capital, environmental and sustainability reporting, public sector economics, as well as social audit insights and the integration process into standard reporting practices, as well as information disclosure.

Silvia Romero is an associate professor in the Department of Accounting and Finance at Montclair State University, Feliciano School of Business. She holds BS and MA degrees from the University of Buenos Aires and an MBA and a PhD from Rutgers the State University of New Jersey. Prior to joining the PhD programme, she spent 15 years in professional practice as comptroller, internal and external auditor and consultant. Her research interests are in the areas of corporate social responsibility, as well as accounting information systems and its links to financial reporting and auditing. She has articles published in the *Journal of Business Ethics*, *CSR and Environmental Management, Accounting and Finance*, the *CPA Journal*, the *Journal of Cleaner Production*, the *ISACA Journal* and *Internal Auditing*, among others.

Maria Roszkowska-Menkes is an assistant professor in the Department of Management Theory in Warsaw School of Economics (SGH). In research and teaching, she specialises in issues of innovation management, corporate social responsibility and public relations, in particular with regard to new media and networked economy. Maria is the author of one of the first studies on open innovation in Poland. She is a business trainer in the area of corporate social responsibility, public relations and innovation management. She is also the author of numerous research and business press articles concerning CSR, crowdsourcing and social media. She was formerly a promotion coordinator in Responsible Business Forum—the oldest and the biggest Polish NGO providing in-depth focus on the concept of CSR.

Antonella Silvestri is a senior lecturer in accounting and assistant professor teaching accounting. She is department member at the Department of Business Administration and Law, University of Calabria. Her main research interests are related to the value relevance of accounting and extra-accounting information, to international accounting standards applied to insurance companies and to systems of measurement and reporting of intellectual capital. On these research themes, she has published books, book chapters and journal articles, and she has presented papers to national and international congresses.

Maria-Teresa Speziale, PhD in business economics, is University Teacher in Accounting at the Sheffield University Management School in Sheffield (UK). In 2013, she was awarded the Italian National Scientific Habilitation for the role of associate professor in business economics.

Since 2004, she has collaborated in teaching and research at the Alma Mater Studiorum—University of Bologna (Italy). In 2010–2011, she was a visiting researcher for 12 months at Brown University (USA), Boston University (USA) and Boston College (USA). She gave lectures and seminars as a visiting professor at the Kaunas University of Technology, Lithuania (2014), and at the Sheffield University Management School, UK (2016–2017). She is a member of the European Accounting Association (EAA), British Accounting and Finance Association (BAFA) and Centre for Social & Environmental Accounting Research (CSEAR). She is also a Certified Management Consultant (CMC—ICMCI). Since 2008, she has been Court-appointed Technical Consultant and Appraiser in many procedures at the Court of Bologna (Italy).

Adriana Tiron-Tudor is a full professor and PhD supervisor in accounting and auditing at Babeş-Bolyai University, Faculty of Economics and Business Administration in Cluj-Napoca (Romania). Her research interest is in corporate and public sector entities reporting including integrated reporting (financial and non-financial reporting). She also published articles in accounting history, accounting profession and auditing.

Stefania Veltri is a senior lecturer in accounting and assistant professor teaching business economics. She is department member at the Department of Business Administration and Law, University of Calabria. Her main research interests are related to the value relevance of accounting and extra-accounting information and the systems of measurement, management and reporting of intellectual capital. All these arguments are pursued employing both quantitative and qualitative methods. On these research themes, she has published books, book chapters and journal articles (such as *Journal of Intellectual Capital*, *Corporate Communications*, *Journal of Management and Governance*), and she has presented papers to national and international congresses.

Chapter 1
Coping with Integrated Reporting: An Overview of Financial and Social Reporting Using the Integrated Approach

Samuel O. Idowu and Mara Del Baldo

Corporate entities of our time are expected to demonstrate that they are responsible in all areas of their operational activities and existence including ensuring that all their actions demonstrate glaringly to all and sundry that they care about the wellbeing of all the inhabitants of planet earth—both animate and inanimate and that they are economically, socially and environmentally responsible. What does this really mean is perhaps a reasonable question to ask. Being responsible in the aforemention areas may not happen naturally or even by accident if the organisation in question were to fail to take a concerted effort to work towards responsibility in them. Corporate social responsibility has provided some of the ingredients which tantamount to societal responsibility. These are activities and actions needed by modern corporate entities to help them in the quest to tread the paths of responsibility; many of these issues and activities that are at the core of the area were not apparent to us some fifty or so years ago. CSR has similarly reoriented how we operate, perceive and understand the world we now live in. This has meant that corporations of the twenty-first century are faced with a number of interesting challenges and choices on how they must conduct their operational activities to ensure that they are looked at by their stakeholders as either being socially responsible or socially irresponsible. It has been argued times and times again that a number of strategic and social benefits will result from being perceived as being socially responsible. An entity that wishes to survive and be at a competitive advantage in its sector, needless to say; would in our view take these issues seriously. The way these modern entities report on their financial and non-financial activities has a number of

S. O. Idowu (✉)
London Metropolitan University, London, UK
e-mail: s.idowu@londonmet.ac.uk

M. Del Baldo
University of Urbino Carlo Bo, Urbino, Italy
e-mail: mara.delbaldo@uniurb.it

S. O. Idowu, M. Del Baldo (eds.), *Integrated Reporting*, CSR, Sustainability, Ethics & Governance, https://doi.org/10.1007/978-3-030-01719-4_1

consequences on how they are understood by readers of their financial and social reports, it also has some connotations on how they are placed in the league table of responsibility.

The King's Report of South Africa—(the origins of integrated reporting) has often been rightly credited by scholars and practitioners worldwide with providing the first impetus and direction on *Integrated Reporting*. Chapter 10 of the King Report III (2009) on Corporate Governance in South Africa made some recommendations on *Integrated Reporting and Disclosure*. The chapter recommends the integration of the financial information and other company information for disclosure purposes. The King Report III 2009 based its recommendations on Integrated Reporting on three principles, namely that:

- the board should "*ensure the integrity*" of the integrated report
- the financial report should be "*integrated*" with social reporting and disclosure
- the social reporting should be "*independently assured*".

The King Report III notes that the information the report contains should be forward looking and it should be "**complete**, **timely**, **relevant**, **accurate**, **honest** and **accessible** and **comparable** with past performance of the company". In other words, King was recommending that the information the report provides should not only be reliable but users of the information must have confidence in what it is telling them for decision making and other purposes.

It has been noted in both practical and theoretical terms some of the problems that could ensue from bad and ineffective corporate governance. There are a number of practical examples globally of the consequences and results of this—see a number of corporate failures and collapses in the literature and the media on some of the consequences of bad corporate governance in both the advanced and emerging economies worldwide. Without naming any particular corporations that failed or experienced serious difficulties as a result of bad and unacceptable system of corporate governance, it should hopefully not be too difficult to convince our readers that good governance is tantamount to good practice. Any action that is put in place to encourage good governance by corporate entities should be celebrated worldwide. This in our view explains why the different versions of King Report on Corporate Governance in the Republic of South Africa have been congratulated worldwide, Integrated Reporting is one of the many spin-offs from the different versions of the King Report.

Integrated reporting in its current form has been part of corporate reporting for nearly one decade. As implied above, this is another welcomed development in corporate social responsibility. Having said this, a number of interesting questions which our readers might want to ask have come to our minds, let's put forward some of them. Are we now the less or more wiser since these better reporting practices came on to the global corporate scene? What does integrated reporting really entails? How are CSR scholars and practitioners taking forward the requirements of integrated reporting around the globe? What are those corporations which have adopted the use of integrated reporting saying about it? What can we learn from organizations that have pioneered the integrated report? How does the integrated reporting process influence the corporate business model and enhance the quality of disclosure? How is non-financial information assured in the context of the <IR> Framework? What are stakeholders saying about integrated reporting? The intention of the book was to add to the discourse on Integrated Reporting by exploring different aspects and perspectives on Integrating Reporting and integrated report based on the authors' critical perspectives, knowledge, expertise and ideas, both from fully theoretical and empirical approaches. Accordingly, these and many other relevant questions are amongst the questions many of the 18 chapters of this book provide answers to.

The 18 chapter book is divided into three parts, each part focusing of different issues relating to integrated reporting. Part I on *Theoretical Insights and Outlooks on Integrated Reporting,* is made up of four Chapters. Part II on *Critical Perspectives on Integrated Reporting in Theory and Practice* is in eight chapters. Part III on *Implementing Integrated Reporting in Different Countries and Organisations: Issues, Benefits and Challenges* contains five Chapters.

In the very second chapter of the book from ltaly entitled '*Theoretical Perspectives on Purposes and Users of Integrated Reporting: A Literature Review*' where Speziale an Italian scholar explores the literature on integrated reporting to elucidate a number of relevant issues on integrated reporting. The chapter provides a critical narrative literature review with the aim of answering a particular research question—on the main theoretical perspectives in the academic literature on integrated reporting. She analyses the academic papers published in international accounting and finance journals for a period of 17 years, focusing on more than 60 articles that explicitly refer to one or more theories. In her review of the literature on the subject, she found that 28 different theories are mentioned especially the stakeholder theory, institutional theory, legitimacy theory, impression management theory and agency theory. Speziale was also able to present an embryonic draft of a conceptual model, structured in four components: theories of the firm, purposes and users, types of integrated reporting, and value creation and distribution. It is a must read chapter to fully understand the issues the chapter explored.

In the third Chapter of the book from Malta, on "*Theoretical Insights on Integrated Reporting: Valuing the Financial, Social and Sustainability Disclosures*" Mark Camilleri notes that organisations are increasingly disclosing information on financial and non-financial performance as they are becoming more accountable and transparent to the providers of capital and other interested parties. Camilleri argues that most of these organisations are specifying their environmental, social and governance (ESG) content, materiality and assurance mechanisms in their corporate disclosures. Very often, Camilleri notes, the organisations integrate financial, social and sustainability reporting. The chapter provides a critical review of key theoretical underpinnings that have anticipated the development of the corporations' integrated disclosures. The chapter also describes the International Integrated Reporting Council's <IR> Framework and its guiding principles. This contribution posits that there are potential tensions and challenges for those organisations that intend using the <IR> Framework. The chapter concludes by outlining future avenues as it identifies knowledge gaps in the realms of integrated reporting.

Ioana Dragu a Romanian scholar in the fourth chapter of the book on the "*Evolution of Corporate Sustainability and Corporate Social Responsibility towards the common goal of Integrated Reporting*" traces back the origins of integrated reporting to the Brundtland Report 1987. Dragu argues that by integrating sustainability and CSR information in the annual report, companies can develop successful reporting models in which financial and non-financial information is combined in a single report.

Mara Del Baldo an excellent Italian scholar of repute in Chap. 5 on *"Small and Medium Sized Organisations: Why and How Do the Implement IR?"* delves into the implementation of integrated reporting by small and medium sized enterprises in Italy. Del Baldo as at the point of her studies for the chapter notes the absence in the literature both theoretical arguments and empirical research about SMEs' implementation of integrated reporting, which was her motivation for the chapter on this topic. She uses a case study which in her view represents the best practice within the Italian context to investigate why and how Italian SMEs embarked on integrated reporting journey. The results of her study reveal the positive impact and the importance of a number of some enabling conditions.

Part II of the book considers the critical perspectives on integrated reporting in its theoretical and practical dimensions.

The first of the eight chapters' part II—Chap. 6 by Federica Casonato, Federica Farneti (from Italy) and John Dumay (from Australia), three excellent scholars of repute, *"From Sustainability to Integrated Reporting*, to explore the evolution of reporting in a financial institution and examine whether or not implementing the Integrated Reporting Framework enhance the quality of corporate disclosures. The chapter uses a case study on the sustainability reporting and integrated reporting practices of CBD an Australian financial services company, and its journey of disclosure. Their analysis suggests that the approach proposed by the IIRC does not support disclosure but rather legitimises a company's activities for the purposes of restoring its reputation. The 'accountingisation' of the process makes information more comparable from a financial point of view, but not necessarily from a social or environmental standpoint note the three scholars. They argue that the concept of materiality, highlighted in the IIRC's Framework, has shifted its focus from stakeholders to shareholders. The research shows that CBD has reached the last stage of a process highlighted in previous studies: definitive acceptance of, and compliance with, a new set of accounting rules that shifts focus from stakeholders to shareholders with the aim of standardising reporting output. Casonato et al argue that possible future changes in the evolution of the integrated reporting framework might influence a company's approach to reporting.

The seventh Chapter by a prolific author and scholar from Romania—Adriana Schiopoiu Burlea entitled *"The impact of Triple Bottom Dispersal of Actions on Integrated Reporting: A critical Perspective"* Schiopoiu Burlea commences her chapter by arguing that integrated reporting is considered in the literature as a practice that brings many benefits to organizations. Schiopoiu Burlea chapter provides some background information on the triple bottom dispersal of actions as a core element of Integrated Reporting. The triple bottom dispersal of actions she argues is a theory explaining the source of the problems and the crisis at an individual, organizational, national and international level. Schiopoiu Burlea states that the aim of the study she carried out for the chapter is to use integrated reporting to explain how organizations adopt the triple bottom dispersal of actions in order to reduce the amount of pressure caused by the increasing number of relevant information that needs to be disclosed. Accordingly, the chapter presents a critical analysis of the relationship between principles of integrated reporting and Triple bottom dispersal of actions and concludes with a set of reflections on how IR diminishes the triple bottom dispersal of actions.

In Chap. 8 entitled *"The mimicry of integrate reporting: An analysis of the principle—based approach"* by Tiron-Tudor, Oprisor and Zanellato three reputable scholars from Romania note that integrated reporting system is not a brand new reporting practice, but a restructuring output of existing reporting components. In this regard, these authors' underlying cause for investigation is that some of the framework's key components are inspired and transposed from other reporting frames. Their chapter analyzes the principles of integrated reporting, using content analysis on the conceptual framework, as well as existing sources from the literature and in order to establish the extent to which the set of principles embedded in this primary guideline document match with the principles from other major reporting frames. The valuable contributions of the paper in line with its main findings consist of tracking the principles' sources of inspiration (and pinpointing new and upgraded instances, such as the strategic focus and future orientation), emphasizing the mimicry given by the integrated reporting system and, ultimately, showing the usefulness of this set of principles in the fuller output.

Antonella Silvestri and Stefania Veltri, two Italian scholars in Chap. 9 on *"Integrated Report: Is it a strong or weak accountability tool?"* use their chapter to develop a specific research template which has the tool to support researchers in analyzing the quality of an organizational integrated report, in terms of degree of accountability towards the stakeholders. The chapter makes an original contribution to the IR literature by simplifying the elements to consider in an attempt to evaluate the degree of accountability followed by an organization in using IR to disclose its financial and non-financial information. Silvestri and Veltri note that the research template they used introduces subjective elements, both in the selection of the key dimensions included in the research framework and their observations and conclusions which are based on their own analysis under an interpretative approach.

In the tenth Chapter of the book on Integrated reporting, integrated thinking and gaming companies: Myths and Paradoxes, by Maria-Gabriella Baldarelli, an Italian scholar with a high expertise in social and environmental accounting, looks at integrated reporting in the gambling sector. The chapter looks at some of the integrated tools of measurement and accountability used in the sector in the attempt to establish how companies in the sector have moved from weak to strong sustainability in terms of eco-justice.

In Chap. 11, entitled *"Assurance on Integrated Reporting: A critical perspective" by* Miriam Corrado, Paola Demartini and John Dumay, these excellent scholars investigate how non-financial information is assured in the context of the Integrated

Reporting Framework. The chapter sets out to understand through the lens of the legitimacy theory, whether the assurance process is a tool that enhances user trust. These scholars research investigates the challenges of developing assurance through semi-structured interviews with a group of experts. The respondents to their study are practitioners, academics and investors directly involved in integrated reporting assurance debates. The interviews outline different opinions in relation to the role of <IR> assurance, consistent with legitimacy theory most practitioners agree assurance adds value to the reporting process. However, other academics and investors are skeptical about whether assurance reduces agent-principal gaps. Their main concerns regard the difficulty in assuring qualitative information, and the discretion <IR> grants to management in the reporting process. Surprisingly, our research shows how assurance is a tool for improving managers' confidence in their own non-financial reporting, rather than a credibility-enhancing instrument for users. They note that despite the evolution of assurance practice in CSR over the last decade, there remain many difficulties in understanding and developing suitable approaches for assuring non-financial information. Specifically, <IR> assurance is at an early stage of development, and very little research on <IR> assurance practices is available. This study contributes to this topic by interviewing key informant actors directly involved in the <IR> debate, including practitioners, academics and investors.

The twelfth Chapter of the book by a famous Bulgarian scholar Ninel Nesheva Kiosseva entitled *"Value-oriented and rental approach in the Integrated Reporting of Water Utility Companies"* who argues that the aims to explore theoretically the possibility of using the value-oriented and rental approach in estimating the performance of water utility companies (WUC) in their integrated reporting. The first part of the chapter deals with the issue of value and valuation of water in relation to the characteristics of water in the International Framework of IR, 2013 and other international "water documents". Based on this the chapter attempts to specify the capital structure of water utility companies (WUC) by integrating water as a natural capital to its full value. The second part of the chapter examines the possibility of using the rental approach in assessing the profits of companies and clarifying their business model. Nesheva-Kiosseva notes that the two approaches are important for assessing the added value, business model, strategy and reputation created by these companies, as they are key elements of integrated reporting.

In the final chapter of Part II the thirteenth in the book by three reputable Romanian scholars Tiron-Tudor, Martin and Farcas entitled "*Corporate Reporting Practices Concerning Non-Financial Aspects" who* argue that new trend in corporate reporting is the integrated reporting. Integrated reporting they note is a reporting practice which integrates financial information with the entity's information on the environment, society and governance (ESG) in a single report. The chapter answers a critical question on how disclosures around social, environmental, human, social and natural capital differ between the IIRC-type of integrated reports and the GRI-type sustainability reports.

The final Part of the book focuses on "Implementing integrated reporting in different countries and organizations: Issues, Benefits and Challenges".

In the first chapter of the part III—Chap. 14 by De Cristofaro and Gulluscio two Italian reputable scholars of CSR and Accounting History in a chapter entitled "*Integrated Reporting and Materiality Process Disclosure in European Sustainability Oriented Companies*" the two scholars note that materiality is a critical principle in the context of companies that are embracing Integrated Reporting (IR), the link between companies' IR orientation and their disclosure about their materiality determination process (MDP) is still under-investigated. Starting out from this premise, the chapter contributes to the empirical branch of IR research focusing on materiality in IR contexts. The study verifies the extent to which companies' IR orientation is associated with their attitude towards disclosing information about their MDP. In the attempt to increase our understanding in this area, De Cristofaro and Gulluscio analysed websites and reports of 108 European top companies included in the 2016 RobecoSAM Sustainability Yearbook. After the selection of two sets of predictors—i.e. regarding respectively IR orientation and MDP disclosure—the z-scores of the extracted principal components for each set were tested with Chi Square and Cramer's V associations. Although the analysis returned interesting results for IR (e.g. the prominence of sustainability-based reporting) and MDP (e.g. generalised lack of materiality in tables of contents) as well as on variables not associated, the most striking result that came out of the study is the unexpected very weak connection between IR orientation and MDP disclosure attitude. Both these results and the study's limitations suggest further research on the evolution of the link addressed, De Cristofaro and Gulluscio concluded.

In Chap. 15 by three scholars of repute from Spain—Romero, Riz-Blanco and Fernandez-Feijoo entitled *"Are Integrated Reports really integrated in Spain?"* who set out to meet three objectives namely to provide an understanding of the evolution of integrated reporting in Spain, to assess the integrated thinking of companies through the analysis of the CEO presentation pages of the Integrated Report and to study the extent to which integrated reports are committed to sustainability and are really integrated. They explored these issues using text-mining and content analysis techniques, besides the descriptive analysis of the documents from 2015. Spain was selected because it is a leading country in CSR reporting. The results of their study show a disparity between the underlying integrated thinking shown in the presentation letters/messages and the reference to the capitals proposed by the IIRC. They note that IR seems to be a tool to send signals to specific stakeholders: employees and financial capital providers.

In Chap. 16 titled "*Early Adopters of Integrated Reporting: The Practical Evidence from Warsaw Stock Exchange Companies*" by Aluchna and Roszkowska-Menkes two Polish well published scholars of CSR and Corporate Governance. Using the case study of five Polish listed companies which are the only quoted companies in Poland that have adopted the use of integrated reporting. They observed the five companies' organisational, sectorial characteristics, disclosure dynamics and evolution as well as the practice of integrated reporting. They also added to knowledge by observing the companies' size, standards and the quality of non-financial information provided by them. They were able to see the challenges and success factors of the companies integrated reporting process.

The penultimate chapter of the book—Chap. 17 on *"King Codes on Corporate Governance and Environment, Social and Governance (ESG) Performance: Evidence from FTSE/JSE All Share Index"*, by Federica Doni, Antonio Corvino and Silvio Bianchi Martini three Italian reputable scholars set out in their chapter to assess the main content of the South African Corporate Governance Code, King III, and to explore the relationship between environmental/social performance and corporate governance practices in a sample of South African listed companies following its adoption. The chapter explores the main objective of King III Report, and the more recent King IV Report, and note that both reports aim to promote good corporate governance as a driver for ethical and effective leadership at board level. In supporting companies, these codes provide recommendations for a culture of sustainability and environmental performance, adequate control by the governing body and for promoting trust in the organization's reputation and legitimacy. These principles of "good governance" Doni et al. argue can be evaluated through the process of Integrated Reporting, which became a mandatory listing requirement for South African listed companies from 2010 onwards. This requirement is consistent with the approach suggested by the Global Governance Principles, which include the recommendation that boards should produce integrated reports. In this context, Doni et al. empirical analysis is focused on the Integrated Reports drawn up by a sample of companies listed on the Johannesburg Stock Exchange (JSE), in an attempt to evaluate whether the recent adoption of corporate governance codes enhances environmental and social performance and improves corporate governance. Doni et al's period of analysis runs from 2012, 2 years after the introduction of King III (2010), to 2016, immediately after the publication of the draft version of King IV. They went on to investigate whether the release of corporate governance codes can influence environmental and social performance and improve corporate governance practices. In this regard, Doni et al. found that King III does exert an influence on environmental, social and governance (ESG) performance. Moreover, the presence of an active board, able to allocate tasks and responsibilities among different committees, along with the creation of an all-encompassing vision aimed at integrating financial and non-financial perspectives—an integrated thinking approach—are positively associated with environmental and social performance. Therefore, this empirical study can contribute to stakeholders' understanding of how mandatory adoption of King III and Integrated Reporting affect ESG factors and boards' attitude to sustainability issues. The chapter discusses a number of managerial implications, offering suggestions for managers, directors, consultants, investors and policymakers.

In the very final chapter of the book—Chap. 18, from India by a young Indian Professor based in Kolkata—Sumona Ghosh entitled *"Integrated Reporting in India: Research and Insights"* who argue that even though integrated reporting has become popular globally but there has been a limited number of research studies conducted with respect to IR in India. Hence the study investigated the extent to which Indian companies are integrating financial and non-financial data in their reporting. Ghosh's study employed Longitudinal Qualitative Document Analysis for the period 2010–2016 using an Integrated Reporting Index (IRI) in examining annual reports of 102 most valuable companies ranked on the basis of market capitalization. Based on the score range of 100 to 0 being the maximum and minimum respectively, the study revealed an average IRI of 70%. Ghosh observed that governance and performance were the two most well reported content elements throughout. In contrast, information about business model was the least reported content element. Ghosh also observed that financial and human capital were the two most well reported capital form disclosed by the companies and the two least reported ones were manufactured and intellectual capital. From the study it was observed that the selected companies fell into four levels of integration namely high integration, low integration, progressive integration and moderate integration. 38.24% of the companies fell into the category of high integration and 26.47% of the companies fell into the category of low integration. Hence Indian companies are nearly well equipped to adopt and implement "integrated reporting" as required by the IIRC framework Ghosh notes, but preparation of "Integrated Report" needs its promotion amongst such companies in the Indian context so that it encourages them to step forward and adopt this method of reporting, she concludes.

A careful read through of all the issues explored in this introductory chapter to each of the seventeen chapters that comprise this book should hopefully reveal that *Integrated Reporting* is reshaping corporate disclosure practices worldwide, integrated reporting has indeed come to stay in our corporate reporting world. Companies operating in different political settings around the globe are engaging themselves in a number of activities to improve their disclosure practices with regard to both their financial and non-financial areas of their operations. Professor Judge Mervyn King the author of King III and IV Reports in South Africa has drastically reshaped how corporate entities of this era perceive, understand and go about doing what the should do in this regard. Stock Exchanges and Standard setting organizations worldwide are also driving forward activities on how companies should deal with the matter. We can only become the wiser as we move forward in reporting both financial and non-financial aspects of the organisation, readers of these reports can only find them more useful in their working day.

Part I
Theoretical Insights and Outlooks on Integrated Reporting

Chapter 2
Theoretical Perspectives on Purposes and Users of Integrated Reporting: A Literature Review

Maria-Teresa Speziale

2.1 Introduction

Integrated reporting seems to be the ultimate tool to provide a holistic representation of an organisation's financial and non-financial performance. Nevertheless, the debate on this complex and delicate subject is lively and heated, involving different streams of research mainly related to corporate governance, accounting, and sustainability. In particular, the nature of integrated reporting as a sustainability tool is very controversial, being interpreted in entirely different ways and even with opposite meanings by different authors.

This chapter was born from an intuition. We observed that different scholars and frameworks attribute different purposes and users to integrated reporting and the reason is that they assume different theoretical perspectives. Therefore, our insight is that bringing this link to light is essential to orientate integrated reporting towards the desired purposes and users, even developing and adopting alternative theories, if needed. Starting from this intuition, we try to answer the following research question: What are the main theoretical perspectives in the academic literature on integrated reporting?

As mentioned by Stent and Dowler (2015), Parker (2005) distinguishes between two main groups of theories. On the one side, 'augmentation' theories treat social and environmental accounting (SEA) as addendums to conventional accounting (e.g. stakeholder theory, economic agency theories, decision–usefulness theories, legitimacy theory, accountability theory). On the other side, 'heartland' theories focus on the role of information in relationships between the organisation and society (e.g. political economy accounting theory, deep green ecological perspectives, eco-feminist approach, communitarian-based theories). Stakeholder theory seems

M.-T. Speziale (✉)
Sheffield University Management School, University of Sheffield, Sheffield, UK
e-mail: m.speziale@sheffield.ac.uk

S. O. Idowu, M. Del Baldo (eds.), *Integrated Reporting*, CSR, Sustainability, Ethics & Governance, https://doi.org/10.1007/978-3-030-01719-4_2

to be the prevailing approach in the studies on social and environmental disclosures, because stakeholders need to be informed of the social and environmental impact of corporate performance, beyond the information provided in financial statements (Gray et al. 1995; Gray et al. 1987).

However, even though integrated reporting is a quite recent phenomenon, the related academic literature draws upon multiple theories (mostly stakeholder theory, institutional theory, and legitimacy theory), which represent partial views enriching the understanding of the nature and impact of this emerging tool.

Therefore, we undertook a critical narrative literature review of papers published in international academic journals, with the aim of providing an overview of the main theories, identifying gaps and potential for other theories, and contributing to the development of a conceptual model to support future academic researches, policies, and practices.

This chapter is structured as follows. Section 2.2 describes research methodology, by illustrating the steps involved in conducting literature review and qualitative content analysis. After showing the paper distribution over time and across journals, Sect. 2.3 summarises the contents of the selected papers, classified into theoretical perspectives that are briefly presented. Section 2.4 critically discusses the primary findings from literature review, by proposing an early draft of a conceptual model. Finally, Sect. 2.5 concludes the chapter, by identifying the main limitations of our study and offering some cues for further research.

2.2 Research Methodology

We present a critical narrative literature review of the academic papers published in international accounting and finance journals from January 2000 to September 2017 (Baumeister and Leary 1997; Cronin et al. 2008; Denney and Tewksbury 2013; Hart 1998; Jesson and Lacey 2006; Van Wee and Banister 2016). First of all, to identify relevant literature on integrating reporting, we searched five well-known academic databases and collections (Scopus, ProQuest Central, Elsevier ScienceDirect, Emerald, and Wiley) by keyword 'integrated reporting' (or 'integrated report'), introduced in the field called 'article title'.

Then, we selected all types of sources (articles, articles in press, books, book chapters, reviews, conference papers, and notes), written in English, related to the following subject fields: business, management, and accounting; social sciences; economics, econometrics, and finance; decision science. We found and imported in Mendeley 164 studies, 119 of which are free downloadable from the above-mentioned databases and collections, whereas the other 45 papers are mostly books and book's chapters.

We conducted a qualitative content analysis of all available papers (articles, articles in press, reviews, and conference papers) with the support of Nvivo 11 software package, having imported all collected sources from Mendeley. We made a query in Nvivo 11, by searching all sources by keyword 'theory', and we obtained a 'word tree' that was extremely helpful for identifying all the theories mentioned in the selected papers. Thus, we made a query for each theory searching for the theory's name (e.g. shareholder, stakeholder, institutional, legitimacy) enclosed in brackets with each of the following terms: 'theory', 'theories', 'approach', 'approaches', 'view', 'views', 'perspective', and 'perspectives' (all these couples of words are searched for in the same query). By saving each query as a node, we created 30 nodes corresponding to 30 theories. These nodes include a total of 85 papers belonging to one or more nodes since they explicitly refer to one or more theories, but we did not analyse all of them in this chapter, because some papers just cite theories in text or references, without grounding their work on them.

Finally, we selected and thoroughly examined 66 papers that are deemed relevant for our goals, by mainly focusing on the single theory or on the combination of theories mentioned in each paper. We also read the previous literature reviews on integrated reporting (six additional papers), so we analysed a total of 72 papers and we identified their structures, methodologies, and purposes, in order to differentiate our review and to add value to the current academic debate. Table 2.1 shows the 30 theoretical perspectives identified in our review and the number of papers that explicitly refer to each theory (considering only the 72 analysed papers).

In our selection of papers, the five theories mentioned by the higher number of contributions are stakeholder theory, institutional theory, legitimacy theory, impression management theory, and agency theory.

The literature review provided by de Villiers et al. (2014) synthesises insights from accounting and accountability research on integrated reporting, by focusing in particular on the other papers published in the same special issue of Accounting, Auditing and Accountability Journal, and proposes a comprehensive agenda for

Table 2.1 Theoretical perspectives and number of analysed papers that explicitly refer to each theory

Theories (Nodes)	N. Papers	Theories (Nodes)	N. Papers
Stakeholder theory	40	Theory of proprietary cost	3
Institutional theory	27	Contractual theory	3
Legitimacy theory	24	Practice theory	2
Agency theory	23	Contingency theory	2
Impression management theory	23	Organisational theory	2
Signalling theory	6	Pragmatic constructivism theory	2
Voluntary disclosure theory	5	Theory of institutional reality	2
Systems theory	4	Behavioural decision theory	1
Political cost theory	4	Resource dependency theory	1
Shareholder theory	4	Decision-usefulness theory	1
Luhmannian theory	4	Regulation theory	1
Diffusion of innovation theory	4	Accountability theory	1
Sociology of worth theory	4	Social theory of trust	1
Stewardship theory	3	Cognitive cost theory	1
Resource-based theory	3	Media richness theory	1

future research. After illustrating the antecedents, the pioneers and the recent developments of integrated reporting, they summarise the research findings of some early academic studies in this field, by highlighting many areas where further robust evidence is needed to support improvements in regulation, policies, and practices. De Villiers et al. (2017a) present a narrative review of the key findings from the papers published in a special issue of Meditari Accountancy Research, with the aim of developing a conceptual model for examining the evolution of integrated reporting and identifying areas for future research. De Villiers et al. (2017b) provide an overview of the integrated reporting literature by classifying the studies into themes and discussing broad research approaches and questions, measurement and research design issues, and research opportunities in integrated reporting. Dumay et al. (2016) undertake a structured literature review to provide insights and critique of the previous research and to offer guidelines for future studies, by referring to the four stages identified in the research on intellectual capital. Perego et al. (2016) conduct a literature review of 17 studies, by distinguishing the papers that analyse the key institutional drivers or firm-level determinants of integrated reporting (antecedents) from those focused on the effects generated by the diffusion of integrated reporting (consequences). Velte and Stawinoga (2017) analyse the current state of empirical research on integrated reporting, by evaluating 44 empirical studies on integrated reporting that have been mostly published after the adoption of the IIRC Framework (IIRC 2013).[1] They create a research agenda based on three different levels of analysis (market level, organisational level, and individual/group decision-making level) by considering the fundamental theories, research methods, and main input factors, which contribute to the implementation and quality of integrated reporting.

2.3 Theoretical Perspectives on Integrated Reporting

2.3.1 Paper Distribution Over Time and Across Journals

Looking at the distribution over time of the 72 analysed papers, we are not surprised to observe a growing trend from January 2010 to September 2017, given the emergence of this topic in the academic literature (Fig. 2.1). In our selection of papers, we notice the absence of publications in 2010 and 2011, a tiny number of

[1] *"The International Integrated Reporting Council (IIRC) is a global coalition of regulators, investors, companies, standard setters, the accounting profession and NGOs. The IIRC's mission is to establish integrated reporting and thinking within mainstream business practice as the norm in the public and private sectors. The IIRC's vision is to align capital allocation and corporate behaviour to wider goals of financial stability and sustainable development through the cycle of integrated reporting and thinking"* (https://integratedreporting.org/the-iirc-2/).

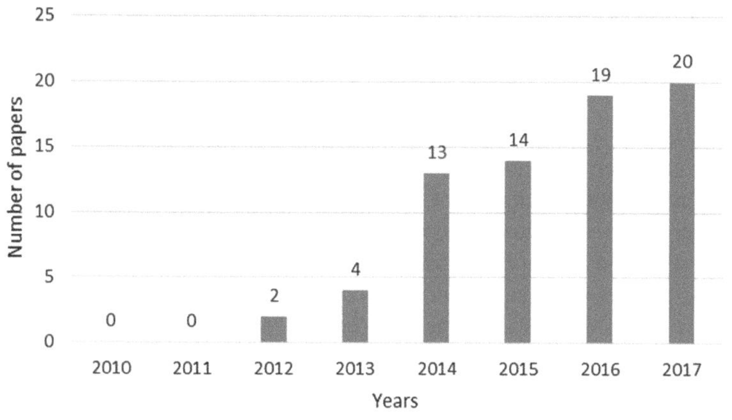

Fig. 2.1 Paper distribution over time (January 2010–September 2017)

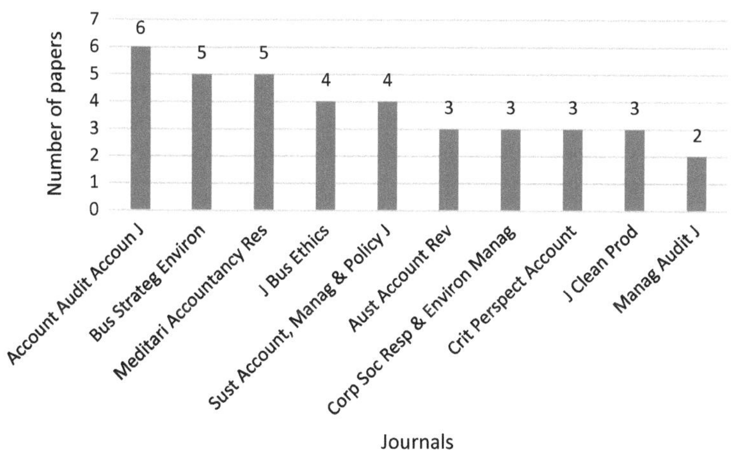

Fig. 2.2 Paper distribution across the top ten journals according to number of papers included in our literature review

papers in 2012 and 2013 (six papers in 2 years), and a sharp increase from 2014 to 2017, after the publication of IIRC Framework.

Moreover, we notice that the 72 analysed papers are distributed across 40 academic journals. By focusing on the top ten journals (25%) according to number of papers included in our literature review, we observe that 38 papers (about 53%) are published in the top ten journals and 24 papers (1 out of 3 papers) are published in the top five journals (Fig. 2.2).

2.3.2 Shareholder (or Capitalistic) Theory and Stakeholder Theory

Over the last few years, the global financial crisis, corporate scandals, and lack of trust in companies have put under discussion the capitalistic theory of the firm, determining a higher request of accountability and transparency and a more stakeholder-inclusive approach to corporate governance. Indeed, shareholder theory conceives the firm as an entity owned by shareholders, who supply its financial capital and appoint its management, obliged to run the firm in their interests. In other words, this theory assumes that the most critical factor of production is capital and capitalists seek the highest possible return on their investment. Stakeholder theory considers too simplistic that the firm's principal aim is to maximise profits and recognises that value creation springs from the relationships between an organisation and many individuals and groups having a stake in its activities and performance (Donaldson and Preston 1995; Enderle 2004; Freeman 1984; Freeman and Reed 1983; Freeman et al. 2007; Freeman et al. 2010).

According to a broad definition, stakeholders are '*any group or individual who can affect (or is affected by) the achievement of the firm's objectives*' (Freeman 1984, p. 46), whereas in a narrower interpretation, '*a primary stakeholder group is one without whose continuing participation the corporation cannot survive as a going concern*" (Clarkson 1995, p. 106).

Since the support from stakeholders is crucial to the long-term survival of companies, managers must disclose information on their programs and activities through corporate social responsibility (CSR) reporting, to foster the dialogue with a range of stakeholders (Perrini and Tencati 2006; Roberts 1992) and to demonstrate that the company is meeting their expectations (Deegan 2002; Gray et al. 1995). Therefore, any organisation must balance the conflicting requirements of the various stakeholders and communicate the results of its activities through public reports, whose purpose is to provide all stakeholders (users) with the information needed to form their judgements, to take their decisions, and to define their behaviours (Chapman et al. 2009).

Donaldson and Preston (1995) identify three contrasting/combining aspects of stakeholder theory: descriptive, instrumental, and normative. Indeed, stakeholder theory is used to describe, and sometimes to explain, specific corporate characteristics and behaviours, but also to explore, from an instrumental perspective, the connections, or lack of connections, between stakeholder management and the achievement of traditional corporate objectives. Further, the normative approach interprets the corporation's function, including the identification of moral or philosophical guidelines for the operation and management of corporations. Based on Immanuel Kant's thinking, normative stakeholder theory claims that stakeholders should be treated as human beings and their rights respected (Evan and Freeman 1988), whereas instrumental stakeholder theory argues that the firm has no interest in the welfare of its stakeholders, but it will have a competitive advantage if it takes proper account of their interests. Jones and Wicks (1999) propose '*converging the social science (instrumental) and ethics (normative) components of stakeholder theory to arrive at a normative theory*' (McWilliams and Siegel 2001, p. 118).

In his critique to the International Integrated Reporting Framework (IIRC 2013), Flower (2015) argues that the IIRC's concept of integrated reporting is founded on instrumental stakeholder theory, which is utterly consistent with the capitalistic theory of the firm, looking at stakeholders as means to maximise shareholder value.

From this theoretical perspective, the purpose of integrated reporting is to improve the information provided to shareholders, who are its primary users, by considering the objectives related to different stakeholders and multiple capitals in an instrumental way to achieve traditional corporate objectives (e.g. profitability, growth). In case of conflict of interests between different stakeholders, the information needs of capital providers to allocate capital efficiently take priority.

Flower (2015) adopts a critical perspective in tracing the history of the IIRC since its formation in 2010, at the initiative of the Prince of Wales Accounting for Sustainability (A4S) Project in collaboration with the Global Reporting Initiative (GRI). He demonstrates that, on its foundation, the IIRC's principal objective was the promotion of sustainability accounting, whereas later in the Framework (IIRC 2013) the IIRC has abandoned this goal. The author attributes the abandoning of sustainability accounting to the composition of the IIRC's governing council, dominated by the accountancy profession and multinational enterprises that seem determined to control an initiative that would undermine the capitalistic theory of the firm and threaten their established position in the field of corporate reporting ('regulatory capture').

Flower argues that the word 'value' can be interpreted in different ways: as 'value to investors' (shareholder theory), 'value to stakeholders' (stakeholder theory), 'value to society' (social and environmental accounting), and 'value to present and future generations' (sustainability). He observes that the IIRC Discussion Paper (IIRC 2011) lists six different categories of 'capital' (financial capital, manufactured capital, human capital, intellectual capital, natural capital, and social capital), by adopting the concept of 'value for stakeholders' and, more broadly, 'value for society'. Conversely, the IIRC Framework (IIRC 2013) adopts the balance sheet approach to the measure of performance and focuses on 'value to investors', even though it recognises that the value created by an organisation over time manifests itself in increases, decreases or transformations of the capitals caused by the organisation's business activities and outputs. Flower confidently concludes that the IIRC Framework refers to 'value' as 'value to the firm' and more precisely as 'value to investors' from the user need perspective. He deems that the values of other capitals (e.g. the quality of the workforce or the value of customer relations) are assessed solely on their contribution to the firm's profit-making activities, in order to achieve a more accurate estimate of future cash flows. Thus, the IIRC's investor orientation essentially determines the content of integrated report, in deciding whether a matter is material and hence should be reported, following the principle that '*an integrated report should disclose information about matters that substantially affect the organization's ability to create value over the short, medium and long-term*' (IIRC 2013, paragraph 3.17).

As highlighted by Thomson (2015), Flower's critique draws upon conventional theories of financial reporting, regulatory theory, agency theory, Kantian ethics, decision-usefulness, stakeholder theory, capitalism, and political economy. By supporting Flower's criticism and conclusions, Thomson argues that, even though integrated reporting and thinking produce some positive social and environmental changes, its current format does not allow for any substantive redistribution of power towards real sustainability. He expresses the need for an authentic effort of corporations to meet the urgent challenges posed by our unsustainable world and to avoid the risk of silencing 'the radical elements of the sustainability'.

Tweedie and Martinov-Bennie (2015) claim that IIRC Framework is double-edged from a sustainability perspective. They carry out a qualitative content analysis of public documents from four leading non-financial reporting organisations: the IIRC, the Accounting for Sustainability Project (A4S), the Global Reporting Initiative (GRI), and the King Committee on Corporate Governance in South Africa. The authors observe that integrated reporting has moved away from three key tenets of prior CSR reporting by privileging: (i) communication over holding organisations accountable; (ii) organisational over social sustainability and (iii) providers of financial capital over other stakeholders. Nonetheless, they conclude that integrated reporting could be a sustainability tool not only by serving the interests of a wide range of stakeholders but also by contributing to a broader social transformation in corporations and financial markets through incentives to reward the long-term investment horizon.

In a conceptual paper based on a comprehensive review of international research, Haller and van Staden (2014) propose to present a 'value-added statement' to report on the monetary effects of different types of capital included in integrated reporting. Indeed, stakeholder theory underlies the traditional measure of 'value-added', conceiving a firm's value creation as the central precondition of the going concern assumption and the result of a collective effort, although with considerable conflicts of interests between different groups of stakeholders (Shaoul 1998). The value-added statement discloses the success of an enterprise in generating and distributing value-added to stakeholders and shows the company's contribution to the national economy, regarding wealth increase (Maunders 1985; Schreuder 1979).

By studying a substantial and diversified international sample of firms, Mervelskemper and Streit (2017) apply the accounting-based valuation model, developed by Ohlson (1995), to investigate the value-relevance of environmental, social, and corporate governance (ESG) performance. They highlight that integrated report seems to be qualitatively superior to stand-alone report (Eccles and Krzus 2012a, b) since it combines shareholder theory and stakeholder theory by offering information that is considered to influence firm value and describing how this value is generated. According to them, a relatively high ESG performance is valued more strongly by capital market investors when firms publish an ESG report, irrespective of its type (stand-alone or integrated). Finally, they conclude that integrated report is preferable since it can even further enhance market valuation of a firm's composite ESG performance to an economically and statistically significant extent at no additional cost. Conversely, the overall effect on market value may be negative for firms with relatively low ESG performance scores.

Some studies analyse the first experiences of integrated reporting, with the aim of mapping the early adoption of the IIRC Discussion Paper and Framework, at an international level, and the King Report (King III) on Corporate Governance (Institute of Directors Southern Africa 2009),[2] at a national level.

Parrot and Tierney (2012) discuss why integrated reporting and stakeholder engagement are central to the success of an American electricity utility and how this company seeks to integrate financial, environmental, and social factors in its investment decisions. They claim that management must daily balance the myriad of stakeholders' interests, maximising the company's long-term value. However, they observe that 'enlightened value maximisation theory' (Jensen 2001), which views stakeholder engagement as an integral part of how companies generate long-term market value, does not tell management how to weigh or balance conflicting interests. They conclude that sound decision-making requires managers to optimise long-term market value and relational value creation while maintaining the highest ethical standards.

García-Sánchez et al. (2013) ground their study on stakeholder theory to examine the influence of the national cultural system (Hofstede 2001; Hofstede and Hofstede 2005), which is representative of the diverse values, norms, practices, and expectations of local stakeholders on integrated reporting. By analysing a non-balanced sample of 1590 companies from 20 countries for the period 2008–2010, they find that companies located in countries with a collectivist, feminist cultural system place greater emphasis on sustainable issues, ethics and good governance and are more likely to publish integrated report. Moreover, observing, at a firm level, that firm size, industry and profitability are associated with the extent of holistic disclosure adopted, the authors identify the need to develop further research on these business characteristics.

Grounded on stakeholder theory, the study of Rensburg and Botha (2014) explores financial communication in South Africa, through a national online survey on a sample of the general public, with a particular focus on those consuming financial information (421 respondents). They find that annual and interim financial reports are still the primary source of financial and investment information, whereas stakeholders currently seldom use integrated reporting and they would increasingly prefer using the Internet for financial information.

Vorster and Marais (2014) illustrate the case of a State-owned electricity utility in South Africa. By analysing the content of the 2012 integrated report and adopting stakeholder theory, they find that the company does not identify and map its

[2]The King III applies to companies listed on the Johannesburg Stock Exchange and requires them to present an integrated report in accordance with the "apply or explain" principle (companies are required to apply King III or provide reasons for not doing so). The King Committee published the King IV Report on Corporate Governance for South Africa 2016 on 1 November 2016. King IV replaces King III in its entirety and is effective in respect of financial years commencing on or after 1 April 2017.

stakeholder groups with their respective expectations, but the report meaningfully discloses stakeholder engagement, showing significant moves towards a stakeholder-inclusive approach to corporate governance. According to them, the primary purpose of integrated report should not be necessarily the disclosure of stakeholder management, but rather the combination of traditionally 'silo'-based financial and non-financial information into 'a coherent whole' explaining their joint impact on value creation in the short, medium, and long-term.

Stubbs and Higgins (2015) conduct an exploratory study on stakeholder's perspectives on the role of voluntary and regulatory approaches to integrated reporting in Australia. Their research is based on 22 in-depth semi-structured interviews with managers of sustainability, finance and communications across 15 organisations in varying stages of implementing integrated reporting. By undertaking a content analysis with qualitative coding techniques, they find that the majority of users of non-financial reporting prefers voluntary approaches, as they think that it will become the reporting norm over time. Their results also show that half of the investors support mandatory integrated reporting and this tool privileges financial value creation over stewardship, following a weak sustainability paradigm. Boonlua and Phankasem (2016) mention the transition from shareholder view to stakeholder view in investigating the adoption of the IIRC framework and promoting the improvement of integrating reporting in Thailand.

Smith (2016) observes that organisations embracing the so-called conscious capitalism model, such as benefit and flexible purpose corporations, have achieved superior financial results versus both the marketplace at large as well as competitors (Simpson et al. 2013). Therefore, by underlining the emergence of stakeholder theory both in academic studies and in the marketplace, he promotes the improvement of systems and processes associated with auditing and forensics, which should be more integrated and evolve into a more strategic function to meet the information needs of increasingly critical non-traditional stakeholders.

By analysing the integrated reports of a sample of 21 industrial companies in 2013, Ruiz-Lozano and Tirado-Valencia (2016) show that strategic approach to information and the relationship between capitals in the process of value creation seem to have a high level of monitoring. However, they bring to light the need for greater stakeholder engagement to identify material issues and to assure the information, making integrated reporting more reliable and credible. Moreover, they notice that the comparability of information is still difficult, as the KPIs are not homogeneous, the principle of 'conciseness' is disregarded, and the attention given to the principles seems not influenced by the environment.

A fundamental issue of stakeholder theory is to understand what stakeholders are legitimate and deserve consideration in the decision-making processes of each company (Freeman et al. 2010). Grounded on stakeholder salience theory (Deegan 2006), the study of Gianfelici et al. (2016) focuses on the managerial issues related to the selection of salient stakeholders, the prioritisation of their requests, and the adoption of effective engagement practices (Donaldson 1999; Freeman et al. 2010; Laplume et al. 2008). By applying the content analysis method to the integrated reports included by the IIRC in its Emerging Integrated Reporting Database (2013), they argue that the process of stakeholder selection presents some criticalities mainly for two reasons. On the one hand, different stakeholder categories may cooperate in the pursuit of common ends or might be in conflict (Greenley et al. 2004; Sen et al. 2006). On the other hand, the same individual may belong to different stakeholder categories at the same time (Neville and Menguc 2006). The authors refer to the framework proposed by Mitchell et al. (1997) for detecting stakeholder salience, by identifying seven categories based on three primary attributes: power (coercive, utilitarian, and normative), legitimacy, and urgency. They find that some business characteristics, such as industry membership, may affect stakeholder salience and therefore the content of integrated reports, regarding the representation of the impact of corporate social and environmental responsibilities on the economic performance. Other factors, such as firm's nationality, seem not to be differentiating elements, but further research on this topic is needed.

2.3.3 Institutional Theory

A comprehensive system of political, financial, educational, cultural, and economic institutions can exert pressure on companies (economic units) through three types of institutionalisation mechanisms: coercive isomorphisms, normative pressures, and mimetic processes (DiMaggio and Powell 1983). Coercive isomorphisms are mainly based on setting rules, norms or laws, monitoring their application, and finally giving reward and punishment. Normative pressures are related to non-imposed values and norms and to the impact that educational or professional authorities have on companies by setting standards. Finally, mimetic processes consist of imitating the strategies of successful organisations or competitors, regarded as best practice.

Several studies use institutional theory to explain the adoption and diffusion of different CSR strategies (Aguilera and Jackson 2003; Brammer et al. 2012; Campbell 2007; Delmas and Toffel 2004; Jackson and Apostolakou 2010; Meyer and Rowan 1977; Oliver 1991; Wild and van Staden 2013). Jensen and Berg (2012) conduct empirical research on 309 corporations selected from various sources (GRI reports list, GRI dataset with best sustainability reports, CRRA Reporting Award 2010). They adopt the Matten and Moon (2008) institutional model with the aim of providing empirical evidence of potential country-level determinants for sustainability reporting in contrast with integrated reporting. Their results show that integrated reporting correlates especially with investor and employment protection laws, the intensity of market coordination and ownership concentration, the level of economic, environmental and social development, the degree of national corporate responsibility, and the value system of the country of origin. They suggest that institutional theory, combined with contingency theory, may be applied to analyse internal processes that lead companies to publish integrated reports, also considering firm-level characteristics such as size, profit, multinational presence, identification of the most critical stakeholders, and sustainability performance.

Frías-Aceituno et al. (2013a) analyse a non-balanced sample of 750 international companies for the years 2008–2010, by showing that companies located in civil law countries, and where indices of law and order are high, are more likely to publish a broad range of integrated reports.

Dragu and Tiron-Tudor (2013a) explore the integrated reports issued by a sample of 58 IIRC Pilot Program companies for the period 2010–2012, through content analysis, disclosure index computation, and SPSS data processing. They observe that the external political, cultural, and economic factors influence the voluntary adoption of integrated reporting, but these emergent factors have a small influence of 8.1% on integrated reporting disclosure. In another article, the same authors (Tudor-Tiron and Dragu 2014) utilise the same sample to compute a disclosure index for each company/report, finding that the disclosure level of integrated reporting is higher in common law countries than in civil law countries.

Stubbs and Higgins (2014) investigate the internal mechanisms used by early adopters of integrated reporting, through content analysis with qualitative coding techniques of 23 semi-structured interviews in 15 Australian organisations. They show that integrated reporting is limited to incremental changes in sustainability reporting, rather than being a revolutionary transformation of the existing reporting practices, and that companies try to create cross-functional teams, instead of shifting the ownership of the reporting process from the sustainability or communications groups to the finance or strategy areas. According to them, even though more radical processes may take place over time, a lack of comprehensive standards may impact negatively on the diffusion of integrated reporting.

By conducting semi-structured interviews with 23 managers of Australian early adopters, Higgins et al. (2014) identify two narratives that emerge simultaneously in the adoption of integrated reporting and conform to the IIRC's rationales, but can create tensions leading to potentially different integrated reports, especially regarding materiality judgements and management responsibilities. On the one hand, integrated reporting is a tool to demonstrate the ability of managers to develop strategies addressing business challenges while protecting shareholder interests. On the other hand, integrated reporting is a disclosure mechanism that requires conformance to a new global reporting framework due to institutional pressures. According to this study, the implementation of integrated reporting will naturally evolve, by combining strategy with external expectations and developing the stakeholder engagement processes.

2.3.4 Legitimacy Theory

According to legitimacy theory, the existence of an organisation is justified if its stakeholders perceive it as legitimate because it appears to conform to social norms, values, and expectations within a kind of implicit social contract (Ashforth and Gibbs 1990; Deegan 2002; Dowling and Pfeffer 1975; Shocker and Sethi 1973). Legitimacy is a condition necessary to attract resources and gain the support of constituents for the going concern and long-term survival of an entity. The communication with the stakeholders (Suchman 1995) and especially the publication of accounting reports as social, political and economic documents (Chu et al. 2013; Guthrie and Parker 1990) are suitable to manage organisational legitimisation as a continuous process. Legitimacy theory is one of the most employed theoretical framework in social responsibility studies (Owen 2008; Parker 2005) and, more generally, is useful to explain organisational reactions to changes in the business environment and societal values (Deegan et al. 2000, 2002; Vourvachis et al. 2016). Indeed, every kind of change in societal values, norms, and expectations can create a legitimacy gap, when organisational actions and practices are not congruent with the expectations of relevant stakeholders (Lindblom 1994). A legitimacy gap can be addressed by implementing various kinds of legitimation strategies (Fernando and Lawrence 2014).

On the one hand, 'substantive management' strategy implies organisational change, seems more effective in managing stakeholder perceptions (Kim et al. 2007), and results in 'consequential legitimacy' or 'procedural legitimacy', both classified as 'moral legitimacy' (Suchman 1995). On the other hand, 'symbolic management' strategy focuses on portraying the organisation as consistent with social values and expectations (Ashforth and Gibbs 1990), resulting in 'dispositional legitimacy' (Suchman 1995). Lindblom (1994) illustrates the following legitimation strategies: influencing the relevant public's perception about the organisation's activities, deflecting the relevant public's attention from the issue at hand, and changing the relevant public's expectations. Organisations can combine different legitimation strategies, their choice depending both on whether they are attempting to gain, maintain, repair, or defend legitimacy (O'Donovan 2002), and on the level of CSR or environmental/social sensitivity (Cuganesan et al. 2010; Hrasky 2012). Moreover, the differences in organisational legitimacy-seeking behaviours have an impact on the extent of disclosure of social, environmental and intellectual capital information. Even though some papers investigate the application of legitimacy theory to sustainability disclosure, accounting researchers highlight the need for further studies.

By extending the work of Solomon and Maroun (2012), Setia et al. (2015) analyse the content of the annual/integrated reports of the top 25 JSE-listed companies (2009/2010 and 2011/2012) and the change in the extent of disclosure of non-financial capitals, moving from an unregulated to a regulated environment for integrated reporting. They test two alternative research propositions (substantive management and symbolic management), finding that the mandatory introduction of integrated reporting in South African listed companies has led to an increase in the extent of disclosure of human, social, relational, natural, and intellectual capitals. Moreover, the increment in the disclosure of social and relational capital is statistically significantly higher than the increment in the disclosure of other capitals. Therefore, the authors conclude that JSE-listed companies are adopting a symbolic legitimation strategy, aiming at maintaining their legitimacy.

Lai et al. (2016) demonstrate that Bloomberg ESG disclosure ratings are significantly higher for adopters (a sample of 52 IIRC Pilot Program companies) than for non-adopters of integrated reporting (a matching sample of 52 firms), rejecting the hypothesis that firms adopt integrated reporting as a response to a poor rating. Additionally, they show that other proxies of legitimacy pressures (size, leverage, profitability, industry) do not seem to play a role in explaining integrated reporting adoption. Thus, they conclude that corporate engagement in integrated reporting is not a strategy to manage corporate legitimacy, as already found for stand-alone CSR reports or similar forms of sustainability reports (Clarkson et al. 2008; Mahoney et al. 2013).

Beck et al. (2017) present the case study of a large Australian corporation in the financial service sector, by exploring its journey from the pragmatic adoption of integrated reporting to repair external legitimation (threatened by a crisis in public confidence in 2004) to the publication of integrated reporting to gain strategic legitimacy. By drawing upon Suchman's (1995) evaluation of institutional and strategic legitimacy, their work investigates the emergence of non-financial reporting as an interactive part of the overall strategic posture of an organisation, not constrained by the promulgation of voluntary guidelines. They suggest that the definition of stakeholders should be broader than the definition provided by the IIRC and that report preparers should understand that adopting integrated reporting would positively affect capital flows.

2.3.5 *Impression Management Theory*

According to impression management theory, developed in the psychology field and embedded in legitimacy theory, managers can employ narrative disclosures as an impression management tool (Brennan et al. 2009) to '*facilitate the construction of a new and different image of the company*' (Hopwood 2009, p. 437). Impression management is defined as a reporting bias introduced by the opportunistic behaviour of managers in selecting a style of presentation and content that is beneficial to them. This concept assumes a conscious and deliberate managerial disclosure strategy (Bowen et al. 2005; Merkl-Davies and Brennan 2007), aimed at presenting information in a manner that distorts readers' perceptions of corporate image and achievements (Hooghiemstra 2000; Leary and Kowalski 1990; Neu 1991; Neu et al. 1998; Riess et al. 1981; Schlenker 1980; Schneider 1981). The accounting literature shows a variety of techniques to manipulate the information presented in corporate reporting (Brennan et al. 2009; Merkl-Davies and Brennan 2007; Merkl-Davies et al. 2011). Neu et al. (1998) observe that managers prefer accounting narratives rather than financial or other quantifiable information because the former can be designed and customised to manage public impressions, by biasing the amount of information disclosed, the scope of topics, and the verbal tone of disclosures (Merkl-Davies and Brennan 2007).

Atkins and Maroun (2015) carry out detailed interviews with 20 institutional investors to explore the initial views of this particular user group on the first sets of South African integrated reports produced by JSE-listed companies. Their study confirms the findings of earlier research, which identify a significant presence of impression management in the proliferation of ESG disclosure in South African integrated reports (Carels et al. 2013; Solomon and Maroun 2012), in social and environmental investor engagement, and in corporate governance reform (Tremblay and Gendron 2011). Therefore, the authors identify the impression management by corporate preparers as one of the main obstacles to high-quality integrated reporting, because it encourages the maintenance of the status quo to the detriment of integrated thinking and sustainability (Gray 2013).

Melloni (2015) analyses all the reports available on the IIRC's website, with the aim of assessing the quality of intellectual capital disclosure (ICD) offered in integrated reporting. This research shows that ICD tends to be more optimistic than non-ICD, is focused on relational capital with limited quantitative and forward-looking information, and is used as an impression management strategy (the positive tone of ICD is significantly associated with declining performance, bigger size, and high level of intangibles). In another article, Melloni et al. (2016) focus on 'thematic manipulation' as an impression management strategy. By analysing all the reports available on the IIRC's website, they find that a positive tone of business model disclosure is significantly associated with weak corporate governance, bad performance and low verifiability of the disclosure itself. Furthermore, by examining the integrated reports issued by a sample of 74 IIRC Pilot Program companies in 2013 and 2014, Melloni et al. (2017) explore a selection of performance determinants to identify the factors associated with conciseness, completeness, and balance in integrated reports. The results show that in the presence of a firm's weak financial performance, the integrated report tends to be significantly longer, less readable, and more optimistic. They additionally find that firms with worse social performance provide reports that are foggier and with less information on their sustainability performance. This evidence implies that early adopters employ both quantity and syntactical reading ease manipulation, and thematic content and verbal tone manipulation. The results also suggest that such impression management strategies depend not only on the level of firms' performance but also on the type of performance (financial versus nonfinancial/sustainability).

Haji and Hossain (2016) analyse 246 integrated reports of large South African companies over a 3-year period (2011–2013), with the aim of understanding how the adoption of integrated reporting, and the embedded multiple capitals framework, has influenced organisational reporting practice. The authors find that companies are increasingly conforming to reporting language espoused in existing guidelines and multiple capitals frameworks over time. Indeed, they are recognising that the capitals are increased, decreased, and transformed through organisational activities, and there are interdependencies and potential trade-offs between multiple capitals, or components of capital. However, checking the utilisation of five impression management techniques (rhetorical manipulation, thematic manipulation, selectivity, emphasis in visual presentation, and performance comparisons), they conclude that the disclosures are generic, rather than company-specific, and lack substance.

By analysing all the reports available on the IIRC's website, the study of Stacchezzini et al. (2016) reveals that firms not only provide limited forward-looking and quantitative disclosure of their actions to achieve sustainability outcomes but also avoid providing information about their sustainability performance when their social and environmental results are poor. Thus, they argue that integrated reporting is not able to encourage the integrative management and to communicate managerial aspects of corporate sustainability.

McNally et al. (2017) identify impression management as an obstacle/challenge to preparing an integrated report. By conducting detailed interviews with 26 preparers at nine South African organisations, they highlight that integrated reporting is not seen as a natural part of the business process, despite the relevance of multiple types of capital for organisations' business models. They suggest considering non-financial issues as strategically relevant, clearly defining sustainability performance targets, linking them to specific performance indicators, and developing management control systems and accounting infrastructure for monitoring sustainability performance.

2.3.6 Agency Theory

Agency theory observes that the relationship between managers and investors is conflicting because of informational asymmetry, which results in adverse selection and moral hazard problems (Berle and Means 1932; Fama and Jensen 1983; Jensen 1986; Jensen and Meckling 1976). According to this theory, the principal delegates to the agent an activity of interest and monitors its achievement, then managers should pursue not only their interests but also shareholder interests (Fama and Jensen 1983; Kiel and Nicholson 2003). An appropriate disclosure will provide shareholders and other investors with the information useful for supporting decision-making processes and supervising managerial actions, thus reducing informational asymmetry. These functions are extended to other stakeholders under stakeholder–agency theory (Hill and Jones 1992). Indeed, stakeholder theory provides the basis for broadening the principal-agent relationship to other interested parties, especially in the public sector.

Some studies find that an increase in company size determines a greater need for external funds and, consequently, a higher likelihood of conflicts of interest and a rise of agency costs, without any impact on the cost of voluntary information disclosure (Lárran and Giner 2002). Other research highlights that managers of profitable companies tend to increase voluntary information disclosure, to ensure the stability of their position and to increase their level of remuneration.

Pavlopoulos et al. (2017) examine a sample consisting of 82 international firms in the 2011–2015 period, finding that a higher number of independent and non-executive board members on the nomination committee tend to display higher integrated reporting disclosure quality. Aligned with the earning quality literature, they also indicate that firms presenting high-quality information tend to adopt milder earnings management techniques and show lower agency costs than those with low integrated reporting disclosure quality. Grounded on agency theory, the study of

García-Sánchez and Noguera-Gámez (2017a) investigates an unbalanced sample of 995 international companies listed for the years 2009–2013, obtaining a total of 3294 observations from 27 countries. They show that the adoption of integrated reporting can help mitigate agency problems, facilitate corporate decision-making, and improve the information among investors. They also observe that the reduction effect on information asymmetry is more statistically significant in countries with stronger investor protection and in firms with higher financial reporting quality.

2.3.7 Other Theories

2.3.7.1 Voluntary Disclosure Theory, Signalling Theory, Cognitive Cost Theory, Political Cost Theory, Proprietary Cost Theory, Theory of Institutional Reality, and Media Richness Theory

Voluntary disclosure helps improve the information environment of companies by enhancing analysts' understanding of companies' prospects (Beyer et al. 2010). Empirical studies reveal a positive relationship between disclosure quality and analysts' earnings forecasting properties, such as lower forecast error and dispersion (Barron et al. 1999; Barth et al. 2001; Hope 2003; Lang and Lundholm 1993; Plumlee 2003). This theory is also applied to non-financial information, assuming that this further information is value-relevant (Orlitzky et al. 2003). Some research shows that the benefit of additional disclosure depends on the company's information environment (Diamond and Verrecchia 1991; Lambert et al. 2007; Verrecchia 1983). Moreover, the reduction in information asymmetry determines an increase both in investors' trust and in financial capital inflow, which has the potential to lower the capital cost (Healy and Palepu 2001; Verrecchia 1983).

Simnett and Huggins (2015) mention voluntary disclosure theory in an archival analysis of the responses to the IIRC's public consultation phases, by providing insights into arguments for and against salient aspects of the framework, and identifying issues that would benefit from future research. Zhou et al. (2017) examine a sample consisted of 443 company-year observations listed on the JSE from 2009 to 2012 (132 unique companies) for the analyst forecast error and dispersion analysis and 430 company-year observations (130 unique companies) for the cost of equity capital analysis. This study provides evidence that an increase in the level of alignment with the IIRC Framework is associated with a decrease in the analyst forecast error and, for some companies, with a reduction in the cost of equity capital. Hoque (2017) refers to voluntary disclosure theory with the aim of discussing the several roles of integrated reporting and providing some insights on the main benefits of its adoption.

Contrary to the findings of impression management theory, some studies based on signalling theory (or incremental information theory) posited in voluntary disclosure theory, indicate that companies with a superior performance (e.g. high profitability and return on investments) use information disclosure as a signal conveyed to the market to show their success. This companies' behaviour seems capable of reducing information asymmetries, optimising financing costs, and increasing the value of the firm (Baiman and Verrecchia 1996; Watson et al. 2002). Indeed, a lower cost of capital enables the company to raise its level of investment (Elliott and Jacobson 1994).

The study of Reimsbach et al. (2017) uses a comprehensive framework based on psychological research. It refers, in particular, to cognitive cost theory, as described in the Maines and McDaniel (2000) model and its extension (Hodge et al. 2010) based on proximity compatibility principle (Wickens and Carswell 1995). By analysing a sample of 104 professional analysts and fund managers in Germany, they find that voluntary assurance of sustainability information positively affects professional investors' evaluation of a firm's sustainability performance, results in a higher weighting of this information, and leads to higher investment-related judgments. This assurance effect is weaker in the case of integrated reporting compared to separate reporting, probably because of a cognitive bias ('halo effect') in decision-making when assured financial performance and non-assured sustainability performance are presented in the same report.

Political cost theory (Watts and Zimmerman 1978) states that companies voluntarily disclose information to reduce political costs (e.g. taxes, fees) and to obtain certain benefits, such as subsidies or government actions in favour of the company. Therefore, companies operating in countries with a higher level of regulation, nationalisation and confiscation tend to disclose more information to reduce these costs. Conversely, proprietary costs theory (or cost of ownership theory) identifies two types of cost associated with information disclosure. Indeed, in addition to the cost of processing data, companies can incur the cost related to the competitive disadvantage caused by external users (e.g. dissenting shareholders, employees and competitors) who may use the information on strategy, operations and technology in a way that can compromise the company's future (Elliott and Jacobson 1994). However, some empirical studies show that firms operating in competitive industries provide less disclosure because management fears the loss of competitive advantage or bargaining power, whereas other research indicates that companies operating in less competitive industries are less likely to disclose segment reporting information.

Grounded on both voluntary disclosure theory and theories of proprietary disclosure costs, the study of Lee and Yeo (2016) examines a sample of 822 firm-year observations in South Africa for the period 2010–2013, covering approximately 73% of the 2013 market capitalisation of all listed firms in JSE. They observe that the positive association between firm valuation and integrated reporting is stronger in the firms with higher organisational complexity (e.g. high intangible assets, firms with multiple business segments and large firms) and with higher external financing needs. Further, they find that firms with high integrated reporting outperform those with low integrated reporting both in terms of stock market and accounting performance. Their findings suggest that integrated reporting improves the information environment in complex firms and reduces the information asymmetry between corporate insiders and capital providers. Therefore, the authors conclude that firm valuation is positively associated with integrated reporting disclosures and the benefits of integrated reporting exceed on average its costs.

With regard to how organisations apply the IIRC's materiality guidelines in preparing integrated reporting, Lai et al. (2017) adopt the theory of institutional reality proposed by the philosopher John R. Searle to describe the 'construction' of social institutions (Searle 1995, 2010). Searle argues that social institutions exist and influence the social world only to the degree that their meaning is shared across their 'audience'. Drawing upon an interpretation of materiality as a 'social construction', the study of Lai et al. (2017) applies the framework proposed by Eccles and Krzus (2014), in line with Searle's theory, to analyse materiality with reference to who defines materiality and to whom the report is addressed. Indeed, they develop a case study to explore these two aspects, by conducting in-depth interviews with report preparers. They find that, according to preparers, capital providers are the primary audience of material information and the meaning of materiality corresponds with the company strategy (integrated report describes strategic priorities and related actions and results) and is intimately connected to the function that preparers assign to the integrated report, which is not only a tool of impression management.

Lodhia and Stone (2017) refer to media richness framework having its origins in media richness theory (Daft and Lengel 1984, 1986), which differentiates between rich media, appropriate to transmit potentially ambiguous information, and lean media, suitable to convey well-defined and unambiguous information. In previous research, some authors formulated and tested media richness theoretical perspectives to investigate the potential benefits offered by web-based media for corporate reporting (Cho and Roberts 2010; Lodhia 2010), also with a particular focus on the small business sector (Stone 2011, 2012). Drawing upon media richness framework, the paper of Lodhia and Stone (2017) indicates that Internet-based technologies, including social media platforms, have a potentially critical role in enhancing the communication aspect of stakeholder-inclusive integrated reporting.

2.3.7.2 Diffusion of Innovation (DOI) Theory

Diffusion of innovation theory aims at explaining how, why and what rate innovations are adopted and integrated into standard practice (Green et al. 2009; Lapsley and Wright 2004; Mellett et al. 2009; Rogers 2003). This theory tries to overcome the limitations of institutional theory, which does not explain why some organisations adopt radical changes while others do not (assuming similar institutional pressures), by focusing more on the processes and problems encountered by the adopters in implementing institutional innovations (Poole and Van de Ven 2004). Diffusion can be defined as a process by which all kinds of innovations are communicated through certain channels over time among members of a social system (Rogers 2003). Indeed, independently from the type of innovation, adopter, place or culture, innovation diffusion is a general process composed of four main elements: innovation, communication channels, time, and social system. As observed by Ax and Bjørnenak (2005), most changes in accounting are the direct or indirect consequences of diffusion processes, so it is important to analyse thoroughly this processes.

The study of Robertson and Samy (2015), based on a content analysis of 22 UK FTSE 100 annual and sustainability reports across industries and semi-structured interviews with ten senior managers, adopts diffusion of innovation theory to identify the individual company-level determinants. The analysis reveals that the lack of clear linkages between the majority of annual and sustainability reports in the sample significantly limits their ability to allow the stakeholders to assess an organisation's performance, strategy, and prospects. Senior managers perceive integrated reporting as having a relative advantage over existing practices. Furthermore, the framing of integrated reporting within the complex international regulatory arena, the cross-functional communication and integrated thinking, and mandatory compliance with integrated reporting seem to be the main factors of innovation determining the likelihood of more widespread diffusion of integrated reporting. The authors suggest to develop further research on capital measurement concept under a complex multiple capital model, to extend this study in different national settings, and to explore the other DOI elements (communication channels, time, and social system) with the aim of identifying other factors likely to impact on integrated reporting diffusion.

Gunarathne and Senaratne (2017) adopt diffusion of innovation theory to examine how and why integrated reporting is diffused in Sri Lanka, an emerging South Asian nation, from an expansion diffusion perspective, and to analyse the various stages of the diffusion process based on both demand-pull and supply-push perspectives. This research includes semi-structured interviews to gather data from both the demand and supply sides of DOI, content analysis of the annual reports published by the companies listed in the Colombo Stock Exchange from 2010 to 2014, and tracking of the material published on integrated reporting. They find that the country was in the diffusion stage of integrated reporting (32 adopters in 2014), with the early adoption mainly driven by the efficient-choice perspective and the diffusion stage mainly driven by fashion setting. This study also reveals the evolution of integrated reporting through the incremental changes in sustainability reporting and the risk that integrated reporting becomes a mere reporting mechanism if companies do not internalise its innovative principles.

2.3.7.3 Systems Theory

Some authors ground their research on systems thinking, mainly developed in General Systems Theory (von Bertalanffy 1969) and Senge's 'fifth' discipline (Senge 1997). This is a widely accepted multidimensional approach to problem-solving, aimed at understanding and solving problems by framing them within a larger system of parts and relationships. The IIRC Framework defines integrated thinking as '*the active consideration by an organisation of the relationships between its various operating and functional units and the capitals that the organisation uses or affects. Integrated thinking leads to integrated decision-making and actions that consider the creation of value over the short, medium and long term*' (IIRC 2013, p. 33). As highlighted by Stent and Dowler (2015), systems thinking and integrated thinking are very similar concepts, which have a great potential for avoiding the possible mistakes and damages of considering problems in isolation. Eccles and Krzus (2012a, b) define integrated reporting as 'One Report' promoting integrated thinking rather than 'silo' thinking for a truly sustainable strategy.

Alexander and Blum (2016) adopt Luhmann's Complex Systems Theory (or Luhmannian Systems Theory), developed by the German sociologist Luhmann, who applies to ecology the description of the world as a set of complex systems. Their study illustrates that the IIRC Framework's final draft has little relevance to either sustainability or ecology because of 'regulatory capture'. Further, they provide guidance and direction for establishing a broader framework of integrated reporting that consider the complex set of systems and sub-systems involved in the multi-capital, multi-measurement-unit, multi-stakeholder, and multi-motivated content of the sustainability issue.

2.3.7.4 Practice Theory

By undertaking document analysis and semi-structured interviews with managers, Lodhia (2015) explores the development of integrated reporting in an Australian customer-owned bank, aiming at identifying the transition's drivers that differentiate this company from its competitors and other organisations. This research adopts practice theory (Schatzki 2001, 2002, 2005; Whittington 2006, 2011), a perspective suitable for contextualising the functioning of systems within organisations, by suggesting that the elements of the practice need to be in place and should function appropriately in order for practices to develop. The author emphasises that integrated reporting is a complex process involving a sequence of activities, which taken together constitute this novel practice, rather than merely an outcome in the form of an integrated report. He observes that the main drivers to undertake effectively a comprehensive integrated reporting are the recognition of the potential value of integrated reporting and thinking, the availability of basic rules and guidelines, the formulation of strategies embedding economic, social and environmental aspects, and the alignment of the entirety of organisational ethical values, goals, structures and business operations to these strategies.

2.3.7.5 Sociology of Worth (SOW) Theory

Van Bommel (2014) utilises a novel theoretical approach—the Boltanski and Thévenot's sociology of worth (SOW) framework (Boltanski and Thévenot 1999, 2006)—to explore the multiplicity of views on integrated reporting in the Netherlands and to consider the possibility of, and impediments to, reconciling these multiple rationales ('orders of worth' or 'economies of worth'), in order to gain legitimacy through a compromise. He analyses the content of 64 semi-structured in-depth interviews with a wide range of relevant actors and various kind of additional sources, such as documents, reports, articles, and notes. This study suggests that integrated reporting combines the different logics of valuation or orders of worth—industrial, market, civic, and green—which need to be dynamically reconciled in a durable legitimate compromise, in order to avoid the risk that integrated reporting gets 'captured' by investors and accountants.

2.3.7.6 Pragmatic Constructivism Theory

Through document analysis, archival research and an interview with the project manager of the CSR department, Lueg et al. (2016) analyse how standards and guidelines for corporate social responsibility can help a Danish carpet manufacturer in its integrated reporting. They adopt the perspective of pragmatic constructivism (Jakobsen et al. 2011), a philosophical school of thought, which is mainly applied to the field of critical management (accounting) studies and focuses on the integration of four dimensions: facts, possibilities, values, and communication. By investigating the motivations of diverse stakeholders (including shareholders), the authors find that the company follows a strategy of 'enlightened shareholder value' that fostered integrated reporting with guidelines and standards for CSR.

2.3.8 Combination of Theories

In exploring the early adoption of integrated reporting, many authors combine multiple theories to increase their explanatory power of corporate disclosure.

Wulf et al. (2014) refer to stakeholder theory, stewardship theory and decision-usefulness theory to present a descriptive analysis of the development towards integrated reporting and its impact on corporate governance. After clarifying the difference between shareholder-oriented and stakeholder-oriented corporate governance, they observe that the obligation of German listed groups to apply IFRS implies the prevalence of shareholders, decision-usefulness and relevance of information on stakeholders, stewardship and reliability of information. Thus, the authors propose a two-dimensional financial statement for capital-market-oriented companies that have to meet both of the two accounting purposes: decision-usefulness and stewardship. By comparing the well-established German management commentary and integrated reporting, they conclude that, if properly implemented, integrated reporting provides even more decision-useful information and it affects several traditional structures and business processes, so it could contribute to strengthening the Continental European stakeholder approach to corporate governance and forcing adequate management control systems.

Dragu and Tiron-Tudor (2014) investigate the leading theories about integrated reporting, claiming that there are three theoretical circles: (1) stakeholder theory versus shareholder theory, (2) new institutionalism and legitimacy theory and

(3) innovation diffusion theory. They find that institutional theory, focusing on the factors affecting integrated reporting, and diffusion and adoption theory, focusing on the spread of this new practice, represent solid research base for this field of studies.

Frías-Aceituno et al. (2014) examine the determinant factors of integrated reporting, by taking into account the postulates of four main theories of information disclosure: stakeholder–agency theory, signalling theory, political cost theory, and proprietary cost theory. Their research, which analyses a non-balanced sample of 1590 international companies in 20 countries for the years 2008–2010, shows that industry concentration has a negative influence on the likelihood of integrated reporting to be produced and published. Company size and profitability seem to have a positive impact, whereas business growth opportunities and industry does not appear significant in this respect. In previous research, the same authors focus on the role played by some features of the Board of Directors on the development of integrated reporting (Frías-Aceituno et al. 2013b). Specifically, by grounding their study mainly on stakeholder theory, but also mentioning agency theory, they examine a non-balanced sample of 568 leading non-financial multinational firms from 15 countries, for the period 2008–2010. Their results show that bigger size, larger growth opportunities and boards with a greater number of directors, and a variety of expertise and gender diversity have a material positive impact on the adoption and diffusion of integrated reporting. Steyn (2014) refers to stakeholder theory and legitimacy theory to examine the perceived benefits and implementation challenges to implementing mandatory integrated reporting in South African listed companies. This study brings to light a disconnect between the perceived audience of integrated reporting, according to senior managers, and the intended audience of integrated reporting as envisioned by the IIRC. Indeed, managers are more motivated by the legitimising aspect of advancing corporate reputation and broader stakeholder needs then by merely satisfying investor needs.

Dumitru et al. (2015) refer to stakeholder theory and impression management theory to analyse both quantitative and qualitative data extracted from the integrated reports published in 2014 by a sample of 95 IIRC Pilot Program companies. Their research aims at investigating how integrated reports are used as marketing tools for the communication of information regarding the value creation and distribution. Differently from most marketing research that focuses on studying the relationship between an organisation and its consumers, perceived as privileged regarding legitimacy, this study takes into consideration a variety of stakeholder groups. Their results show that organisations have only partially and selectively observed the IIRC Framework and most of them report primarily for shareholders and customers. There is no precise definition of value-added and quantitative information is poorly disclosed in comparison with the abundance of qualitative and especially narrative information (e.g. the amounts of the value distributed to different groups of stakeholders are not clearly specified). It seems that many organisations adopt an opportunistic behaviour to gain a competitive advantage through the affiliation to the IIRC Pilot Program, even if the main integrated reporting principles are not yet fully incorporated in their reporting practices.

By drawing upon economic-based theories (e.g. agency theory) and socio-political theories (e.g. stakeholder, legitimacy and political economy perspectives), Haji and Anifowose (2016a) develop rival hypotheses to predict the role of the audit committee function in integrated reporting practice. Consistently with the predictions of agency theory (Beasley et al. 2009), the authors find that the overall effectiveness of the audit committee function (and especially its authority and meetings) has a strong positive association with the extent and quality of integrated reporting practice. However, as implied by legitimacy theory (Beasley et al. 2009; Brennan and Solomon 2008; Michelon and Parbonetti 2012), they do not find a significant association between the key aspects of the audit committee function, such as audit committee independence and financial expertise, and integrated reporting practice. By analysing 246 integrated reports of 82 large South African JSE-listed companies over a 3-year period (2011–2013) after the introduction of an 'apply or explain' integrated reporting requirement, the same authors (Haji and Anifowose 2016b) observe the trend of integrated reporting practice. They mainly refer to legitimacy theory and impression management theory to illustrate that organisations undertake symbolic or substantive legitimation strategies, and in some cases, combination of both, to respond to changes in the business environment (Berrone et al. 2009; Oliver 1991; Soobaroyen and Ntim 2013; Vourvachis et al. 2016). Despite the significant increase in the extent and quality of integrated reporting practice and the evidence of both symbolic and substantive strategies, the authors conclude that the current practices are far from being accountability mechanisms and show a ceremonial nature, as symbolic conformity with integrated reporting guidelines, peppered with impression management techniques.

Stent and Dowler (2015) refer to systems thinking as well as accountability and stakeholder theory to provide early assessments of the gap between integrated reporting and current corporate reporting. Accountability can be defined as '*the duty to provide an account (by no means necessarily a financial account) or reckoning of those actions for which one is held responsible*' (Gray et al. 1996, p. 38). Moreover, Deegan (2009) argues that this notion of accountability may be derived from the ethical and normative perspective of stakeholder theory. By analysing the 2011 annual reports and related online reporting practices for four New Zealand 'best practice reporting entities', Stent and Dowler (2015) find that current reporting processes lack the integration. Thus the stated aims of integrated reporting, sustainability and financial stability, seem to be unachieved.

Reuter and Messner (2015) analyse the comment letters issued towards the IIRC Discussion Paper with the aim of shedding light on the characteristics of lobbying parties and the determinants of their lobbying behaviour towards the IIRC. They adopt Sutton's (1984) rational-choice model of lobbying and combine regulation theory, agency theory, and political cost theory. According to regulation theory, the auditors tend to pursue their benefit, even in case of conflict with the preparers' preferences (Puro 1984), whereas agency theory states that auditors act in the interest of their clients (Georgiou 2002, 2004). Moreover, as suggested by political cost theory, larger organisations may have a higher incentive to take lobbying action as they attract more public attention than smaller firms and they are put under pressure by interest groups, such as labour unions and environmental activists. This study finds that comment letters are mainly written by large multinational firms and by preparers. Sustainability service firms and professional bodies seem quite critical towards the emphasis on investor needs. Further concerns are related to the scope of the audience of integrated reporting, materiality issues, and the relationship with other existing reporting frameworks.

Stefanescu et al. (2016) combine agency theory, stakeholder theory, and legitimacy theory for developing an assessment tool, based on integrated reporting approach, with the aim of enhancing public accountability throughout greater transparency and holistic view in reporting.

Adams et al. (2016) deem that stewardship theory and specific dimensions of institutional theory (namely isomorphism and isopraxism) are potentially helpful for understanding why corporate reporting approaches may converge (or not) over time towards integrated reporting. To be more precise, isomorphism assumes that there is little or no reinvention of an idea or innovation as it travels between adopters (homogeneity), whereas isopraxism considers that ideas are translated and modified as they travel (heterogeneity). The authors argue that, following stewardship theory, management tends to be more integrated, collectivist, and long-term oriented. Further, integrated reporting, which requires stewardship of multiple capitals (IIRC 2013), is expected to increase the level of accountability towards society and the environment. They also highlight that stewardship theory seems helpful for explaining that organisations undertake CSR activities believing that it is the right thing to do. Institutional theory and legitimacy theory seem more suitable for explicating that organisations implementing CSR activities seek to do what it is good for business. By examining the evolution of business reporting of four large multi-national corporations towards more integrated approaches from 2009 to 2013, their study finds in some companies significant progress towards more human-centred value creation. They also observe improvements in the articulation of purposes and outcomes of social investments and in their association with longer-term notions of progress, risk and strategy.

Vaz et al. (2016) combine institutional theory and stakeholder theory, given that in previous research institutional theory has often been used to analyse the effects of pressure from supra-level structures on organisational practices and company behaviour, whereas stakeholder theory has been mainly used to analyse value creation at a firm level. Indeed, a double level of analysis is needed, since some determinants both at a country level (e.g. legal, political, economic, and cultural characteristics) and at a company level (e.g. size, profitability, and industry) may explain the decision to present an integrated report. By analysing a sample of 1449 companies registered in the GRI database in 13 countries, the authors find that more than half of the variability in adopting integrated reporting depends on country-level determinants. They conclude, in agreement with the results of other authors (García-Sánchez et al. 2013; Sierra-García et al. 2015), that companies in countries adopting a 'comply or explain' integrated reporting regulation and having a less individualistic orientation are more likely to publish integrated report. Therefore, their research shows that coercive and normative institutional mechanisms exert pressure to present integrated reports, whereas the mimetic mechanism appears to be ineffective, even though, being integrated reporting quite a new approach to corporate reporting, it might be that this institutionalisation mechanism will be observed in future.

Rivera-Arrubla and Zorio-Grima (2016) claim that the Internet and digital technologies provide new channels of communication and interaction through social media, as indicated by the IIRC within the context of 'connectivity' of information. By emphasising the concept of integrated thinking, the IIRC Framework illustrates the connectivity between all kinds of internal and external information. According to these authors, not only legitimacy theory and stakeholder theory but also reputation risk management (Bebbington et al. 2008) and multistakeholder co-creation approach (Killian and O'Regan 2016; Singaraju et al. 2016) provide suitable frameworks. Indeed, integrated reporting informs on reputation and risk management, aiming at multiple stakeholders, especially capital providers. Therefore, social media make available effective platforms to disseminate a wide variety of integrated reporting contents or announcements and to bring stakeholders' knowledge into the firm's strategy, facilitating stakeholder engagement and integrated reporting transparency. From this theoretical perspective, they analyse 78 integrated reports for the year 2012 of IIRC Pilot Program companies, finding that the level of integrated reporting connectivity (or its absence) can be due to the combination of three types of elements: information quality, corporate characteristics, and communication factors.

Chaidali and Jones (2017) mainly refer to social theory of trust (Sztompka 1999), focusing on trust in social relationships, together with the Luhmannian and the impression management theories, to explore the views and opinions of integrated reporting preparers few months after the publication of the final version of the IIRC Framework. By following on from the findings of Robertson and Samy (2015), who performed a similar analysis prior to the publication of the IIRC Framework (2013), they conduct semi-structured interviews with 15 senior managers from a sample of FTSE 100 companies. Their research highlights trust's problems: interviewed preparers are often suspicious of the motives of the IIRC professionals and express concerns about the credibility, performance, appearance, benefits, and beneficiaries of integrated report.

Buitendag et al. (2017) adopt signalling theory and agency theory to analyse the integrated reports of the top 100 JSE-listed entities for the years 2013–2015. They show that the type of industry, the size and profitability of the entity, as well as the composition of the board of directors, affect and can predetermine the integrated report quality. More precisely, they find that entities whose business has an effect on the environment, larger entities (greater resources), more profitable entities (greater cash flow), entities with more females (Frías-Aceituno et al. 2013a), more directors of colour and fewer executive directors, and entities with less public shareholding (this result contradicts findings reported in earlier literature) tend to produce a better integrated report. However, by adopting the same four theories utilised by Frías-Aceituno et al. (2014), García-Sánchez and Noguera-Gámez (2017b) investigate an unbalanced sample of 995 international companies listed for the years 2009–2013, obtaining a total of 3294 observations from 27 countries. They conclude that some empirical evidence show a positive influence of profitability on the degree of voluntary disclosure, whereas other studies find no statistically significant relationship.

Trébucq and Magnaghi (2017) mention stakeholder theory, organisational theory, resource-based view theory, contingency theory, and legitimacy theory to check how EFQM model could help managers understand the connectivity between the six capitals suggested by the IIRC Framework and manage intangibles, improving the quality of narratives in future reports. They propose considering this model as a complementary management control tool for improving strategic thinking and implementing integrated reporting.

Fasan and Mio (2017) utilise stakeholder theory, agency theory, and legitimacy theory for formulating the research hypotheses to study the determinants of materiality disclosure among the IIRC Pilot Program companies. Their results show that industry and some firm-level characteristics (board size and diversity) are the main determinants, whereas the legal environment does not play a significant role. They also find that the IIRC Pilot Program companies disclose their materiality to a greater extent when compared with their competitors that are not part of the Program.

Tweedie et al. (2017) unpack and critically evaluate the core features of the IIRC's business model concept, by conducting a structured comparative analysis of the IIRC's model against eight influential alternatives in prior business model research. They identify tensions both in the internal coherence of IIRC's business model and in the consistency of this business model with the IIRC's broader reporting objectives. Their study mentions stakeholder theory to distinguish three different scopes of business models (narrow, broad, and eco-system approaches) referring to a narrower or broader set of stakeholders (Evan and Freeman 1988; Freeman and Reed 1983). They also observe that, although business models lack a single overarching theory, purpose or structure, there is a significant shift towards more systems-focused theories and open-ended relationships between model elements.

2.4 Discussion of Findings Towards a Conceptual Model

Our analysis provides an overview of the main theoretical perspectives adopted by the academic literature on integrated reporting, an emergent tool that seems currently limited to incremental changes, rather than being a revolutionary transformation in existing reporting practices (Gunarathne and Senaratne 2017; Stubbs and Higgins 2014).

Some authors explore country-level and firm-level determinants for integrated reporting, adopting stakeholder theory, institutional theory, diffusion of innovation theory or a combination of theories (Adams et al. 2016; Buitendag et al. 2017; Dragu and Tiron-Tudor 2013a, 2014; Fasan and Mio 2017; Frías-Aceituno et al. 2013a, b, 2014; García-Sánchez and Noguera-Gámez 2017b; García-Sánchez et al. 2013;

Jensen and Berg 2012; Robertson and Samy 2015; Tudor-Tiron and Dragu 2014; Vaz et al. 2016).

Legitimacy theory helps explain why organisations adopt substantive or symbolic legitimation strategies, depending both on whether they are attempting to gain, maintain, repair or defend legitimacy and on the level of CSR or environmental/ social sensitivity (Haji and Anifowose 2016b; Setia et al. 2015). Further, one of the main obstacles to high-quality integrated thinking and reporting for sustainability is represented by impression management strategies (Atkins and Maroun 2015; Chaidali and Jones 2017; Dumitru et al. 2015; Haji and Hossain 2016; Melloni 2015; Melloni et al. 2016, 2017; Stacchezzini et al. 2016), which can be seen as a kind of legitimation strategy. However, some empirical evidence shows that managers are motivated by the legitimising aspect of advancing corporate reputation (Steyn 2014), whereas other evidence indicates that integrated reporting does not seem a strategy to manage corporate legitimacy (Lai et al. 2016).

Nevertheless, the findings of our literature review show not only criticism towards the concept of integrated reporting according to the IIRC Framework (2013), but also some significant steps towards a more system-focused and stakeholder-inclusive approaches to corporate governance and integrated reporting (Boonlua and Phankasem 2016; Smith 2016; Tweedie et al. 2017; Vorster and Marais 2014). Adopting systems theory, some contributions highlight that systems thinking and integrated thinking are very similar concepts, thus integrated reporting can promote integrated thinking for a truly sustainable strategy (Stent and Dowler 2015). A research based on Luhmann's complex systems theory proposes a broader framework of integrated reporting that consider the complex set of systems and sub-systems involved in the multi-capital, multi-measurement-unit, multi-stakeholder, and multi-motivated content of the sustainability issue (Alexander and Blum 2016). Another study, grounded on sociology of worth (SOW) theory, suggests reconciling in a durable legitimate compromise the different logics of valuation or orders of worth (industrial, market, civic, and green) within integrated reporting (van Bommel 2014).

Several authors claim that organisations should consider integrated reporting as a natural part of business strategy, processes (McNally et al. 2017) and practices (Lodhia 2015). Some studies underline the need to clearly define sustainability performance targets, link them to specific performance indicators, and develop management control systems, accounting infrastructures, and auditing processes for monitoring sustainability performance (Beck et al. 2017; Haji and Anifowose 2016a; Ruiz-Lozano and Tirado-Valencia 2016; Stefanescu et al. 2016; Trébucq and Magnaghi 2017; Wulf et al. 2014). From this perspective, integrated report should describe strategic priorities and related actions and results, thus materiality should be identified by means of stakeholder selection and engagement (Gianfelici et al. 2016; Higgins et al. 2014) and correspond with the company strategy (Lai et al. 2017). Some works illustrate that the Internet and digital technologies play a critical role in the development of stakeholder-inclusive integrated reporting (Lodhia and Stone 2017; Rensburg and Botha 2014), as stated by the IIRC within the context of connectivity of information (Rivera-Arrubla and Zorio-Grima 2016).

Agency theory (García-Sánchez and Noguera-Gámez 2017a; Pavlopoulos et al. 2017), voluntary disclosure theory (Hoque 2017; Lee and Yeo 2016; Simnett and Huggins 2015; Zhou et al. 2017), cognitive cost theory (Reimsbach et al. 2017), and all the other theories mentioned in this chapter help analyse the main benefits and challenges to implementing integrated reporting.

Finally, some studies emphasise the importance of maintaining high ethical standards, also by observing that some multi-national corporations adopting integrated reporting seem progressing towards more human-centred value creation (Adams et al. 2016) and that organisations embracing 'conscious capitalism' model show superior financial results (Smith 2016).

Starting from the findings of our literature review, we outline a conceptual model structured in the following components: theories of the firm (nature and purposes of the firm), purposes and users of integrated reporting, and types of integrated reporting (Fig. 2.3).

According to shareholder theory, the primary purpose of a firm is to maximise profit for shareholders in the short-term. Thus, the companies adopting the narrow business case perspective do not need integrated reporting because financial reporting is suitable to communicate the value creation and distribution to shareholders. In a broader perspective, instrumental stakeholder theory supports the purpose of maximising shareholder value in the long-run, by meeting the needs of some key stakeholders instrumentally to achieve this purpose (Lueg et al. 2016; Mervelskemper and Streit 2017; Parrot and Tierney 2012). This theory seems consistent with the concept of integrated reporting assumed by the IIRC Framework (2013); nevertheless, some authors highlight the difficulty of going beyond a weak sustainability paradigm that privileges financial stakeholders (Alexander and Blum 2016; Brown and Dillard 2014; de Villiers et al. 2017a, b; Flower 2015; Reuter and Messner 2015; Stubbs and Higgins 2015; Thomson 2015; Tweedie and Martinov-Bennie 2015). Conversely, normative stakeholder theory supports the purpose of the firm to create and distribute value for all the stakeholders and, in a broader sense,

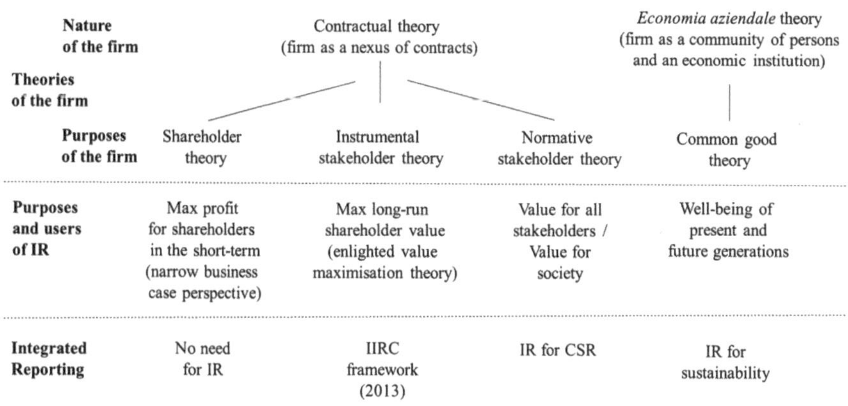

Fig. 2.3 Conceptual model on purposes and users of integrated reporting (first draft)

even for the society as a whole (Haller and van Staden 2014). Therefore, adopting this theoretical perspective, integrated reporting should be considered as a tool of Corporate Social Responsibility (CSR). However, both shareholder theory and (instrumental and normative) stakeholder theory assume the contractual theory, conceiving the firm as a 'nexus of contracts' between individuals or groups of individuals.

Given that the self-legitimating behaviour of corporations has a big responsibility in causing the ongoing global crisis, some scholars are advocating the central role of firms in creating value for sustainability, with the aim of ensuring the well-being of present and future generations. Our research highlights a great potential for the adoption of other theories of the firm, on which to ground the integrated reporting as a sustainability tool. In particular, we deem that it could be helpful to refer to common good theory, which can be seen as a broader development of stakeholder theory (Argandoña 1998). The notion of 'common good' is a universal in Catholic Social Teaching[3] (Wallace 1977) and a general condition to everyone's advantage (Rawls 1971; O'Brien 2009), which has been developed over centuries by Plato, Aristotle, St. Augustine, St. Thomas of Aquinas, St. Francis of Assisi, Locke, and many other philosophers and business ethicists. The common good has been defined as '*that order of society in which every member enjoys the possibility of realising his true self by participating in the effects of the cooperation of all*' (Messner 1965, p. 118). Created and shared by all community members, common good should be aimed at human flourishing (Del Baldo 2017; Finnis 2011; Melé 2002, 2012; Sison 2007).

The firm is much more than a mere 'nexus of contracts' focused on consumer preferences and shareholder value or, more generally, on the needs of stakeholders as 'contractors'. The firm oriented to common good is a 'community of persons' and an institution designed to meet real human needs. We could say that the firm should meet the needs of presently living human beings without compromising the ability of future people to meet their own needs, consistently with the definition of sustainable development provided by the Brundtland Report (UNWED 1987).

This notion of firm can draw upon the concept of *azienda*, developed by the best tradition of Italian *economia aziendale.*[4] An *azienda* has been defined as an 'economic coordination in action, instituted and run by human beings for the satisfaction of human needs' (Zappa 1920, 1929), giving a central emphasis to a complex and synergic system of interdependencies and relationships of varying natures between

[3]Catholic Social Teaching is the branch of Catholic Moral Theology concerned with social ethics and primarily focused upon the integral development of each human person. In his first Encyclical Letter, Pope Saint John Paul II explained the need to consider "*man in the full truth of his existence, of his personal being and also of his community and social being*" (Pope Saint John Paul II 1979, 14).

[4]"*Economia aziendale*, defined by its founder Gino Zappa as *'the science that focuses on the conditions of existence and the manifestations of life of aziende*" (Zappa 1927, p. 30), encompasses three disciplines—accounting, organisation and operations—in a unitary whole and a systemic vision.

several stakeholder-subjects and components (Signori and Rusconi 2009). Later, an *azienda* has been conceived as an 'economic institution designed to continue its existence through time' (Zappa 1957)[5] and as the 'tightly economic order of an institution' (Masini 1974). In other words, a group or community of persons (Mintzberg 2009) constitutes an *azienda* to carry out an economic activity aimed at satisfying human needs, thus serving the common good of its members and the broader common good of society (Costa and Ramus 2012). Indeed, having being influenced by Catholic Social Teaching since its foundation, the best tradition of *economia aziendale* is consistent with the notion of common good and provides many significant contributions for implementing common good in practice.

According to *economia aziendale*, which adopts a systems perspective, the concept of *azienda* is suitable for all types and sizes of organisations (private/public sector; profit/non-profit orientation; big/medium/small) and the *azienda* has to be accountable to the entire human community, with strong implications for corporate social responsibility and sustainability (Caldarelli et al. 2011).

Furthermore, the concept of *azienda* would better flourish within a context of 'civil economy', a model of socio-economic organisation that emphasises the value of community and reciprocity and promotes human and social well-being through a greater and more systematic interplay between State, market and civil society, according to the principle of circular subsidiarity. Originated in the Kingdom of Naples and the Parthenopean Republic in the second half of the eighteenth century (Genovesi 1965-70), for many historical reasons civil economy did not enter mainstream economic thought. Nevertheless, assuming a different anthropological perspective, civil economy offers a helpful paradigm to inspire new solutions for overcoming the limits and problems of global individualistic financial capitalism, making the socio-economic context more human and sustainable (see Bruni 2006; Bruni and Zamagni 2004, 2016; Zamagni 1998, 2013).

2.5 Conclusions

In this chapter, we undertook a critical narrative literature review, by starting from the intuition that it is essential to shed light on the different theoretical perspectives adopted by the academic literature on integrated reporting, with the aim of improving the awareness of this aspect and orienting integrated reporting consciously towards the desired purposes and users. We analysed the academic papers published in international accounting and finance journals from January 2000 to September 2017, by focusing on 66 papers that explicitly refer to one or more theories. We found that 30 different theories are utilised in our selection of papers and the most

[5]"*The system of azienda operates in conditions of long-lasting economic-financial balance. This is a condition necessary for the continuity of azienda as a basic economic unit, in other words for its survival as a cell of the general economic tissue*" (Paganelli 1976, p. 11).

used are stakeholder theory, institutional theory, legitimacy theory, impression management theory, and agency theory.

The originality and value added of our literature review consist in analysing a large number of papers that adopt different methodologies and theories, utilising literature databases and collections partially unexplored in previous reviews, and focusing on theoretical perspectives, with the aim of clarifying their implications for integrated reporting. Nevertheless, the main limitations of our work are related to the selection of literature databases and collections, the identification of keywords ('integrated report' and 'integrated reporting'), the choice of searching databases by keywords just in the article title, the availability of a limited number of free downloadable papers, the focus only on academic literature in English, and the setting of queries in Nvivo 11 software package.

We also outlined an early draft of a conceptual model to be developed as a comprehensive model to support future academic researches, practices, and policies.

This draft identifies four distinct perspectives on the nature and purposes of firms, and consequently on the nature and purposes of integrated reporting, according to four different theories: shareholder theory, instrumental stakeholder theory, normative stakeholder theory, and common good theory. Moreover, we highlight that all the other theories found in our literature review provide a valuable contribution to the research on integrated reporting.

Finally, this chapter intends to inspire both conceptual and empirical studies, by promoting interdisciplinary approach and openness to alternative theories to foster the ongoing debate on integrated reporting. In particular, further research could investigate how common good theory, *economia aziendale* theory, and the concrete application of civil economy can collaborate in achieving sustainability from a perspective of 'integral ecology' (Pope Francis 2015, 156). Indeed, we deem that it is worth trying to realise the dream of an integrated reporting as a sustainability tool that elaborates and communicates to the entire human community the system of values and other information summarising the contribution of an *azienda* to sustainable development and common good for the well-being of present and future generations.

References

Abeysekera, I. (2013). A template for integrated reporting. *Journal of Intellectual Capital, 14*(2), 227–245.

Adams, C. A. (2015). The international integrated reporting council: A call to action. *Critical Perspectives on Accounting, 27*, 23–28.

Adams, S., & Simnett, R. (2011). Integrated reporting: An opportunity for Australia's not-for-profit sector. *Australian Accounting Review, 21*(3), 292–301.

Adams, C. A., Potter, B., Singh, P. J., & York, J. (2016). Exploring the implications of integrated reporting for social investment (disclosures). *The British Accounting Review, 48*(3), 283–296.

Alexander, D., & Blum, V. (2016). Ecological economics: A Luhmannian analysis of integrated reporting. *Ecological Economics, 129*, 241–251.
Ambachtsheer, K. P. (2016). Pension organizations and integrated reporting. In K. P. Ambachtsheer (Ed.), *The future of pension management: Integrating design, governance, and investing* (pp. 87–92). Hoboken, NJ: Wiley.
Atkins, J., & Maroun, W. (2015). Integrated reporting in South Africa in 2012: Perspectives from South African institutional investors. *Meditari Accountancy Research, 23*(2), 197–221.
Baboukardos, D., & Rimmel, G. (2016). Value relevance of accounting information under an integrated reporting approach: A research note. *Journal of Accounting and Public Policy, 35*(4), 437–452.
Balashova, N. N., Šilerová, E., & Melikhov, V. A. (2015). Developing the methodology to form integrated reporting of Agroholdings in the Russian Federation. *Agris On-Line Papers in Economics and Informatics, 7*(4), 19–29.
Bartocci, L., & Picciaia, F. (2013). Towards integrated reporting in the public sector. In C. Busco, M. L. Frigo, A. Riccaboni, & P. Quattrone (Eds.), *Integrated reporting: Concepts and cases that redefine corporate accountability* (pp. 191–204). Cham: Springer.
Beck, C., Dumay, J., & Frost, G. (2017). In pursuit of a "single source of truth": From threatened legitimacy to integrated reporting. *Journal of Business Ethics, 141*(1), 191–205.
Bernardi, C., & Stark, A. W. (2016). Environmental, social and governance disclosure, integrated reporting, and the accuracy of analyst forecasts. *The British Accounting Review, 50*, 16–31.
Boonlua, S., & Phankasem, S. (2016). Engagement in integrated reporting: Evidence from the international integrating reporting council adoption framework. *Journal of Business and Retail Management Research, 10*(3), 126–136.
Bouten, L., & Hoozée, S. (2015). Challenges in sustainability and integrated reporting. *Issues in Accounting Education, 30*(4), 373–381.
Brown, J., & Dillard, J. (2014). Integrated reporting: On the need for broadening out and opening up. *Accounting, Auditing & Accountability Journal, 27*(7), 1120–1156.
Buitendag, N., Fortuin, G. S., & De Laan, A. (2017). Firm characteristics and excellence in integrated reporting. *South African Journal of Economic and Management Sciences, 20*(1), 1–8.
Burke, J. J., & Clark, C. E. (2016). The business case for integrated reporting: Insights from leading practitioners, regulators, and academics. *Business Horizons, 59*(3), 273–283.
Busco, C., Frigo, M. L., Quattrone, P., & Riccaboni, A. (2013a). Towards integrated reporting: Concepts, elements and principles. In C. Busco, M. L. Frigo, A. Riccaboni, & P. Quattrone (Eds.), *Integrated reporting: Concepts and cases that redefine corporate accountability* (pp. 3–18). Cham: Springer.
Busco, C., Frigo, M. L., Riccaboni, A., & Quattrone, P. (2013b). *Integrated reporting: Concepts and cases that redefine corporate accountability*. Cham: Springer.
Buys, P. W., & Van Niekerk, E. (2014). The South African financial services industry's integrated reporting compliance with the global reporting initiative framework. *Banks and Bank Systems, 9*(4), 107–115.
Caliskan, A. O., & Esen, E. (2016). Inseparable parts of sustainability: Business, climate change, and integrated reporting. In M. A. Gonzalez-Perez & L. Leonard (Eds.), *Climate change and the 2030 corporate agenda for sustainable development, advances in sustainability and environmental justice* (Vol. 19, pp. 25–43). Bingley: Emerald Group Publishing.
Carels, C., Maroun, W., & Padia, N. (2013). Integrated reporting in the South African mining sector. *Corporate Ownership and Control, 11*(1), 991–1005.
Cavazzoni, C., & Orlandi, F. (2013). The relationship between multinational enterprises and territory in the integrated reporting. In C. Busco, M. L. Frigo, A. Riccaboni, & P. Quattrone (Eds.), *Integrated reporting: Concepts and cases that redefine corporate accountability* (pp. 171–190). Cham: Springer.
Chaidali, P., & Jones, M. J. (2017). It's a matter of trust: Exploring the perceptions of integrated reporting preparers. *Critical Perspectives on Accounting, 48*, 1–20.

Cheng, M., Green, W., Conradie, P., Konishi, N., & Romi, A. (2014). The international integrated reporting framework: Key issues and future research opportunities. *Journal of International Financial Management & Accounting, 25*(1), 90–119.

Cho, C. H., & Roberts, R. W. (2010). Environmental reporting on the internet by America's toxic 100: Legitimacy and self-presentation. *International Journal of Accounting Information Systems, 11*(1), 1–16.

Churet, C., & Eccles, R. G. (2014). Integrated reporting, quality of management, and financial performance. *Journal of Applied Corporate Finance, 26*(1), 56–64.

Clayton, A. F., Rogerson, J. M., & Rampedi, I. (2015). Integrated reporting vs. sustainability reporting for corporate responsibility in South Africa. *Bulletin of Geography, 29*(29), 7–17.

Correa Ruiz, C. (2013). A commentary on "integrated reporting: A review of developments and their implications for the accounting curriculum". *Accounting Education, 22*(4), 360–362.

da Cunha Pinto, T., & Bandeira, A. M. (2013). Sustainability reporting and financial reporting: The relevance of an integrated reporting approach. In H. E. Muga & K. D. Thomas (Eds.), *Cases on the diffusion and adoption of sustainable development practices* (pp. 167–194). Hershey, PA: IGI Global.

de Villiers, C., Rinaldi, L., & Unerman, J. (2014). Integrated reporting: Insights, gaps and an agenda for future research. *Accounting, Auditing & Accountability Journal, 27*(7), 1042–1067.

de Villiers, C., Hsiao, P.-C. K., & Maroun, W. (2017a). Developing a conceptual model of influences around integrated reporting, new insights and directions for future research. *Meditari Accountancy Research, 25*, 450–460. https://doi.org/10.1108/MEDAR-07-2017-0183.

de Villiers, C., Venter, E.R., & Hsiao, P.-C. K. (2017b). Integrated Reporting: Background, Measurement Issues, Approaches and an Agenda for Future Research. Accounting & Finance, 57(4), 937-959.

Dey, C., & Burns, J. (2010). Integrated reporting at a Novo Nordisk. In A. Hopwood, J. Unerman, & J. Fries (Eds.), *Accounting for sustainability: Practical insights* (pp. 215–232). Oxford: Earthscan.

Dragu, I.-M., & Tiron-Tudor, A. (2013a). The integrated reporting initiative from an institutional perspective: Emergent factors. *Procedia - Social and Behavioral Sciences, 92*, 275–279.

Dragu, I.-M., & Tiron-Tudor, A. (2013b). GRI compliance and prerequisites of integrated reporting for Asian-Pacific companies. *Annales Universitatis Apulensis Series Oeconomica, 15*(2), 432–442.

Dragu, I.-M., & Tiron-Tudor, A. (2014). Research agenda on integrated reporting: New emergent theory and practice. *Procedia Economics and Finance, 15*, 221–227.

Dumay, J., Bernardi, C., Guthrie, J., & Demartini, P. (2016). Integrated reporting: A structured literature review. *Accounting Forum, 40*(3), 166–185.

Dumitru, M., Glavan, M. E., Gorgan, C., & Dumitru, V. F. (2013). International integrated reporting framework: A case study in the software industry. *Annales Universitatis Apulensis Series Oeconomica, 15*(1), 24–39.

Dumitru, M., Guşe, R. G., Feleagă, L., Mangiuc, D. M., & Feldioreanu, A. I. (2015). Marketing communications of value creation in sustainable organizations. The practice of integrated reports. *Amfiteatru Economic, 17*(40), 955–976.

Eccles, R. G., & Krzus, M. P. (2012a). *One report: Integrated reporting for a sustainable strategy*. Hoboken, NJ: Wiley.

Eccles, R. G., & Krzus, M. P. (2012b). United Technologies Corporation's first integrated report. In R. G. Eccles & M. P. Krzus (Eds.), *One report: Integrated reporting for a sustainable strategy* (pp. 29–50). Hoboken, NJ: Wiley.

Fasan, M. (2013). Annual reports, sustainability reports and integrated reports: Trends in corporate disclosure. In C. Busco, M. L. Frigo, A. Riccaboni, & P. Quattrone (Eds.), *Integrated reporting: Concepts and cases that redefine corporate accountability* (pp. 41–57). Cham: Springer.

Fasan, M., & Mio, C. (2017). Fostering stakeholder engagement: The role of materiality disclosure in integrated reporting. *Business Strategy and the Environment, 26*(3), 288–305.

Feng, T., Cummings, L., & Tweedie, D. (2017). Exploring integrated thinking in integrated reporting – An exploratory study in Australia. *Journal of Intellectual Capital, 18*(2), 330–353.

Fiori, G., Di Donato, F., & Izzo, M. F. (2016). Exploring the effects of corporate governance on voluntary disclosure: An explanatory study on the adoption of integrated report. In M. J. Epstein, F. Verbeeten, & S. K. Widener (Eds.), *Performance measurement and management control: Contemporary issues. Studies in managerial and financial accounting* (Vol. 31, pp. 83–108). Bingley: Emerald Group Publishing.

Flower, J. (2015). The international integrated reporting council: A story of failure. *Critical Perspectives on Accounting, 27*, 1–17.

Frías-Aceituno, J. V., Rodríguez-Ariza, L., & García-Sánchez, I. M. (2013a). Is integrated reporting determined by a country's legal system? An exploratory study. *Journal of Cleaner Production, 44*, 45–55.

Frías-Aceituno, J. V., Rodríguez-Ariza, L., & García-Sánchez, I. (2013b). The role of the board in the dissemination of integrated corporate social reporting. *Corporate Social Responsibility and Environmental Management, 20*(4), 219–233.

Frías-Aceituno, J. V., Rodríguez-Ariza, L., & García-Sánchez, I. M. (2014). Explanatory factors of integrated sustainability and financial reporting. *Business Strategy and the Environment, 23*(1), 56–72.

Garanina, T., & Dumay, J. (2017). Forward-looking intellectual capital disclosure in IPOs: Implications for intellectual capital and integrated reporting. *Journal of Intellectual Capital, 18*(1), 128–148.

García-Sánchez, I.-M., & Noguera-Gámez, L. (2017a). Integrated reporting and stakeholder engagement: The effect on information asymmetry. *Corporate Social Responsibility and Environmental Management, 24*, 395–413.

García-Sánchez, I.-M., & Noguera-Gámez, L. (2017b). Institutional investor protection pressures versus firm incentives in the disclosure of integrated reporting. *Australian Accounting Review, 28*, 199–219.

García-Sánchez, I.-M., Rodríguez-Ariza, L., & Frías-Aceituno, J.-V. (2013). The cultural system and integrated reporting. *International Business Review, 22*(5), 828–838.

Gelmini, L., Bavagnoli, F., Comoli, M., & Riva, P. (2015). Waiting for materiality in the context of integrated reporting: Theoretical challenges and preliminary empirical findings. In L. Songini & A. Pistoni (Eds.), *Sustainability disclosure: State of the art and new directions. Studies in managerial and financial accounting* (Vol. 30, pp. 135–163). Bingley: Emerald Group Publishing.

Gianfelici, C., Casadei, A., & Cembali, F. (2016). The relevance of nationality and industry for stakeholder salience: An investigation through integrated reports. *Journal of Business Ethics, 150*, 541–558.

Gonzalbez, J. M., & Rodriguez, M. M. (2012). XBRL and integrated reporting: The Spanish accounting association taxonomy approach. *The International Journal of Digital Accounting Research, 12*, 59–91.

Gunarathne, N., & Senaratne, S. (2017). Diffusion of integrated reporting in an emerging South Asian (SAARC) nation. *Managerial Auditing Journal, 32*(4/5), 524–548.

Gupta, S. K. (2015). Integrated reporting – Walk the talk. *The Management Accountant, 50*(5), 27–28.

Gupta, M. C. (2016). Disclosures in integrated reporting: A review. *Splint International Journal of Professionals, 3*(1), 38–40.

Haji, A. A., & Anifowose, M. (2016a). Audit committee and integrated reporting practice: Does internal assurance matter? *Managerial Auditing Journal, 31*(8/9), 915–948.

Haji, A. A., & Anifowose, M. (2016b). The trend of integrated reporting practice in South Africa: Ceremonial or substantive? *Sustainability Accounting, Management and Policy Journal, 7*(2), 190–224.

Haji, A. A., & Anifowose, M. (2017). Initial trends in corporate disclosures following the introduction of integrated reporting practice in South Africa. *Journal of Intellectual Capital, 18*(2), 373–399.
Haji, A. A., & Hossain, D. M. (2016). Exploring the implications of integrated reporting on organisational reporting practice: Evidence from highly regarded integrated reporters. *Qualitative Research in Accounting and Management, 13*(4), 415–444.
Haller, A., & van Staden, C. (2014). The value added statement – An appropriate instrument for integrated reporting. *Accounting, Auditing & Accountability Journal, 27*(7), 1190–1216.
Hamid, F. Z. A., Shafie, R., & Othman, Z. (2015). The "cognitive experience, area of business, responsiveness and engagement" conceptual framework for integrated reporting. *Advanced Science Letters, 21*(6), 1791–1793.
Havlová, K. (2015). What integrated reporting changed: The case study of early adopters. *Procedia Economics and Finance, 34*, 231–237.
Higgins, C., Stubbs, W., & Love, T. (2014). Walking the talk(s): Organisational narratives of integrated reporting. *Accounting, Auditing & Accountability Journal, 27*(7), 1090–1119.
Hindley, T., & Buys, P. W. (2012). Integrated reporting compliance with the global reporting initiative framework: An analysis of the South African mining industry. *The International Business & Economics Research Journal (Online), 11*(11), 1249–1260.
Hoque, M. E. (2017). Why company should adopt integrated reporting? *International Journal of Economics and Financial Issues, 7*(1), 241–248.
Institute of Directors Southern Africa. (2009). King code of governance for South Africa 2009. *SAICA Legislation Handbook 2010/2011*, vol. 3.
Ito, K., & Iijima, M. (2017). Integrated reporting and its impact on organisational change. *International Journal of Human Resources Development and Management, 17*(1/2), 73–88.
James, M. L. (2013a). Sustainability and integrated reporting: A case exploring issues, benefits and challenges. *Journal of the International Academy for Case Studies, 19*(7), 89–95.
James, M. L. (2013b). Sustainability and integrated reporting: Opportunities and strategies for small and midsize companies. *Entrepreneurial Executive, 18*, 17–28.
James, M. L. (2014). The benefits of sustainability and integrated reporting: An investigation of accounting majors' perceptions. *Journal of Legal, Ethical and Regulatory Issues, 17*(2), 93–113.
James, M. L. (2015). Accounting majors' perceptions of the advantages and disadvantages of sustainability and integrated reporting. *Journal of Legal, Ethical and Regulatory Issues, 18*(2), 107–123.
Jensen, J. C., & Berg, N. (2012). Determinants of traditional sustainability reporting versus integrated reporting. An institutionalist approach. *Business Strategy and the Environment, 21*(5), 299–316.
Jhunjhunwala, S. (2014). Beyond financial reporting-international integrated reporting framework. *Indian Journal of Corporate Governance, 7*(1), 73–80.
Kamp-Roelands, N. (2013). A commentary on "integrated reporting: A review of developments and their implications for the accounting curriculum". *Accounting Education, 22*(4), 357–359.
Kaspina, R. G., & Molotov, L. A. (2016). Corporate integrated reporting: An efficient tool of economic security management. *Academy of Strategic Management Journal, 15*(1), 225–232.
Kaspina, R. G., Molotov, L. A., & Kaspin, L. E. (2015). Cash flow forecasting as an element of integrated reporting: An empirical study. *Asian Social Science, 11*(11), 89–94.
Kiran, V. U., & Goud, M. M. (2015). A study of integrated reporting in Indian banks. *The Management Accountant, 50*(5), 29–33.
Knauer, A., & Serafeim, G. (2014). Attracting long-term investors through integrated thinking and reporting: A clinical study of a biopharmaceutical company. *Journal of Applied Corporate Finance, 26*(2), 57–64.
Köhler, K., & Hoffmann, C. P. (2017). Integrated reporting: Bridging investor relations and strategic management. In A. V. Laskin (Ed.), *The handbook of financial communication and investor relations* (pp. 209–219). Chichester: Wiley.

Kuzina, R. W. (2014). Integrated reporting as a mechanism of increasing business value. *Actual Problems of Economics, 158*(8), 385–392.
Lai, A., Melloni, G., & Stacchezzini, R. (2016). Corporate sustainable development: Is "integrated reporting" a legitimation strategy? *Business Strategy and the Environment, 25*(3), 165–177.
Lai, A., Melloni, G., & Stacchezzini, R. (2017). What does materiality mean to integrated reporting preparers? An empirical exploration. *Meditari Accountancy Research, 25*, 533–552.
Lambooy, T., Hordijk, R., & Bijveld, W. (2014). Communicating about integrating sustainability in corporate strategy: Motivations and regulatory environments of integrated reporting from a European and Dutch perspective. In R. Tench, W. Sun, & B. Jones (Eds.), *Communicating corporate social responsibility: Perspectives and practice. Critical studies on corporate responsibility, governance and sustainability* (Vol. 6, pp. 217–255). Bingley: Emerald Group Publishing.
Laptes, R., & Sofian, I. (2016). A new dimension of the entities' financial reporting: Integrated reporting. *Bulletin of the Transilvania University of Brasov. Economic Sciences. Series V, 9*(2), 239–250.
Lee, K.-W., & Yeo, G. H.-H. (2016). The association between integrated reporting and firm valuation. *Review of Quantitative Finance and Accounting, 47*(4), 1221–1250.
Lodhia, S. (2015). Exploring the transition to integrated reporting through a practice lens: An Australian customer owned bank perspective. *Journal of Business Ethics, 129*(3), 585–598.
Lodhia, S., & Stone, G. (2017). Integrated reporting in an internet and social media communication environment: Conceptual insights. *Australian Accounting Review, 27*(1), 17–33.
Lueg, K., Lueg, R., Andersen, K., & Dancianu, V. (2016). Integrated reporting with CSR practices: A pragmatic constructivist case study in a Danish cultural setting. *Corporate Communications: An International Journal, 21*(1), 20–35.
Maniora, J. (2017). Is integrated reporting really the superior mechanism for the integration of ethics into the core business model? An empirical analysis. *Journal of Business Ethics, 140*(4), 755–786.
Maroun, W. (2017). Assuring the integrated report: Insights and recommendations from auditors and preparers. *The British Accounting Review, 49*(3), 329–346.
McNair-Connolly, C. J., Silvi, R., & Bartolini, M. (2013). Integrated reporting and value-based cost management: A natural union. In C. Busco, M. L. Frigo, A. Riccaboni, & P. Quattrone (Eds.), *Integrated reporting: Concepts and cases that redefine corporate accountability* (pp. 147–157). Cham: Springer.
McNally, M.-A., Cerbone, D., & Maroun, W. (2017). Exploring the challenges of preparing an integrated report. *Meditari Accountancy Research, 25*, 481–504.
Melloni, G. (2015). Intellectual capital disclosure in integrated reporting: An impression management analysis. *Journal of Intellectual Capital, 16*(3), 661–680.
Melloni, G., Stacchezzini, R., & Lai, A. (2016). The tone of business model disclosure: An impression management analysis of the integrated reports. *Journal of Management and Governance, 20*(2), 295–320.
Melloni, G., Caglio, A., & Perego, P. (2017). Saying more with less? Disclosure conciseness, completeness and balance in integrated reports. *Journal of Accounting and Public Policy, 36*, 220–238.
Mervelskemper, L., & Streit, D. (2017). Enhancing market valuation of ESG performance: Is integrated reporting keeping its promise? *Business Strategy and the Environment, 26*(4), 536–549.
Meyrick, J. (2016). Telling the story of culture's value: Ideal-type analysis and integrated reporting. *The Journal of Arts Management Law and Society, 46*(4), 141–152.
Mio, C., & Fasan, M. (2014). Beyond financial reporting: A journey from sustainability towards integrated reporting. *Journal of Environmental Accounting and Management, 2*(3), 1–14.
Mio, C., Marco, F., & Pauluzzo, R. (2016). Internal application of IR principles: Generali's internal integrated reporting. *Journal of Cleaner Production, 139*, 204–218.

Mmako, N., & Van Rensburg, M. J. (2017). Towards integrated reporting: The inclusion of content elements of an integrated annual report in the chairmen's statements of JSE-listed companies. *South African Journal of Business Management, 48*(1), 45–54.
Morros, J. (2016). The integrated reporting: A presentation of the current state of art and aspects of integrated reporting that need further development. *Intangible Capital, 12*(1), 336–356.
Nathuramka, S. (2016). Integrated reporting – Accountants redefining the role of management. *The Management Accountant, 51*(11), 49–54.
Needles Jr., B. E. (2013). Capacity building and integrated reporting: A framework for development. *Independent Business Review, 6*(2), 1–25.
Needles Jr., B. E., Frigo, M. L., Powers, M., & Shigaev, A. (2016). Integrated reporting and sustainability reporting: An exploratory study of high performance companies. In M. J. Epstein, F. Verbeeten, & S. K. Widener (Eds.), *Performance measurement and management control: Contemporary issues. Studies in managerial and financial accounting* (Vol. 31, pp. 41–81). Bingley: Emerald Group Publishing.
Oprisor, T. (2015). Auditing integrated reports: Are there solutions to this puzzle? *Procedia Economics and Finance, 25*(15), 87–95.
Oshika, T., & Saka, C. (2017). Sustainability KPIs for integrated reporting. *Social Responsibility Journal, 13*(3), 625–642.
Owen, G. (2013a). A rejoinder to commentaries on "integrated reporting: A review of developments and their implications for the accounting curriculum". *Accounting Education, 22*(4), 363–365.
Owen, G. (2013b). Integrated reporting: A review of developments and their implications for the accounting curriculum. *Accounting Education, 22*(4), 340–356.
Parrot, K. W., & Tierney, B. X. (2012). Integrated reporting, stakeholder engagement, and balanced investing at American electric power. *Journal of Applied Corporate Finance, 24*(2), 27–37.
Paternostro, S. (2013). The connectivity of information for the integrated reporting. In C. Busco, M. L. Frigo, A. Riccaboni, & P. Quattrone (Eds.), *Integrated reporting: Concepts and cases that redefine corporate accountability* (pp. 59–77). Cham: Springer.
Pavlopoulos, A., Magnis, C., & Iatridis, G. E. (2017). Integrated reporting: Is it the last piece of the accounting disclosure puzzle? *Journal of Multinational Financial Management, 41*, 23–46.
Perego, P., Kennedy, S., & Whiteman, G. (2016). A lot of icing but little cake? Taking integrated reporting forward. *Journal of Cleaner Production, 136*, 53–64.
Pistoni, A., & Songini, L. (2015). New trends and directions in CSD: The integrated reporting. In L. Songini & A. Pistoni (Eds.), *Sustainability disclosure: State of the art and new directions. Studies in managerial and financial accounting* (Vol. 30, pp. 81–105). Bingley: Emerald Group Publishing.
Pozzoli, M., & Gesuele, B. (2016). From theory to practice: First adoption of integrated reporting by the Italian public utilities. In P. Spagnoletti, M. De Marco, N. Pouloudi, D. Te'eni, J. vom Brocke, R. Winter, & R. Baskerville (Eds.), *Lecture notes in information systems and organisation (LNISO)* (Vol. 14, pp. 121–132). Cham: Springer.
Pratama, A. (2017). Clustering Indonesian companies' annual reports: Preliminary assessment of the implementation of integrated reporting by Indonesian listed companies. *International Journal of Globalisation and Small Business, 9*(1), 46–54.
Pucci, S., Cenci, M., Tutino, M., & Luly, R. (2014). Intangible assets: Current requirements, social statements, integrated reporting, and new models. In M. Russ (Ed.), *Value creation, reporting, and signaling for human capital and human assets: Building the foundation for a multi-disciplinary, multi-level theory* (pp. 179–211). London: Palgrave Macmilian.
Rambaud, A., & Richard, J. (2015). The "triple depreciation line" instead of the "triple bottom line": Towards a genuine integrated reporting. *Critical Perspectives on Accounting, 33*, 92–116.
Rambe, P., & Mangara, T. B. (2016). Influence of integrated reporting ratings, CEO age, and years of experience on the share price of top 106 JSE listed companies. *Problems and Perspectives in Management, 14*(3/1), 216–231.

Ramin, K., & Reiman, C. (2013). *IFRS and XBRL: How to improve business reporting through technology and object tracking* (pp. 525–539). Hoboken, NJ: Wiley.

Reimsbach, D., Hahn, R., & Gürtürk, A. (2017). Integrated reporting and assurance of sustainability information: An experimental study on professional investors' information processing. *European Accounting Review, 27*, 559–581.

Rensburg, R., & Botha, E. (2014). Is integrated reporting the silver bullet of financial communication? A stakeholder perspective from South Africa. *Public Relations Review, 40*(2), 144–152.

Reuter, M., & Messner, M. (2015). Lobbying on the integrated reporting framework: An analysis of comment letters to the 2011 discussion paper of the IIRC. *Accounting, Auditing & Accountability Journal, 28*(3), 365–402.

Rivera-Arrubla, Y. A., & Zorio-Grima, A. (2016). Integrated reporting, connectivity, and social media. *Psychology & Marketing, 33*(12), 1159–1165.

Rivera-Arrubla, Y. A., Zorio-Grima, A., & García-Benau, M. A. (2017). Integrated reports: Disclosure level and explanatory factors. *Social Responsibility Journal, 13*(1), 155–176.

Robertson, F. A., & Samy, M. (2015). Factors affecting the diffusion of integrated reporting – A UK FTSE 100 perspective. *Sustainability Accounting, Management and Policy Journal, 6*(2), 190–223.

Ruiz-Lozano, M., & Tirado-Valencia, P. (2016). Do industrial companies respond to the guiding principles of the integrated reporting framework? A preliminary study on the first companies joined to the initiative. *Revista de Contabilidad, 19*(2), 252–260.

Serafeim, G. (2015). Integrated reporting and investor clientele. *Journal of Applied Corporate Finance, 27*(2), 34–51.

Setia, N., Abhayawansa, S., Joshi, M., & Huynh, A. V. (2015). Integrated reporting in South Africa: Some initial evidence. *Sustainability Accounting, Management and Policy Journal, 6*(3), 397–424.

Sierra-García, L., Zorio-Grima, A., & García-Benau, M. A. (2015). Stakeholder engagement, corporate social responsibility and integrated reporting: An exploratory study. *Corporate Social Responsibility and Environmental Management, 22*(5), 286–304.

Signori, S., & Rusconi, G. (2009). Ethical thinking in traditional Italian Economia Aziendale and the stakeholder management theory: The search for possible interactions. *Journal of Business Ethics, 89*, 303–318.

Simnett, R., & Huggins, A. L. (2015). Integrated reporting and assurance: Where can research add value? *Sustainability Accounting, Management and Policy Journal, 6*(1), 29–53.

Smith, S. S. (2016). Integrated reporting & the future of auditing. *Journal of Accounting and Finance, 16*(1), 140–145.

Sofian, I., & Dumitru, M. (2017). The compliance of the integrated reports issued by European financial companies with the international integrated reporting framework. *Sustainability (Switzerland), 9*(8), 1319.

Soyka, P. A. (2013). The international integrated reporting council (IIRC) integrated reporting framework: Toward better sustainability reporting and (way) beyond. *Environmental Quality Management, 23*(2), 1–14.

Stacchezzini, R., Melloni, G., & Lai, A. (2016). Sustainability management and reporting: The role of integrated reporting for communicating corporate sustainability management. *Journal of Cleaner Production, 136*, 102–110.

Stefanescu, A. C., Oprisor, T., & Sntejudeanu, M. A. (2016). An original assessment tool for transparency in the public sector based on the integrated reporting approach. *Accounting and Management Information Systems, 15*(3), 542–564.

Stent, W., & Dowler, T. (2015). Early assessments of the gap between integrated reporting and current corporate reporting. *Meditari Accountancy Research, 23*(1), 92–117.

Steyn, M. (2014). Organisational benefits and implementation challenges of mandatory integrated reporting: Perspectives of senior executives at South African listed companies. *Sustainability Accounting, Management and Policy Journal, 5*(4), 476–503.

Strong, P. T. (2015). Is integrated reporting a matter of public concern?: Evidence from Australia. *The Journal of Corporate Citizenship, 60*, 81–100.
Stubbs, W., & Higgins, C. (2014). Integrated reporting and internal mechanisms of change. *Accounting, Auditing & Accountability Journal, 27*(7), 1068–1089.
Stubbs, W., & Higgins, C. (2015). Stakeholders' perspectives on the role of regulatory reform in integrated reporting. *Journal of Business Ethics, 147*, 489–508.
Thomson, I. (2015). "But does sustainability need capitalism or an integrated report" a commentary on "the international integrated reporting council: A story of failure" by flower, J. *Critical Perspectives on Accounting, 27*, 18–22.
Trébucq, S., & Magnaghi, E. (2017). Using the EFQM excellence model for integrated reporting: A qualitative exploration and evaluation. *Research in International Business and Finance, 42*, 522–531.
Tudor-Tiron, A., & Dragu, I. (2014). From sustainability to integrated reporting – The political perspective of institutional theory. *Studia Universitatis Babes-Bolyai, 59*(2), 20–33.
Tweedie, D., & Martinov-Bennie, N. (2015). Entitlements and time: Integrated reporting's double-edged agenda. *Social and Environmental Accountability Journal, 35*(1), 49–61.
Tweedie, D., Nielsen, C., & Martinov-Bennie, N. (2017). The business model in integrated reporting: Evaluating concept and application. *Australian Accounting Review, 28*(3), 405–420.
Ünal, G., & Çoşkun, A. (2014). Disclosure for sustainability: The case of integrated reporting. In U. Akkucuk (Ed.), *Handbook of research on developing sustainable value in economics, finance, and marketing* (pp. 297–305). Hershey, PA: IGI Global.
van Bommel, K. (2014). Towards a legitimate compromise?: An exploration of integrated reporting in the Netherlands. *Accounting, Auditing & Accountability Journal, 27*(7), 1157–1189.
van Zyl, A. S. (2013). Sustainability and integrated reporting in the South African corporate sector. *The International Business & Economics Research Journal, 12*(8), 903–926.
Vaz, N., Fernandez-Feijoo, B., & Ruiz, S. (2016). Integrated reporting: An international overview. *Business Ethics, 25*(4), 577–591.
Velte, P., & Stawinoga, M. (2017). Integrated reporting: The current state of empirical research, limitations and future research implications. *Journal of Management Control, 28*, 275–320.
Veltri, S., & Silvestri, A. (2015). The Free State University integrated reporting: A critical consideration. *Journal of Intellectual Capital, 16*(2), 443–462.
Vorster, S., & Marais, C. (2014). Corporate governance, integrated reporting, and stakeholder management: A case study of Eskom. *African Journal of Business Ethics, 8*(2), 31–57.
Wild, S. (2011). Public sector accountability for cultural assets: An integrated reporting approach. *International Journal of Environmental, Cultural, Economic and Social Sustainability, 7*(5), 379–390.
Wolniak, R. (2013). The role of Grenelle II in corporate social responsibility integrated reporting. *Manager, 18*, 109–119.
Wulf, I., Niemöller, J., & Rentzsch, N. (2014). Development toward integrated reporting, and its impact on corporate governance: A two-dimensional approach to accounting with reference to the German two-tier system. *Journal of Management Control, 25*(2), 135–164.
Zappettini, F., & Unerman, J. (2016). "Mixing" and "bending": The recontextualisation of discourses of sustainability in integrated reporting. *Discourse & Communication, 10*(5), 521–542.
Zhou, S., Simnett, R., & Green, W. (2017). Does integrated reporting matter to the capital market? *Abacus, 53*(1), 94–132.

Further References

Aguilera, R. V., & Jackson, G. (2003). The cross-national diversity of corporate governance: Dimensions and determinants. *The Academy of Management Review, 28*(3), 447–465.

Argandoña, A. (1998). The stakeholder theory and the common good. *Journal of Business Ethics, 17*(9/10), 1093–1102.

Ashforth, B. E., & Gibbs, B. W. (1990). The double-edge of organizational legitimation. *Organization Science, 1*(2), 177–194.

Ax, C., & Bjørnenak, T. (2005). Bundling and diffusion of management accounting innovations – The case of the balanced scorecard in Sweden. *Management Accounting Research, 16*(1), 1–20.

Baiman, S., & Verrecchia, R. (1996). The relation among capital markets, financial disclosure, production efficiency, and insider trading. *Journal of Accounting Research, 34*(1), 1–22.

Barron, O. E., Kyle, C., & O'Keefe, T. B. (1999). MD&A quality as measured by the SEC and analysts' earnings forecasts. *Contemporary Accounting Research, 16*(1), 75–109.

Barth, M. E., Kasznik, R., & McNichols, M. F. (2001). Analyst coverage and intangible assets. *Journal of Accounting Research, 39*(1), 1–34.

Baumeister, R. F., & Leary, M. R. (1997). Writing narrative literature reviews. *Review of General Psychology, 1*(3), 311–320.

Beasley, M. S., Carcello, J. V., Hermanson, D. R., & Neal, T. L. (2009). The audit committee oversight process. *Contemporary Accounting Research, 26*(1), 65–122.

Bebbington, J., Larrinaga, C., & Moneva, J. M. (2008). Corporate social reporting and reputation risk management. *Accounting, Auditing & Accountability Journal, 21*(3), 337–361.

Berle, A. A., & Means, G. C. (1932). *The modern corporation and private property*. New York: Macmillan.

Berrone, P., Gelabert, L., & Fosfuri, A. (2009). *The impact of symbolic and substantive actions of environmental legitimacy*. Working Paper Business School, University of Navarra.

Beyer, A., Cohen, D. A., Lys, T. Z., & Walther, B. R. (2010). The financial reporting environment: Review of the recent literature. *Journal of Accounting and Economics, 50*(2/3), 296–343.

Boltanski, L., & Thévenot, L. (1999). The sociology of critical capacity. *European Journal of Social Theory, 2*(3), 359–377.

Boltanski, L., & Thévenot, L. (2006). *On justification: Economies of worth*. Princeton, NJ: Princeton University Press.

Bowen, R. M., Davis, A. K., & Matsumoto, D. A. (2005). Emphasis on pro forma versus GAAP earnings in quarterly press releases: Determinants, SEC intervention, and market reactions. *The Accounting Review, 80*(4), 1011–1038.

Brammer, S., Jackson, G., & Matten, D. (2012). Corporate social responsibility and institutional theory: New perspectives on private governance. *Socio-Economic Review, 10*(1), 3–28.

Brennan, N. M., & Solomon, J. (2008). Corporate governance, accountability and mechanisms of accountability: An overview. *Accounting, Auditing & Accountability Journal, 21*(7), 885–906.

Brennan, N. M., Encarna, G. S., & Pierce, A. (2009). Impression management: Developing and illustrating a scheme of analysis for narrative disclosures – A methodological note. *Accounting, Auditing & Accountability Journal, 22*(5/7), 789–832.

Bruni, L. (2006). *Civil happiness*. Abingdon: Routledge.

Bruni, L., & Zamagni, S. (2004). *Economia civile: Efficienza, equità, felicità pubblica* [Civil economy: Efficiency, equity, public happiness]. Bologna: Il Mulino.

Bruni, L., & Zamagni, S. (2016). *Civil economy: Another idea of the market*. Newcastle-Upon-Tyne: Agenda Publishing.

Caldarelli, A., Fiondella, C., Maffei, M., Spano, R., & Zagaria, C. (2011). The common good and economia aziendale theory: Insights for corporate social responsibility from the Italian perspective. *Journal of the Asia-Pacific Centre for Environmental Accountability, 17*(4), 197–216.

Campbell, J. L. (2007). Why would corporations behave in socially responsible ways? An institutional theory of corporate social responsibility. *Academy of Management Review, 32*(3), 946–967.

Chapman, C. S., Cooper, D. J., & Miller, P. (Eds.). (2009). *Accounting, organizations and institutions: Essays in honor of Anthony Hopwood*. New York: Oxford University Press.

Chu, C. I., Chatterjee, B., & Brown, A. (2013). The current status of greenhouse gas reporting by Chinese companies: A test of legitimacy theory. *Managerial Auditing Journal, 28*(2), 114–139.

Clarkson, M. E. (1995). A stakeholder framework for analyzing and evaluating corporate social performance. *Academy of Management Review, 20*(1), 92–117.
Clarkson, P. M., Li, Y., Richardson, G. D., & Vasvari, F. P. (2008). Revisiting the relation between environmental performance and environmental disclosure: An empirical analysis. *Accounting, Organizations and Society, 33*(4/5), 303–327.
Costa, E., & Ramus, T. (2012). The Italian Economia Aziendale and Catholic social teaching: How to apply the common good principle at the managerial level. *Journal of Business Ethics, 1*, 103–116.
Cronin, P., Ryan, F., & Coughlan, M. (2008). Undertaking a literature review: A step-by-step approach. *British Journal of Nursing, 17*(1), 38–43.
Cuganesan, S., Guthrie, J., & Ward, L. (2010). Examining CSR disclosure strategies within the Australian food and beverage industry. *Accounting Forum, 34*(3/4), 169–183.
Daft, R. L., & Lengel, R. H. (1984). Information richness: A new approach to managerial behavior and organizational design. *Research in Organizational Behavior, 6*, 191–233.
Daft, R. L., & Lengel, R. H. (1986). Organizational information requirements, media richness and structural design. *Management Science, 32*(5), 554–571.
Deegan, C. (2002). The legitimising effect of social and environmental disclosures: A theoretical foundation. *Accounting, Auditing & Accountability Journal, 15*(3), 282–311.
Deegan, C. (2006). *Financial accounting theory*. North Ryde: McGraw-Hill.
Deegan, C. (2009). *Financial accounting theory* (3rd ed.). Sydney: McGraw-Hill Australia Pty.
Deegan, C., Rankin, M., & Voght, P. (2000). Firms' disclosure reactions to major social incidents: Australian evidence. *Accounting Forum, 24*(1), 101–130.
Deegan, C., Rankin, M., & Tobin, J. (2002). An examination of the corporate social and environmental disclosures of BHP from 1983–1997. A test of legitimacy theory. *Accounting, Auditing & Accountability Journal, 15*(3), 312–343.
Del Baldo, M. (2017). Moral and virtues-based leadership for enhancing integral ecology. *Contributions to conflict management, peace, economics, and development, 26*, 203–228.
Delmas, M., & Toffel, M. W. (2004). Stakeholders and environmental management practices: An institutional framework. *Business Strategy and the Environment, 13*(4), 209–222.
Denney, A. S., & Tewksbury, R. (2013). How to write a literature review. *Journal of Criminal Justice Education, 24*(2), 218–234.
de Villiers, C., & Mähönen, J. (2014). Article 11: Integrated reporting or non-financial reporting? In B. Sjafjell & A. Wiesbrock (Eds.), *The greening of European business under EU law: Taking article 11 TFEU seriously* (pp. 118–143). Abingdon: Routledge.
de Villiers, C., & Sharma, U. (2017). A critical reflection on the future of financial, intellectual capital, sustainability and integrated reporting. *Critical Perspectives on Accounting*. https://doi.org/10.1016/j.cpa.2017.05.003.
Diamond, D. W., & Verrecchia, R. E. (1991). Disclosure, liquidity, and the cost of capital. *The Journal of Finance, 46*(4), 1325–1359.
Dimaggio, P. J., & Powell, W. W. (1983). The iron cage revisited: Institutional isomorphism and collective rationality in organizational fields. *American Sociological Review, 48*(2), 147–160.
Donaldson, T. (1999). Making stakeholder theory whole. *Academy of Management Review, 24*(2), 237–241.
Donaldson, T., & Preston, L. E. (1995). The stakeholder theory of the corporation: Concepts, evidence, and implications. *Academy of Management Review, 20*(1), 65–91.
Dowling, J., & Pfeffer, J. (1975). Organizational legitimacy: Social values and organizational behaviour. *Pacific Sociological Review, 18*(1), 122–136.
Eccles, R. G., & Krzus, M. P. (2014). *The integrated reporting movement: Meaning, momentum, motives, and materiality*. New York: Wiley.
Eccles, R. G., Krzus, M. P., & Ribot, S. (2015a). Meaning and momentum in the integrated reporting movement. *Journal of Applied Corporate Finance, 27*(2), 8–17.
Eccles, R. G., Krzus, M. P., & Ribot, S. (2015b). Models of best practice in integrated reporting 2015. *Journal of Applied Corporate Finance, 27*(2), 103–115.

Elliott, R. K., & Jacobson, P. D. (1994). Costs and benefits of business information disclosure. *Accounting Horizons, 8*(4), 80–96.

Enderle, G. (2004). The ethics of financial reporting. In G. Brenkert (Ed.), *Corporate integrity and accountability* (pp. 87–99). London: Sage Publications.

Evan, W. M., & Freeman, R. E. (1988). A stakeholder theory of the modern corporation: Kantian capitalism. In T. Beauchamp & N. Bowie (Eds.), *Ethical theory and business* (pp. 75–93). Englewood Cliffs, NJ: Prentice Hall.

Fama, E. F., & Jensen, M. C. (1983). Separation of ownership and control. *Journal of Law and Economics, 26*(2), 301–326.

Fernando, S., & Lawrence, S. (2014). A theoretical framework for CSR practices: Integrating legitimacy theory, stakeholder theory and institutional theory. *Journal of Theoretical Accounting Research, 10*(1), 149–178.

Finnis, J. M. (2011). *Human rights and common good: Collected essays*. New York: Oxford University Press.

Freeman, R. E. (1984). *Strategic management: A stakeholder approach*. Boston, MA: Pitman Publishing.

Freeman, R. E., & Reed, D. L. (1983). Stockholders and stakeholders: A new perspective on corporate governance. *California Management Review, 25*(3), 88–106.

Freeman, R. E., Harrison, J. S., & Wicks, A. C. (2007). *Managing for stakeholders: Survival, reputation, and success*. New Haven, CT: Yale University Press.

Freeman, R. E., Harrison, J. S., Wicks, A. C., Parmar, B. L., & De Colle, S. (2010). *Stakeholder theory: The state of the art*. Cambridge: Cambridge University Press.

Genovesi, A. (1965-70). *Lezioni di economia civile* [Lessons of civil economy]. Naples: Cugini Pomba.

Georgiou, G. (2002). Corporate non-participation in the ASB standard-setting process. *European Accounting Review, 11*(4), 699–722.

Georgiou, G. (2004). Corporate lobbying on accounting standards: Methods, timing and perceived effectiveness. *Abacus, 40*(2), 219–237.

Gray, R. (2013). Back to basics: What do we mean by environmental (and social) accounting and what is it for? A reaction to Thornton. *Critical Perspectives on Accounting, 24*(6), 459–468.

Gray, R. H., Owen, D. L., & Maunders, K. T. (1987). *Corporate social reporting: Accounting & Accountability*. Hemel Hempstead: Prentice-Hall.

Gray, R. H., Kouhy, R., & Lavers, S. (1995). Corporate social and environmental reporting: A review of the literature and a longitudinal study of UK disclosure. *Accounting, Auditing & Accountability Journal, 8*(2), 47–77.

Gray, R., Owen, D., & Adams, C. (1996). *Accounting and accountability: Changes and challenges in corporate social and environmental reporting*. London: Prentice Hall.

Green, L., Ottoson, J., Garcia, C., & Hiatt, R. (2009). Diffusion theory and knowledge dissemination, utilization, and integration in public health. *Annual Review of Public Health, 30*, 151–174.

Greenley, G. E., Hooley, G. J., Broderick, A. J., & Rud, J. M. (2004). Strategic planning differences among different multiple stakeholder orientation profile. *Journal of Strategic Management, 12*(3), 163–182.

Guthrie, J., & Parker, L. D. (1990). Corporate social disclosure practice: A comparative international analysis. *Advances in Public Interest Accounting, 3*(1), 159–175.

Hart, C. (1998). *Doing a literature review*. London: Sage Publications.

Healy, P. M., & Palepu, K. G. (2001). Information asymmetry, corporate disclosure, and the capital markets: A review of the empirical disclosure literature. *Journal of Accounting and Economics, 31*(1/3), 405–440.

Hill, C. W., & Jones, T. M. (1992). Stakeholder agency theory. *Journal of Management Studies, 29*(2), 131–154.

Hodge, F. D., Hopkins, P. E., & Wood, D. A. (2010). The effects of financial statement information proximity and feedback on cash flow forecasts. *Contemporary Accounting Research, 27*(1), 101–133.

Hofstede, G. H. (2001). *Culture's consequences, comparing values, behaviors, institutions, and organizations across nations*. Thousand Oaks, CA: Sage Publications.

Hofstede, G., & Hofstede, G. J. (2005). *Cultures and organizations: Software of the mind* (2nd ed.). New York: McGraw Hill.

Hooghiemstra, R. (2000). Corporate communication and impression management – New perspectives why companies engage in corporate social reporting. *Journal of Business Ethics, 27*(1/2), 55–68.

Hope, O. K. (2003). Disclosure practices, enforcement of accounting standards, and analysts' forecast accuracy: An international study. *Journal of Accounting Research, 41*(2), 235–272.

Hopwood, A. (2009). Accounting and the environment. *Accounting, Organisations and Society, 34*(3/4), 433–439.

Hrasky, S. (2012). Carbon footprints and legitimation strategies: Symbolism or action? *Accounting. Auditing & Accountability Journal, 5*(1), 174–198.

IIRC. (2011). *Towards integrated reporting. Communicating value in the 21st century*. International Integrated Reporting Council. http://integratedreporting.org/wp-content/uploads/2011/09/IR-Discussion-Paper-2011_spreads.pdf

IIRC. (2013). *The international IR framework*. International Integrated Reporting Council. http://integratedreporting.org/resource/international-ir-framework/

Jackson, G., & Apostolakou, A. (2010). Corporate social responsibility in Western Europe: An institutional mirror or substitute? *Journal of Business Ethics, 94*(3), 371–394.

Jakobsen, M., Johanson, I.-L., & Nørreklit, H. (2011). *An Actor's approach to management: Conceptual framework and company practices*. Copenhagen: DJØF.

Jensen, M. (1986). Agency costs of free cash flow, corporate finance, and takeovers. *The American Economic Review, 76*(2), 323–329.

Jensen, M. C. (2001). Value maximization, stakeholder theory, and the corporate objective function. *Journal of Applied Corporate Finance, 14*(3), 8–21.

Jensen, M. C., & Meckling, W. H. (1976). Theory of the firm: Managerial behavior, agency costs and ownership structure. *Journal of Financial Economics, 3*(4), 305–360.

Jesson, J., & Lacey, F. (2006). How to do (or not to do) a critical literature review. *Pharmacy Education, 6*(2), 139–148.

Jones, T., & Wicks, A. (1999). Convergent stakeholder theory. *Academy of Management Review, 24*(2), 206–221.

Kiel, G. C., & Nicholson, G. J. (2003). Board composition and corporate governance: How the Australian experience informs contrasting theories of corporate governance. *Corporate Governance, 11*(3), 189–205.

Killian, S., & O'Regan, P. (2016). Social accounting and the cocreation of corporate legitimacy. *Accounting, Organizations and Society, 50*, 1–12.

Kim, J., Bach, S., & Clelland, I. (2007). Symbolic or behavioural management? Corporate reputation in high-emission industries. *Corporate Reputation Review, 10*(2), 77–98.

Lambert, R., Leuz, C., & Verrecchia, R. E. (2007). Accounting information, disclosure, and the cost of capital. *Journal of Accounting Research, 45*(2), 385–420.

Lang, M., & Lundholm, R. (1993). Cross-sectional determinants of analyst ratings of corporate disclosures. *Journal of Accounting Research, 31*(2), 246–271.

Laplume, A. O., Sonpar, K., & Litz, R. A. (2008). Stakeholder theory: Reviewing a theory that moves US. *Journal of Management, 34*(6), 1152–1189.

Lapsley, I., & Wright, E. (2004). The diffusion of management accounting innovations in the public sector: A research agenda. *Management Accounting Research, 15*(3), 355–374.

Lárran, M., & Giner, B. (2002). The use of the internet for corporate reporting by Spanish companies. *International Journal of Digital Accounting Research, 2*(1), 53–82.

Leary, M. R., & Kowalski, R. M. (1990). Impression management: A literature review and two component model. *Psychological Bulletin, 107*(1), 34–47.

Lindblom, C. K. (1994). *The implications of organizational legitimacy for corporate social performance and disclosure*. Paper presented at the Critical Perspectives on Accounting Conference, New York.

Lodhia, S. (2010). Research methods for analysing web based sustainability communication. *Social and Environmental Accountability Journal, 30*(1), 26–36.

Luhmann, N. (1995). *Social systems*. Stanford, CA: Stanford University Press (originally published in German in 1984).

Mahoney, L. S., Thorne, L., Cecil, L., & Lagore, W. (2013). A research note on standalone corporate social responsibility reports: Signalling or greenwashing? *Critical Perspectives on Accounting, 24*(4/5), 350–359.

Maines, L. A., & McDaniel, L. S. (2000). Effects of comprehensive-income characteristics on nonprofessional investors' judgments: The role of financial-statement presentation format. *The Accounting Review, 75*(2), 179–207.

Masini, C. (1974). *Lavoro e Risparmio* [Work and Savings]. Turin: UTET.

Matten, D., & Moon, J. (2008). 'Implicit' and 'explicit' CSR: A conceptual framework for a comparative understanding of corporate social responsibility. *Academy of Management Review, 33*(2), 404–424.

Maunders, K. T. (1985). The decision relevance of value added reports. In F. Choi & G. G. Mueller (Eds.), *Frontiers of international accounting: An anthology* (pp. 225–245). Ann Arbor, MI: UMI Research Press.

McWilliams, A., & Siegel, D. (2001). Corporate social responsibility: A theory of the firm perspective. *The Academy of Management Review, 26*(1), 117–127.

Melé, D. (2002). Not only stakeholders interests: The firm oriented towards the common good. In S. A. Cortright & M. J. Naughton (Eds.), *Rethinking the purpose of business, interdisciplinary essays from Catholic social tradition* (pp. 190–214). Notre Dame: Notre Dame University Press.

Melé, D. (2012). The firm as a "community of persons": A pillar of humanistic business ethos. *Journal of Business Ethics, 106*(1), 89–101.

Mellett, H., Marriott, N., & MacNiven, L. (2009). Diffusion of an accounting innovation: Fixed asset accounting in the NHS in Wales. *European Accounting Review, 18*(4), 745–764.

Merkl-Davies, D. M., & Brennan, N. M. (2007). Discretionary disclosure strategies in corporate narratives: Incremental information or impression management? *Journal of Accounting Literature, 26*, 116–196.

Merkl-Davies, D. M., Brennan, N. M., & McLeay, S. J. (2011). Impression management and retrospective sense-making in corporate narratives: A social psychology perspective. *Accounting, Auditing & Accountability Journal, 24*(3), 315–344.

Messner, J. (1965). *Social ethics: Natural law in the Western World*. Trans, by Doherty J. J. St. Louis, MI: B. Herder Book.

Meyer, J. W., & Rowan, B. (1977). Institutionalized organizations: Formal structures as myth and ceremony. *The American Journal of Sociology, 83*(2), 340–363.

Michelon, G., & Parbonetti, A. (2012). The effect of corporate governance on sustainability disclosure. *Journal of Management and Governance, 16*(3), 477–509.

Mintzberg, H. (2009). L'azienda come comunità. *Harvard Business Review Italia, 9*, 94–99.

Mitchell, R. K., Angle, B. R., & Wood, D. J. (1997). Toward a theory of stakeholder identification and salience: Defining the principle of who and what really counts. *Academy of Management Review, 22*(4), 853–886.

Neu, D. (1991). Trust, impression management and the auditing profession. *Critical Perspectives on Accounting, 2*(4), 295–313.

Neu, D., Warsame, H., & Pedwell, K. (1998). Managing public impressions: Environmental disclosures in annual reports. *Accounting, Organizations and Society, 23*(3), 265–282.

Neville, B. A., & Menguc, B. (2006). Stakeholder multiplicity: Toward an understanding of the interactions between stakeholders. *Journal of Business Ethics, 66*(4), 377–391.

O'Brien, T. (2009). Reconsidering the common good in a business context. *Journal of Business Ethics, 85*(1), 25–37.
O'Donovan, G. (2002). Environmental disclosures in the annual report: Extending the applicability and predictive power of legitimacy theory. *Accounting, Auditing & Accountability Journal, 15*(3), 344–371.
Ohlson, J. A. (1995). Earnings, book values and dividends in equity valuation. *Contemporary Accounting Research, 11*(2), 661–687.
Oliver, C. (1991). Strategic responses to institutional processes. *The Academy of Management Review, 16*(1), 145–179.
Orlitzky, M., Schmidt, F. L., & Rynes, S. L. (2003). Corporate social and financial performance: A meta-analysis. *Organization Studies, 24*(3), 403–441.
Owen, D. (2008). Chronicles of wasted time? A personal reflection on the current state of, and future prospects for, social and environmental accounting research. *Accounting, Auditing & Accountability Journal, 21*(2), 240–267.
Paganelli, O. (1976). *Il sistema aziendale* [The system of azienda]. Bologna: Clueb.
Parker, L. D. (2005). Social and environmental accounting research: A view from the commentary box. *Accounting, Auditing & Accountability Journal, 18*(6), 824–860.
Perrini, F., & Tencati, A. (2006). Sustainability and stakeholder management: The need for new corporate performance evaluation and reporting systems. *Business strategy & the Environment, 15*(5), 296–308.
Plumlee, M. A. (2003). The effect of information complexity on analysts' use of that information. *The Accounting Review, 78*(1), 275–296.
Poole, M., & Van de Ven, A. (2004). *Handbook of organisational change and innovation.* New York: Oxford University Press.
Pope Francis. (2015). *Laudato sì* [Praised Be]. Encyclical Letter. Rome: Libreria Editrice Vaticana.
Pope Saint John Paul II. (1979). *Redemptor Hominis* [The Redeemer of Man]. Encyclical Letter. Rome: Libreria Editrice Vaticana.
Puro, M. (1984). Audit firm lobbying before the financial accounting standards board: An empirical study. *Journal of Accounting Research, 22*(2), 624–646.
Rawls, J. (1971). *A theory of justice*. Cambridge, MA: Harvard University Press.
Riess, M., Rosenfeld, P., Melburg, V., & Tedeschi, J. T. (1981). Self-serving attributions: Biased private perceptions and distorted public descriptions. *Journal of Personality and Social Psychology, 41*(2), 224–231.
Roberts, R. W. (1992). Determinants of corporate social responsibility disclosure: An application of stakeholder theory. *Accounting, Organizations and Society, 17*(6), 595–612.
Rogers, E. M. (2003). *Diffusion of innovations* (5th ed.). New York: Free Press.
Schatzki, T. R. (2001). Introduction, practice theory. In T. R. Schatzki, K. Knorr-Cetina, & E. von Savigny (Eds.), *The practice turn in contemporary theory* (pp. 1–14). London: Routledge.
Schatzki, T. R. (2002). *The site of the social: A philosophical account of the constitution of social life and change*. University Park, PA: Pennsylvania State University Press.
Schatzki, T. R. (2005). Peripheral vision: The sites of organizations. *Organisation Studies, 26*(3), 465–484.
Schlenker, B. R. (1980). *Impression management: The self-concept, social identity, and interpersonal relations*. Monterey, CA: Brooks-Cole Publishing.
Schneider, D. J. (1981). *Tactical self-presentations: Toward a broader conception.* New York: Academic.
Schreuder, H. (1979). Corporate social reporting in the federal republic of Germany: An overview. *Accounting, Organizations and Society, 4*(1/2), 109–122.
Searle, J. R. (1995). *The construction of social reality*. New York: Simon & Schuster.
Searle, J. R. (2010). *Making the social world: The structure of human civilization*. Oxford: Oxford University Press.

Sen, S., Bhattacharya, C., & Korschun, D. (2006). The role of corporate social responsibility in strengthening multiple stakeholder relationships: A field experiment. *Journal of the Academy of Marketing Science, 34*(2), 158–166.
Senge, P. M. (1997). The fifth discipline. *Measuring Business Excellence, 1*(3), 46–51.
Shaoul, J. (1998). Critical financial analysis and accounting for stakeholders. *Critical Perspectives on Accounting, 9*(2), 235–249.
Shocker, A., & Sethi, P. (1973). An approach to incorporating societal preferences in developing corporate action strategies. *California Management Review, 15*(4), 97–105.
Simpson, S., Fischer, B. D., & Rhode, M. (2013). The conscious capitalism philosophy pay off: A qualitative and financial analysis of conscious capitalism corporations. *Journal of Leadership, Accountability & Ethics, 10*(4), 19–29.
Singaraju, S. P., Nguyen, Q. A., Niininen, O., & Sullivan-Mort, G. (2016). Social media and value co-creation in multistakeholder systems: A resource integration approach. *Industrial Marketing Management, 54*, 44–55.
Sison, A. J. G. (2007). Towards a common good theory of the firm: The Tasubinsa case. *Journal of Business Ethics, 74*(4), 471–480.
Solomon, J., & Maroun, W. (2012). *Integrated reporting: The influence of king III on social, ethical and environmental reporting*. London: The Association of Chartered Certified Accountants.
Soobaroyen, T., & Ntim, C. G. (2013). Social and environmental accounting as symbolic and substantive means of legitimation: The case of HIV/AIDS reporting in South Africa. *Accounting Forum, 37*(2), 92–109.
Stone, G. (2011). Let's talk: Adapting accountants' communications to small business managers' objectives and preferences. *Accounting, Auditing, & Accountability Journal, 24*(6), 781–809.
Stone, G. (2012). The effectiveness of newsletters in accountants' client relations with small business managers: An Australian qualitative study. *Qualitative Research in Accounting and Management, 9*(1), 21–43.
Suchman, M. C. (1995). Managing legitimacy: Strategic and institutional approaches. *The Academy of Management Review, 20*(3), 571–610.
Sutton, T. G. (1984). Lobbying of accounting standard-setting in the UK and the USA: A downsian analysis. *Accounting, Organizations and Society, 9*(1), 81–95.
Sztompka, P. (1999). *Trust: A sociological theory*. Cambridge: Cambridge University Press.
Tremblay, M.-S., & Gendron, Y. (2011). Governance prescriptions under trial: On the interplay between the logics of resistance and compliance in audit committees. *Critical Perspectives on Accounting, 22*(3), 259–272.
UNWED. (1987). *Our Common Future*. Report of the World Commission on Environment and Development. http://www.un-documents.net/wced-ocf.htm
Van Wee, B., & Banister, D. (2016). How to write a literature review paper? *Transport Reviews, 36*(2), 278–288.
Verrecchia, R. E. (1983). Discretionary disclosure. *Journal of Accounting and Economics, 5*(1), 179–194.
von Bertalanffy, L. (1969). *General system theory*. New York: George Braziller.
Vourvachis, P., Woodward, T., Woodward, D. G., & Patten, D. M. (2016). CSR disclosure in response to major airline accidents: A legitimacy-based exploration. *Sustainability Accounting, Management and Policy Journal, 7*(1), 22–43.
Wallace, W. A. (1977). *The elements of philosophy: A compendium for philosophers and theologians*. New York: Alba House.
Watson, A., Shrives, P., & Marston, C. (2002). Voluntary disclosure of accounting rations in the UK. *The British Accounting Review, 34*(4), 289–313.
Watts, R., & Zimmerman, J. (1978). Towards a positive theory of the determination of accounting standards. *The Accounting Review, 53*(1), 112–134.
Whittington, R. (2006). Completing the practice turn in strategy research. *Organization Studies, 27*(5), 613–634.

Whittington, R. (2011). The practice turn in organization research: Towards a disciplined trans-disciplinarity. *Accounting, Organisations and Society, 36*(3), 183–186.
Wickens, C. D., & Carswell, C. M. (1995). The proximity compatibility principle: Its psychological foundation and relevance to display design. Human factors. *The Journal of the Human Factors and Ergonomics Society, 37*(3), 473–494.
Wild, S., & van Staden, C. (2013). *Integrated Reporting: Initial analysis of early reporters – An institutional theory approach.* Paper presented at the Seventh Asia Pacific Interdisciplinary Research in Accounting Conference.
Zamagni, S. (Ed.) (1998). *Nonprofit come Economia Civile* [Nonprofit as Civil Economy]. Bologna: Il Mulino.
Zamagni, S. (2013). *Impresa Responsabile e Mercato Civile* [Responsible Firm and Civil Market]. Bologna: Il Mulino.
Zappa, G. (1920-29). *La Determinazione del Reddito nelle Imprese Commerciali. I Valori di Conto in Relazione alla Formazione dei Bilanci* [The Determination of Income in Commercial Enterprises. Accounting Values in Relation to the Formation of Financial Statements]. Rome: Anonima Libreria Romana.
Zappa, G. (1927). *Tendenze Nuove negli Studi di Ragioneria* [New Trends in Accounting]. Opening Speech of the Academic Year 1926–1927 University of Venice. Milan: Istituto Editoriale Scientifico.
Zappa, G. (1957). *Le Produzioni nell'Economia delle Imprese* [The Productions in the Economy of the Firms]. Milan: Giuffrè.

Chapter 3
Theoretical Insights on Integrated Reporting: Valuing the Financial, Social and Sustainability Disclosures

Mark Anthony Camilleri

3.1 Introduction

Academic research is proliferating on the documentation and analysis of non-financial reporting. Many studies are clearly indicating how non-financial reporting and disclosures relating to environmental, social and governance (ESG) performance are increasingly becoming the norm for global multinational corporations (Ioannou and Serafeim 2016; Camilleri 2015a, b; Idowu et al. 2013; Aerts and Cormier 2009). The corporate financial reporting consists of backward-looking financial information (Beck et al. 2017; Camilleri 2017, 2018; Crowther 2016). However, such reported content may not necessarily reveal the whole picture of the organisations' performance (Burritt and Schaltegger 2010). This issue has inevitably led to the development of the integrated reporting guidelines (Adams and Frost 2008; Bhimani and Langfield-Smith 2007). Hedberg and Von Malmborg (2003) reported that there were companies who were already using the GRI guidelines to enhance their visibility and control of their triple bottom line at the corporate level. Subsequently, IIRC has recently formalised its guidelines on financial and non-financial disclosures. Its International Framework for Integrated Reporting <IR> has been promoted as a solution to the shortcomings in corporate financial reporting (Dumay et al. 2016; Cheng et al. 2014; IIRC 2013). Hence, integrated reporting sought to offer a broad picture of the modern organisations by shifting away from stand-alone financial statements, sustainability or social responsibility reports, towards a document that communicated a holistic picture of their value creating activities.

Gone are the days where financial performance could be considered as the only measure of a company's worth. With a wide plethora of possible disclosure formats, the integrated reports have bridged the gap by including non-financial information

M. A. Camilleri (✉)
University of Malta, Msida, Malta
e-mail: Mark.A.Camilleri@um.edu.mt

S. O. Idowu, M. Del Baldo (eds.), *Integrated Reporting*, CSR, Sustainability, Ethics & Governance, https://doi.org/10.1007/978-3-030-01719-4_3

that is relevant to communicating corporate strategy. <IR> combines financial and non-financial disclosures of the organisations' performance in one statement. Therefore, <IR>'s 'integrated thinking' stimulates the businesses and other entities to think about how they could generate value for themselves and for society. The purpose of this framework is to raise awareness on guiding principles and relevant content that could be featured in corporate reports, and to explain the fundamental concepts that underpin them (Dumay et al. 2016). Additionally, the adoption of <IR> is further expected to tackle a number of problems that were (and are still) evident in conventional, stand-alone sustainability reports, such as; the failure to account for all sources of value creation, the complex interconnections between sustainability and financial performance, and the communication of a company's business model (Eccles and Krzus 2010; Eccles et al. 2011).

As integrated reporting focuses on long-term performance from various perspectives; it offers a great opportunity for companies to instill greater confidence among stakeholders as they become more accountable and transparent (Stacchezzini et al. 2016; Lozano and Huisingh 2011). <IR> commends that its integrated reports ought to be concise, reliable and complete, in all material matters, both positive and negative, in a balanced way and without error or bias. <IR>'s online site provides links to exemplary organisations who are already using its guiding framework (IR 2017). The ease of access to IIRC's public data provides scholars with greater knowledge and understanding of the benefits and costs that are associated with the <IR> Framework. Hence, the future development of <IR> should be informed by stakeholder engagement with practitioners and a systematic, rigorous research approach that gathers experts' opinions on the implications for standard setting bodies, report preparers and their users.

This chapter addresses a research gap in academia along two lines of investigation. Firstly, it examines the International Integrated Reporting Council's <IR> guiding principles and content elements. Secondly, it links <IR> with key theoretical underpinnings. The author suggests that the agency, stewardship and institutional theories have contributed to the development of the <IR> field (Ioannou and Serafeim 2012; Brammer et al. 2012; Muth and Donaldson 1998; Davis et al. 1997; Scott 1995; Ness and Mirza 1991; DiMaggio and Powell 1991; Eisenhardt 1989). Indeed, prior research has used legitimacy theory (Beck et al. 2015; Deegan 2002; Suchman 1995) to interpret corporate reporting practices, but it had also focused on disclosure content as it considered the perceived users of corporate reports. Thus, this chapter adds value to the extant literature by exploring the emergence of non-financial reporting within a broader legitimation strategy (Idowu et al. 2013). The author contends that the concepts of isomorphism (Dacin 1997; Deephouse 1996) and isopraxism (Adams et al. 2016) elucidate our interpretation on why corporate reporting approaches are (or are not) converging toward integrated reporting.

The rationale for this contribution is to better understand how the <IR> framework improves the transparency and accountability of financial, social and sustainability disclosures. This research explores *the theoretical insights from social sciences and links key conceptual developments with the integrated reporting of financial and non-financial disclosures.* This study critically appraises <IR>'s framework and discusses about its potential pitfalls and challenges.

3.2 The Conceptual Developments: Paving the Way for Integrated Reporting

The <IR> Framework's broader view of value creation and its multiple capital concept calls for an enhanced stewardship of the organisations' capitals; whilst promoting a better understanding of the interdependencies between the capitals (IIRC 2013, p.8). Relevant theoretical perspectives as well as sound empirical research suggest that the practicing organisations' underlying motive behind their non-financial disclosures is to maximise their financial capital and profit. This argumentation is synonymous with many conceptual theories in academic literature that seek to justify the rationale for voluntary, integrated reporting (Adams et al. 2016; Idowu et al. 2013; Deegan 2002; Suchman 1995; Scott 1995; Eisenhardt 1989).

3.3 The Agency Theory

In the twentieth century, corporations were clearly distinguishing the difference between ownership and control of wealth. The business owners were considered as principals as they employed executives (agents) to manage their firms. The latter executives acted as agents for the principals, and they were morally responsible to maximise their shareholders' wealth (i.e. the prinicipals' wealth). The executives have accepted their agents' status because they perceived the opportunity to maximise their own utility. The agency theory suggests that the company executives and their principals are motivated by opportunities for their own personal gain (Eisenhardt 1989). Rightly so, the principals may invest their wealth in profitable companies and design governance systems in ways that maximise their investments. On the other hand, agents accept the responsibility of managing their principals' undertakings to secure their employment prospects.

However, at times, there may be interest divergence between the managers and their principals. There may be situations where the agents may feel constrained by their principals' imposed structures and controlling mechanisms (Davis et al. 1997). This matter could lead to unproductivity outcomes and will ultimately bring significant losses to the principals themselves. In the event where the agent would have no discretion at all, the firm would be owner-managed. In this case, having a situation where principals are autocratic towards their agents could result in serious repercussions for the businesses' prospects. The crux of the agency theory is that principals are expected to delegate authority to agents to act on their behalf (Ness and Mirza 1991). It is this delegation that at times allows agents to opportunistically build their own utility at the expense of the principals' utility. This happens when there are unaligned objectives; where managers may be motivated by their individualistic, self-serving goals, rather than being stewards for their principals (Eisenhardt 1989).

3.4 The Stewardship Theory

The stewardship theory is the collective-serving model of behaviour that is driven by the organisations' intrinsic values and a genuine desire to do what is best for society and the planet (Donaldson and Davis 1991). The stewardship behaviours benefit principals through the positive effects of profits on corporate dividends and share prices. Consequently, the stewards place higher value on cooperation than defection (these terms are also found in the game theory), because they perceive greater utility in cooperative behaviours. Stewardship theorists assume that there is a strong relationship between successful organisations and their principals' satisfaction. The stewards protect and maximise their shareholders' wealth because by so doing, they maximize their utility functions toward principals.

Stewards who successfully improve their organisational performance will also satisfy other stakeholder groups who have their own vested interests. Therefore, pro-organisational stewards are motivated to maximise organisational performance, whilst satisfying the competing interests of shareholders. The utility that they gain from pro-organisational behaviours is higher than the utility that could be gained through individualistic, self-serving behaviours. This theory suggests that stewards believe that their interests are aligned with those of the corporation that engaged them (Muth and Donaldson 1998). Ideally, the stewards ought to be committed to improve their organisational performance rather than satisfying their personal motivations. This theory's ideals are closely aligned with <IR>'s principles for value creation. IIRC's <IR> Framework emphasises the stewardship of multiple capitals, including; financial, manufactured, intellectual, human, social and natural capital. In the past, the accountability of social and environmental capitals has often been found to be completely lacking in financial reporting (Adams et al. 2016; Muth and Donaldson 1998). In addition, some anecdotal evidence suggests that companies are not always presenting a true and fair view of their negative impacts. On the other hand, there are other organisations who may be reluctant to promote their responsible and sustainable behaviours. This may be due to a lack of awareness on the business case for such activities. The motivations for undertaking stewardship behaviours, including; material ESG initiatives (that may be reported within integrated reports) seem to fall into two increasingly converging camps: doing good practices (this is consistent with the predictions of the stewardship theory) or doing well (this is consistent with both institutional and legitimacy theories).

3.5 The Institutional Theory

Different components of the institutional theory explain how certain processes become established as authoritative guidelines for societal behaviours. Very often, structures and institutions are created, diffused, adopted, and adapted over space and time; and eventually they may also fall into decline and disuse. Unlike the efficiency-based theories which focus on profit maximisation or on the interactions between markets and governments, the institutional theory considers a wider range of variables that could influence the decision-making processes in organisations.

This theory clarifies how firms respond to their institutional environments in which they operate. Stakeholders, including; governments, regulatory authorities, non-governmental organisations (NGOs), and organisations within the supply chain can exert their influence on any business. Scott (1995) held that, in order to survive, organisations must conform to norms and rules that are prevailing in their operating environment. Their compliance with the institutions' formal regulations and ethos will earn them legitimacy among stakeholders (Beck et al. 2015; Dacin 1997; Deephouse 1996; Suchman 1995). The institutional theory's applications have expanded even further; as more research is showing how the institutions effect organisational behaviours, particularly on CSR issues. Historically, the notion of CSR has emerged from the institutionalised forms of social solidarity from liberal market economies. The institutional theory offers promising ways of investigating what lies at the heart of the publics' concern. Therefore, corporations may be influenced by the institutions' voluntary principles, policies and programmes. Their responsible behaviours have often been triggered by socio-political forces and pressure groups. In this case, CSR practice rests on the dichotomy between the corporations' voluntary engagement and their socially binding responsibilities (Brammer et al. 2012). The fact that CSR is 'voluntary' is a clear reflection of the practicing organisations' institutional context. Alternatively, CSR may be driven by legal, customary, religious or other defined institutions.

Undoubtedly, numerous institutions have played a dynamic role, both individually and collectively in the development of integrated reporting. While governments have been the primary force for the promotion of financial reporting standards through security exchange commissions; other institutions like IIRC or GRI have facilitated the growth and diffusion of ESG reporting among practicing organisations. For the time being, it may appear that there is a demand for CSR reporting mechanisms by marketplace stakeholders. For this reason, corporations are communicating their ESG credentials (Camilleri 2015a). This way, they are accountable and transparent about their modus operandi with regulators, industry, and stakeholder groups. Moreover, the corporations' continuous engagement with external institutions, particularly multi-governmental organisations, social and environmental NGOs as well as the standard-setting organisations have brought valuable principles and guidelines in the realms of sustainability reporting (Camilleri 2015a).

Isomorphism has been constructed in conjunction with the applications of the institutional theory (Erlingsdóttir and Lindberg 2005; Dacin 1997; DiMaggio and Powell 1991). This concept has largely been propagated through global cultural and associational processes. Isomorphic developments arise when ideas or innovations travel and are adopted in different contexts (Harding 2012; Dacin 1997; Deephouse 1996). For instance, despite all possible configurations of local economic forces, power relationships, and forms of traditional culture it might consist of, a previously-isolated island society that has made contact with the rest of the globe would quickly take on standardised forms that are similar to a hundred other nation-states around the world (Meyer et al. 1997). Similarly, the notion of isopraxism refers to ideas that are translated and modified by different actors to suit their own needs.

Isomorphism and its related notion, isopraxism are potentially helpful for framing our interpretation of why corporate reporting approaches may converge (or not) over time. For example, the principles-based and non-mandatory <IR> Framework could potentially create explicit and implicit reporting norms that shape the non-financial information of organisations that ought to be communicated through their integrated reporting. In this sense, isomorphism may be useful to understand how and why the disclosures of ESG content can become widely accepted across companies, over time (Adams et al. 2016; Deephouse 1996). In a similar vein, isopraxism has been used to describe instances where identifiable institutional forces lead to new and different actions within specific organisational and social instances. Therefore, isopraxism suggests that organisations may be intrigued to move toward more integrated approaches to reporting. At times, legitimate organisations may be willing to voluntarily disclose their adapted ESG reports, out of their own volition. However, they may not necessarily label them 'integrated', or join the IIRC's <IR> Framework (Erlingsdóttir and Lindberg 2005; Harding 2012).

3.6 The Legitimacy Theory

Very often, the institutional environments provide regulatory frameworks and may be considered as a considerable breath of narratives pertaining to non-financial disclosures, in different jurisdictions. Hence, there is a possibility that responsible organisations will become legitimate if they comply with relevant societal rules that are found in the countries where they operate (Beck et al. 2015; Deegan 2002). The stakeholders perceive that organisations are legitimate when "their actions are desirable, proper, or appropriate within some socially-constructed system of norms, values, beliefs, and definitions" (Suchman 1995, p. 574). This conception suggests that the role of the legitimacy theory is to justify the organisations' behaviour, particularly when they implement and develop social and environmental initiatives. It goes without saying that the stakeholders will recognise those legitimate organisations that are upholding their social contract in accordance with the expectations of society. Therefore, the drivers of institutional legitimacy may be influenced by the organisations' external environment; according to the culturally-defined values and beliefs. On the other hand, stakeholders will severely sanction irresponsible organisations when they do not respect social norms and ethical values.

Suchman (1995) described legitimacy as an operational resource assuming a "high level of managerial control over legitimating processes" (p. 576). Others suggested that legitimacy is strategic as it emanates from recurring conflicts between management and stakeholders (Dacin et al. 2007; Suchman 1995). Organisational legitimacy could be achieved by forging strong relationships with external stakeholders (Camilleri 2017). For this reason, organisations may decide to change and adapt their corporate disclosures according to their stakeholders' expectations to achieve legitimacy. On the other hand, changes in disclosure patterns may be driven by internal decisions on materiality. Corporate reporting could be considered as a mitigating factor that is driven from inside the organisation (Campbell and Cornelia Beck 2004). Therefore, the managers' agenda is to strategically enhance their legitimacy through stakeholder engagement. They may also make financial and ESG disclosures widely available to interested parties to achieve legitimation. This position is consistent with <IR>'s framework. Within this context, the <IR> framework provides significant support to organisations who are willing to disclose their non-financial reports. However, when organisations utilise IIRC's framework for their very first time, they may inevitably have to adapt their financial and ESG reports as per <IR>'s recommended framework. Hence, <IR>'s reporting guidelines provide a passive avenue for institutional legitimisation. It is through the development of such guiding principles that society and external stakeholders are continuously influencing organisations to restore their ethical and social disclosures (Campbell and Cornelia Beck 2004).

The conditions for legitimacy are often constructed by responsible organisational behaviours. For example, relevant research on the legitimacy theory reported that there were organisations who were voluntarily disclosing their non-financial reports. Companies were seeking external legitimation by reporting their environmental performance (Brown and Deegan 1998). Other corporations who decided to follow GRI's reporting guidelines or resorted to the <IR>'s framework were increasingly aligning their internal reflections with external outputs (Beck et al. 2015). Initially, the rationale behind their integrated reporting was to improve their organisations' external legitimation among stakeholders. However, at a later stage they realised that their external reports were informed by their organisation's strategic positioning, and not constrained by the promulgation of the voluntary guidelines (Beck et al. 2015). Evidently, more organisations are conforming to the prevailing definitions of legitimacy through their disclosures of responsible and sustainable actions. Consequently, these responsible organisations' leadership sets the agenda for stakeholder engagement and ESG reporting. The underlying objective is to build or enhance reputation (Aerts and Cormier 2009) that will positively impact on the organisations' capital flows.

3.7 An Appraisal of Integrated Reporting

In the aftermath of the global economic and financial crisis of 2007–2008, many policy makers, regulatory authorities and leading financial institutions were striving in their endeavours to improve their corporate reporting mechanisms (Crowther 2016). At the time, there was an increased awareness on how ESG issues could help to improve corporate reputation and economic value (Camilleri 2017). The ethical behaviours in financial reporting is often equated with the obligation of companies to disclose a true and fair view of their organisational performance (Maniora 2015; Simnett and Huggins 2015). Organisations are accountable and transparent to their stakeholders when they report their financial as well as non-financial information to their stakeholders. PWC (2015) also suggested that CEOs believed that measuring and reporting total non-financial impacts contributes to long term success. Although, historical financial statements are essential in corporate reporting, particularly for compliance purposes, their disclosures do not always provide meaningful information regarding business value (ACCA 2016). Therefore, the communication on value creation should be the next step in the evolution of corporate reporting. The <IR> framework is one of the internationally-accepted frameworks that provides guidance on the reporting of financial information and ESG issues and on how their integrated reporting may lead to the creation of value over the short, medium and long term (IR 2017; de Villiers et al. 2014; Adams and Simnett 2011; Adams and Larrinaga-González 2007). The council has promoted the concept of integrated thinking and reporting as its <IR> framework includes metrics with the aim of communicating all activities and outputs that could potentially create value to the organisations' capitals. The <IR> Framework categorises different stocks of value, including; Financial Capital; Manufactured Capital; Intellectual Capital; Human Capital; Social (and Relationship) Capital; as well as Natural Capital. IIRC has aligned these capital allocations and corporate behaviours with its wider goals of financial stability and sustainable development.

Whilst the end of year financial statements focuses on financial capital, IIRC (2013) has paved the way for the next generation of annual reports that enable stakeholders to make a more informed assessment of the organisation's strategy, governance, performance and prospects. The <IR> Framework relies on resources—such as the expertise of people, intellectual property that was developed through research and development, and interactions with the environment and the societies in which they operate, along with its financial metrics. From this perspective, IIRC's guidelines were developed to address value creating activities that were missing from the corporations' disclosures. Hence, IIRC's (2013) <IR> framework established 'Guiding Principles' and 'Content Elements'. Its content is summarised in Table 3.1.

The 'Guiding Principles' underpin the preparation of an integrated report, whilst, the 'Content Elements' are the key categories of information that ought to be included in an integrated report (according to the <IR> Framework). There are no benchmarking on the above matters and the report is primarily aimed at the private sector; however <IR> could also be adapted to the public sector and to not-for-profit organisations. IIRC has set out a principle-based framework rather than specifying a detailed disclosure and measurement standard. This way each company sets out its own report rather than adopting a checklist approach. Hence, the report acts as a platform which explains what creates value to the business and how management protects this value. This gives the report more business impetus rather than mandating compliance-led approaches.

For the time being, the integrated reporting is not going to replace other forms of reporting but the vision is that large undertakings, including corporations, state-owned entities and government agencies, among others, may be expected to pull together relevant information to explain the key drivers of their non-financial performance. The disclosed information will only be included in the corporate report

Table 3.1 A summary of the <IR> framework's guiding principles and content elements

Guiding principles
• Strategic focus and future orientation—gives an insight of the organisation's strategy;
• Connectivity of information—provides a holistic picture of the combination, inter-relatedness and dependencies between the factors that affect the organisation's ability to create value over time;
• Stakeholder relationships—describes the nature and quality of the organisation's relationships with its key stakeholders;
• Materiality—discloses relevant information about matters that substantively affect the organisation's ability to create value over the short, medium and long term;
• Conciseness—provides sufficient context to understand the organisation's strategy, governance and prospects without being burdened by less relevant information;
• Reliability and completeness—includes all material matters, both positive and negative, in a balanced way and without material error;
• Consistency and comparability—ensures consistency over time and enabling comparisons with other organisations to the extent material to the organisation's own ability to create value.
Content elements
• Organisational overview and external environment—What does the organisation do and what are the circumstances under which it operates?
• Governance—How does an organisation's governance structure support its ability to create value in the short, medium and long term?
• Business model—What is the organisation's business model?
• Risks and opportunities—What are the specific risk and opportunities that affect the organisation's ability to create value over the short, medium and long term, and how is the organisation dealing with them?
• Strategy and resource allocation—Where does the organisation want to go and how does it intend to get there?
• Performance—To what extent has the organisation achieved its strategic objectives for the period and what are its outcomes in terms of effects on the capitals?
• Outlook—What challenges and uncertainties is the organisation likely to encounter in pursuing its strategy, and what are the potential implications for its business model and future performance?
• Basis of preparation and presentation—How does the organization determine what matters to include in the integrated report and how are such matters quantified or evaluated?

Adapted from IIRC (2013)

if it is material for the stakeholders. The term 'materiality' suggests that there are legal connotations that may be related to non-financial reporting. Yet, a few entities, out of their own volition, are already including ESG information in their integrated reports (Adams and Larrinaga-González 2007). In a nutshell, these organisations aim to provide a good insight into their resources, their relationships as they explain how they interact with their external environment to create value for themselves and toward society.

The development of the <IR> framework has brought significant improvements in terms of reporting and integration of financial and non-financial information. The IIRC has developed its very own <IR> framework following multi-stakeholder discussions with international financial accounting standard setters, institutional investors, providers of voluntary guidelines for corporate responsibility disclosures, the national accounting bodies and NGOs, among others. Therefore, IIRC represents key stakeholders that are poised to change the existing duality in corporate reporting of traditionally financial reports and discreet non-financial reporting. Initially, non-financial reporting was part of the Management Report and were filed within the corporations' annual reports. Back in the 1990s, there was the first spike of spurious social responsibility disclosures. Today, many companies are publishing their elaborated CSR Reports, Sustainability Reports, Corporate Citizenship Reports, Creating Shared Value Reports, and the like. These ESG reports have a lengthy tradition in voluntary reporting. The need for more comparable disclosures has led to the development of the reports which integrate financial and non-financial information (Adams et al. 2016; Adams and Frost 2008). The scope of the integrated reporting is to provide a more holistic picture of an entity that encompasses financial and ESG information. IIRC created a globally accepted <IR> framework that elicits material information from organisations about their strategy, governance, performance and prospects in a clear, concise and comparable format. This framework has accelerated the evolution of integrated thinking in corporate reporting (Perego et al. 2016). It has brought interesting developments in financial and ESG disclosures for the benefit of stakeholders, including institutional investors.

Yet, the reporting of non-financial performance remains voluntary. The reporting organisations are frequently providing an incomplete picture of their activities; even when they adopt non-financial guidelines (Adams et al. 2016). This implies that their reporting function ought to be subjected to external acceptance by an externally-recognised framework. It may appear that IIRC is still addressing this potential deficiency as it has provided a decent framework with clear definitions of target audiences, financial resource providers, et cetera (IIRC 2013, 2017). Arguably, it may appear that the standardisation of integrated reporting practices limits confusion as organisations disclose relevant and material information that is easily comparable to the other entities' disclosures. Perhaps, for the time being, the greatest challenge for report bearers is to identify which content should be incorporated within the integrated report. This is the reason why it is important to define the target groups of the report (Adams et al. 2016; Parent and Deephouse 2007). Prospective non-financial reporting that is based on <IR> framework could provide a single source document that gives a good snapshot of both financial and non-financial data. Debatably, while this framework is a significant development for the corporate disclosures of both financial and ESG matters, it has emerged following fruitful and collaborative debates among regulatory institutions, accounting bodies and interested parties.

3.8 Potential Tensions for the Development of Integrated Reporting

Many academic commentators have criticised the development of integrated reporting (Perego et al. 2016). Brown and Dillard (2014) argued that "IR remains an ideologically-closed approach that is more likely to reinforce rather than encourage critical reflection on ‘business as usual’ practices” (p. 1120). In a similar vein, Flower (2015) had voiced serious concerns about IIRC’s approach to sustainability. The author held that the <IR> framework focuses on investors rather than stakeholders, society and the natural environment. These reproaches emphasise that there are some relevant critiques on integrated reporting. Moreover, the accountancy profession has undue power over the institutional processes that were expected to deliver a fundamental shift in framing corporate reporting and sustainability accounting practices (Crowther 2016; Flower 2015; Adams and Larrinaga-González 2007). For instance, Deegan (2007) maintained that environmental reporting is designed to repair organisational legitimacy. Whilst, Adams and Larrinaga-González (2007) held that sustainability accounting is being carried out to conform with institutionalised norms. In a similar way, Hopwood (2009) pointed out that environmental accounting protected the firm’s inner workings from external views. Thus, few papers have attempted to assess the consequences (costs and benefits) of integrated reporting (Stacchezzini et al. 2016; Stubbs and Higgins 2014; O’Dwyer 2003). Previously, Neu et al. (1998) had admitted that environmental disclosures might advance the corporate image; in the absence of corresponding engagement. However, they also contended that socially responsible behaviours and their accounting are not necessarily concomitant. This issue could possibly limit the implementation of integrative reporting.

Stubbs and Higgins (2014) argued that the <IR> framework focuses on the ‘supply side’, namely, the preparers of integrated reporting whilst leaving out the ‘demand side’, i.e. the users’ perspectives on integrated reporting. These authors explored how Australian providers of financial capital were interpreting IIRC’s <IR> framework. They concluded that there was a significant gap between the information that was supplied by reporting companies and information that was sought by the financial markets. The authors also claimed that IIRC’s six capital model was not acknowledged by the Australian investors. Similarly, Perego et al. (2016) hinted that the users of standalone sustainability reports were adjusting their bad ESG valuations to the level of integrated (financial and sustainability) report users. However, they also suggested that none of the standalone reports users were adjusting their valuations following the corporate disclosures about good ESG performance. They concluded that the report preparers made different value judgments when anchoring the effects of ESG information (Perego et al. 2016).

3.9 Discussion and Conclusions

In simple terms, an integrated report is a single report that combines the financial and non-financial disclosures. However, a thorough literature review suggests that the integrated report is more than just a summary of financial, social and sustainability information in corporate disclosures (Aluchna and Idowu 2017; Idowu et al. 2013). The integrated reports constitute a full picture of a company’s overall business performance. Organisations are looking at all aspects of their value-creating capitals, including; financial; manufactured; intellectual; human; social (and relationship); as well as natural capitals (IR 2013). These capitals complement and compete against each other. Therefore, the practitioners who would like to comply with IIRC's <IR> framework will probably experience a dynamic process of adaptation, learning and action to redesign their corporate reporting. They may have to change their internal management systems (Churet and Eccles 2014; Eccles and Krzus 2010), processes, strategies to incorporate ESG issues into their core business model.

Relevant academic literature has yielded many recommendations, ideas, concepts, and other types of inspiration that have surely improved corporate reporting (Crowther 2016). This contribution reported how "integrated thinking" in corporate reporting involves the inclusion of material information on financial and non-financial matters (Adams and Simnett 2011). Moreover, the organisations' integrated reporting could be linked to the conceptual developments in the stewardship, institutional and legitimacy theories, among others. This chapter has indicated that these theoretical insights have focused on the rationale for the inclusion of non-financial information in corporate disclosures (Adams et al. 2016; Eccles and Krzus 2010). Although, there are reasonable arguments in favour and against integrated reporting; the researcher believes that the IIRC's <IR> framework has proved to be a useful instrument for many exemplary organisations who were recognised for their integrated reporting (IIRC 2017). The <IR> framework contains guiding principles and content elements that will enable organisations to disclose a true and fair view of their holistic activities. Conversely, the avoidance of ESG disclosures from their corporate reports can result in a highly-distorted picture of current and future business activities (Camilleri 2017).

This chapter has indicated how the theoretical insights from academic literature have led to the development of integrated reporting. It explained that the organisations' stewardship behaviours, including their 'integrated thinking' can help them improve their legitimacy among stakeholders and institutions. The researcher contended that IIRC's <IR> framework supports organisations in their holistic reporting approaches as it takes into account material information on financial, manufactured, intellectual, human, social and natural capitals.

3.10 Implications for Practitioners

Indeed, the IIRC's <IR> framework was a recent development in corporate reporting. This framework has its inherent limitations that were duly pointed out in this chapter. However this contribution maintains that integrated reporting provides a road map for those organisations who would like to pursue the sustainability path (Dacin et al. 2007). The <IR> framework is based on the general notion that integrated accounting considers both financial and non-financial information to give a true and fair view of the company's overall business performance. When practitioners embed ESG disclosures and "integrated thinking" they help to catalyse behavioural change in integrated reporting (Adams and Simnett 2011). This integrated thinking influences the practitioners' ethical behaviours and their stance on financial and non-financial performance (Camilleri 2015b). The <IR> framework's strategic focus calls for both internalisation and externalisation processes. Internalisation is a process through which human resources adopt the <IR> framework's external ideas, opinions, views or concepts as their own. This process starts with learning what this guiding principle is all about and why its development makes sense to businesses themselves. The internal stakeholders will probably experience a process of adaptation until they finally accept that integrated reporting creates value over time. Thus, internalisation can be understood as a process of acceptance of a new set of norms and working practices.

In the context of integrated reporting, organisations' internal transformation leads to changes in terms of the embeddedness of ESG performance in their operational processes. The responsible organisations commit themselves to disclose externalities that affect stakeholders and other unrelated parties. In other words, through integrated reporting; the internal effects of integrated reporting are finally externalised outside the organisations' boundaries. At times, organisations may intentionally or unintentionally conceal ESG information from stakeholders. Certain unethical practices may result from conscious or unconscious organisational behaviours or simply from misconduct when dealing with extensive information outputs. In this light, <IR> framework is a step in the right direction as integrated reporting leads to the re-evaluation of corporate ethics (The IoDSA and the King Committee, 2009). This framework calls for sustainability accounting of ESG issues. Hence, the IIRC's <IR> framework could be considered as a driver for the creation of an ethically-minded corporate culture. Practicing organisations will therefore bind themselves to disclose material information about financial and ESG matters, both positive and negative that substantively affect their ability to create value over the short, medium and long term. They are also expected to provide sufficient context about their strategy, governance and prospects in a balanced way and without material error (Camilleri 2017). The implementation of integrated reporting represents the companies' ethical obligation in their role of corporate citizens, as they provide a more holistic picture of the company's performance across departments (Eccles and Krzus 2010). Moreover, practitioners could ensure consistency over time as their integrated reporting could facilitate comparisons with other organisations.

3.11 Future Research

This chapter has addressed a gap in the literature as it linked relevant theoretical developments to better understand the rationale for integrated reporting in today's era. Academic literature has often relied on limited publicly available datasets on the diffusion of integrated reporting (Perego et al. 2016). Moreover, past contributions may have only focused on the 'supply side' pertaining to the <IR> framework; without investigating in much depth and breadth which organisational processes are crucial for integrated reporting (Simnett and Huggins 2015; de Villiers et al. 2014). It may appear that to date, there is still scant evidence about the strengths and weaknesses that are associated with the implementation of integrated reporting (Perego et al. 2016; Cheng et al. 2014). Therefore, empirical studies could explore how internal performance measurements and their disclosures may impact integrated reporting. Future qualitative research could identify which organisational processes and practices ought to be reported in integrated disclosures. Undoubtedly, there is potential for further studies that examine the content of integrated reports. Other research might investigate the usefulness of the organisations' integrated approaches to reporting financial and ESG information. In conclusion, the researcher posits that this construct is still emerging among academia; its conceptual development at this early stage is valuable because it exposes how the subject of financial, social and sustainability accounting is evolving among academia and practitioners.

References

ACCA. (2016). *Insights into integrated reporting*. Accessed April 19, 2017, from http://integratedreporting.org/wp-content/uploads/2017/04/ACCAInsights_into_Integrated_Reporting.pdf

Adams, C. A., & Frost, G. R. (2008). Integrating sustainability reporting into management practices. *Accounting Forum, 32*(4), 288–302.

Adams, C. A., & Larrinaga-González, C. (2007). Engaging with organisations in pursuit of improved sustainability accounting and performance. *Accounting, Auditing & Accountability Journal, 20*(3), 333–355.

Adams, S., & Simnett, R. (2011). Integrated reporting: An opportunity for Australia's not-for-profit sector. *Australian Accounting Review, 21*(3), 292–301.

Adams, C. A., Potter, B., Singh, P. J., & York, J. (2016). Exploring the implications of integrated reporting for social investment (disclosures). *British Accounting Review, 48*(3), 283–296.

Aerts, W., & Cormier, D. (2009). Media legitimacy and corporate environmental communication. *Accounting, Organizations and Society, 34*(1), 1–27.

Aluchna, M., & Idowu, S. O. (2017). Responsible corporate governance: An introduction. In *Responsible corporate governance* (pp. 1–7). Heidelberg: Springer.

Beck, C., Dumay, J., & Frost, G. (2015). In pursuit of a 'single source of truth': From threatened legitimacy to integrated reporting. *Journal of Business Ethics, 141*, 1–15.

Beck, C., Dumay, J., & Frost, G. (2017). In pursuit of a 'single source of truth': From threatened legitimacy to integrated reporting. *Journal of Business Ethics, 141*(1), 191–205.

Bhimani, A., & Langfield-Smith, K. (2007). Structure, formality and the importance of financial and non-financial information in strategy development and implementation. *Management Accounting Research, 18*(1), 3–31.

Brammer, S., Jackson, G., & Matten, D. (2012). Corporate social responsibility and institutional theory: New perspectives on private governance. *Socio-Economic Review, 10*(1), 3–28.

Brown, N., & Deegan, C. (1998). The public disclosure of environmental performance information—A dual test of media agenda setting theory and legitimacy theory. *Accounting and Business Research, 29*(1), 21–41.

Brown, J., & Dillard, J. (2014). Integrated reporting: On the need for broadening out and opening up. *Accounting, Auditing & Accountability Journal, 27*(7), 1120–1156.

Burritt, R. L., & Schaltegger, S. (2010). Sustainability accounting and reporting: Fad or trend? *Accounting, Auditing & Accountability Journal, 23*(7), 829–846.

Camilleri, M. A. (2015a). Valuing stakeholder engagement and sustainability reporting. *Corporate Reputation Review, 18*(3), 210–222.
Camilleri, M. A. (2015b). Environmental, social and governance disclosures in Europe. *Sustainability Accounting, Management and Policy Journal, 6*(2), 224–242.
Camilleri, M. A. (2017). *Corporate sustainability, social responsibility and environmental management: An introduction to theory and practice with case studies*. Heidelberg: Springer.
Camilleri, M. A. (2018). Theoretical insights on integrated reporting: The inclusion of non-financial capitals in corporate disclosures. *Corporate Communications: An International Journal, 23*(4), 567–581.
Campbell, D., & Cornelia Beck, A. (2004). Answering allegations: The use of the corporate website for restorative ethical and social disclosure. *Business Ethics: A European Review, 13*(2–3), 100–116.
Cheng, M., Green, W., Conradie, P., Konishi, N., & Romi, A. (2014). The international integrated reporting framework: Key issues and future research opportunities. *Journal of International Financial Management & Accounting, 25*(1), 90–119.
Churet, C., & Eccles, R. G. (2014). Integrated reporting, quality of management, and financial performance. *Journal of Applied Corporate Finance, 26*(1), 56–64.
Crowther, D. (2016). *A social critique of corporate reporting: Semiotics and web-based integrated reporting*. Abington: Routledge.
Dacin, M. T. (1997). Isomorphism in context: The power and prescription of institutional norms. *Academy of Management Journal, 40*(1), 46–81.
Dacin, M. T., Oliver, C., & Roy, J. P. (2007). The legitimacy of strategic alliances: An institutional perspective. *Strategic Management Journal, 28*(2), 169–187.
Davis, J. H., Schoorman, F. D., & Donaldson, L. (1997). Toward a stewardship theory of management. *Academy of Management Review, 22*(1), 20–47.
de Villiers, C., Rinaldi, L., & Unerman, J. (2014). Integrated reporting: Insights, gaps and an agenda for future research. *Accounting, Auditing & Accountability Journal, 27*(7), 1042–1067.
Deegan, C. (2002). Introduction: The legitimising effect of social and environmental disclosures–A theoretical foundation. *Accounting, Auditing & Accountability Journal, 15*(3), 282–311.
Deegan, C. (2007). Organisational legitimacy as a motive for sustainability reporting. In J. Unerman, J. Bebbington, & B. O'Dwyer (Eds.), *Sustainability accounting and accountability* (pp. 127–149). London: Routledge.
Deephouse, D. L. (1996). Does isomorphism legitimate? *Academy of Management Journal, 39*(4), 1024–1039.
DiMaggio, P. J., & Powell, W. W. (Eds.). (1991). *The new institutionalism in organizational analysis* (Vol. 17). Chicago, IL: University of Chicago Press.
Donaldson, L., & Davis, J. H. (1991). Stewardship theory or agency theory: CEO governance and shareholder returns. *Australian Journal of Management, 16*(1), 49–64.
Dumay, J., Bernardi, C., Guthrie, J., & Demartini, P. (2016). Integrated reporting: A structured literature review. *Accounting Forum, 40*(3), 166–185.
Eccles, R. G., & Krzus, M. P. (2010). *One report: Integrated reporting for a sustainable strategy*. Hoboken, NJ: John Wiley & Sons.
Eccles, R. G., Serafeim, G., & Krzus, M. P. (2011). Market interest in nonfinancial information. *Journal of Applied Corporate Finance, 23*(4), 113–127.
Eisenhardt, K. M. (1989). Agency theory: An assessment and review. *Academy of Management Review, 14*(1), 57–74.
Erlingsdóttir, G., & Lindberg, K. (2005). *Isomorphism, isopraxism and isonymism-complementary or competing processes?* (No. 2005/4). Accessed April 19, 2017, from http://www.lri.lu.se/pdf/wp/2005-4.pdf
Flower, J. (2015). The international integrated reporting council: A story of failure. *Critical Perspectives on Accounting, 27*, 1–17.

Harding, T. (2012). How to establish a study association: Isomorphic pressures on new CSOs entering a neo-corporative adult education field in Sweden. *Voluntas: International Journal of Voluntary and Nonprofit Organizations, 23*(1), 182–203.
Hedberg, C. J., & Von Malmborg, F. (2003). The global reporting initiative and corporate sustainability reporting in Swedish companies. *Corporate Social Responsibility and Environmental Management, 10*(3), 153–164.
Hopwood, A. G. (2009). Accounting and the environment. *Accounting, Organizations and Society, 34*(3), 433–439.
Idowu, S. O., Capaldi, N., & Zu, L. (2013). *Encyclopedia of corporate social responsibility*. Heidelberg: Springer.
Ioannou, I., & Serafeim, G. (2012). What drives corporate social performance? The role of nation-level institutions. *Journal of International Business Studies, 43*(9), 834–864.
Ioannou, I., & Serafeim, G. (2016). *The consequences of mandatory corporate sustainability reporting: Evidence from four countries* (Harvard Business School Research Working Paper No. 11-100). Accessed April 14, 2017, from https://papers.ssrn.com/soL3/papers.cfm?abstract_id=1799589
IR. (2013). *International IR Framework, international integrated reporting council*. Accessed May 16, 2017, from http://integratedreporting.org/wp-content/uploads/2013/12/13-12-08-THE-INTERNATIONAL-IR-FRAMEWORK-2-1.pdf
IR. (2017). *Integrated reporting. International integrated reporting council*. Accessed May 19, 2017, from https://integratedreporting.org/the-iirc-2/
Lozano, R., & Huisingh, D. (2011). Inter-linking issues and dimensions in sustainability reporting. *Journal of Cleaner Production, 19*(2), 99–107.
Maniora, J. (2015). Is integrated reporting really the superior mechanism for the integration of ethics into the core business model? An empirical analysis. *Journal of Business Ethics, 140*(4), 755–786.
Meyer, J. W., Boli, J., Thomas, G. M., & Ramirez, F. O. (1997). World society and the nation-state. *American Journal of Sociology, 103*(1), 144–181.
Muth, M., & Donaldson, L. (1998). Stewardship theory and board structure: A contingency approach. *Corporate Governance: An International Review, 6*(1), 5–28.
Ness, K. E., & Mirza, A. M. (1991). Corporate social disclosure: A note on a test of agency theory. *The British Accounting Review, 23*(3), 211–217.
Neu, D., Warsame, H., & Pedwell, K. (1998). Managing public impressions: Environmental disclosures in annual reports. *Accounting, Organizations and Society, 23*, 265–282.
O'Dwyer, B. (2003). Conceptions of corporate social responsibility: The nature of managerial capture. *Accounting, Auditing & Accountability Journal, 16*(4), 523–557.
Parent, M. M., & Deephouse, D. L. (2007). A case study of stakeholder identification and prioritization by managers. *Journal of Business Ethics, 75*(1), 1–23.
Perego, P., Kennedy, S., & Whiteman, G. (2016). A lot of icing but little cake? Taking integrated reporting forward. *Journal of Cleaner Production, 136*, 53–64.
PWC. (2015). *What do 1,300 CEOs worldwide say about today's business challenges and opportunities*? Accessed March 11, 2017, from http://www.pwc.com/gx/en/ceo-agenda/ceosurvey/2015.html
Scott, W. R. (1995). *Institutions and organizations* (Vol. 2). Thousand Oaks, CA: Sage.
Simnett, R., & Huggins, A. L. (2015). Integrated reporting and assurance: Where can research add value? *Sustainability Accounting, Management and Policy Journal, 6*(1), 29–53.
Stacchezzini, R., Melloni, G., & Lai, A. (2016). Sustainability management and reporting: The role of integrated reporting for communicating corporate sustainability management. *Journal of Cleaner Production, 136*, 102–110.
Stubbs, W., & Higgins, C. (2014). Integrated reporting and internal mechanisms of change. *Accounting, Auditing & Accountability Journal, 27*(7), 1068–1089.
Suchman, M. C. (1995). Managing legitimacy: Strategic and institutional approaches. *Academy of Management Review, 20*(3), 571–610.

Chapter 4
The Evolution of Corporate Sustainability and Corporate Social Responsibility Towards the Common Goal of Integrated Reporting

Ioana Dragu

4.1 A Contextual Analysis of Terminology Evolution

We start by defining the terms of *sustainability* and *corporate social responsibility*, analysing their evolution in time. Then, we continue with a research on sustainability and CSR evolution, until they unified in a single reporting trend that was soon to become the *integrated report*. This provides a better understanding of the composition of the integrated reporting, that includes sustainability and corporate social responsibility reporting.

Sustainability and corporate social responsibility had parallel evolutions in time (Fig. 4.1). From a historical point of view, the CSR and sustainability debate debuted in 1950s, when they were two different concepts—one focusing on *social aspects* and the other on *environmental* concerns. The two continued to raise awareness on a global scale between 1950 and 1970, while 10 years later the first concept developed into *corporate social responsiveness*—defined as organizations' impact on society-and the environmental debate lead to the issuance of *World Conservation Strategy* (Loew et al. 2004). The *World Conservation Strategy* represented an intellectual framework and practical guidance for rational development of the resources of our planet prepared by IUCN- International Union for Conservation of Nature and Natural Resources.

In 1984 Freeman was introducing the *stakeholder theory* that argues for the importance of internal and external stakeholders (employees, managers, owners, customers, suppliers, society, government, creditors) as in contrast to the classical shareholder approach that focused on shareholder value maximization only (Freeman 1984). During the period of 1990–1997, many scholars and academics worldwide became advocates of stakeholder theory implementation (Hill and Jones 1992; Donaldson and Preston 1995; Jones 1995; Mitchell et al. 1997).

I. Dragu (✉)
Babes-Bolyai University, Cluj-Napoca, Romania

S. O. Idowu, M. Del Baldo (eds.), *Integrated Reporting*, CSR, Sustainability, Ethics & Governance, https://doi.org/10.1007/978-3-030-01719-4_4

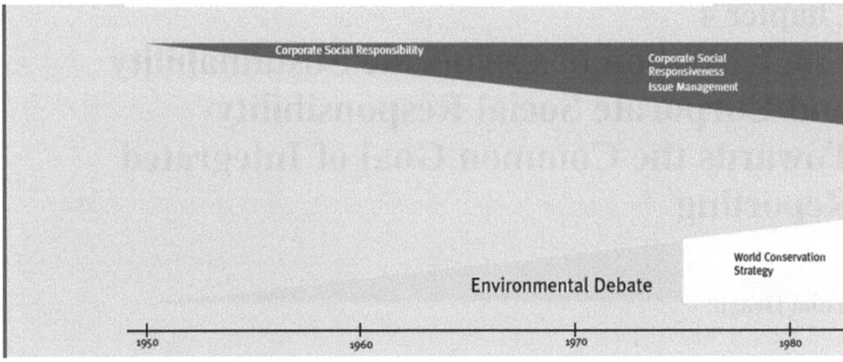

Fig. 4.1 CSR and Sustainability debate between 1950 and 1980. Source: Loew et al. 2004: 8

In the same time, the globalization and economic growth offsetting the natural environment and planet resources, was becoming in the mid of 1990 a big alert for international organizations. This conflict between economic wellbeing and prosperity versus negative effects on environment such as water, air pollution, and resource scarcity was the main reason for the work developed by the Brundtland Commission which culminated with the *Brundtland Report* published by Oxford University Press in 1987. This report mentioned for the first time the concept of *sustainable development.*

Corporate social responsibility and corporate sustainability include various elements, some of them being common for both concepts. In year 2000, corporate social responsibility became known as 'CSR', and companies were measuring their *corporate social performance* using internal analysis, setting targets for the planning process (Loew et al. 2004), introducing pollution indices, financial reports, surveys on company reputation, CSR orientation studies (Foote et al. 2010), or evaluating the performance from stakeholders' perspectives (Ruff et al. 1998). On the other hand, sustainability line was also evolving with the UN Earth Summit organized in Rio de Janeiro in 1992. Rio Conference continued the work on sustainable development initiating an action plan called Agenda 21, that eventually led to the introduction of corporate sustainability as a notion defined by "company's efforts to achieve social, ecological and economic sustainability objectives" (Loew et al. 2004: 13). The Johannesburg Summit from 2002 represented a suitable occasion for discussing the implementation of Agenda 21 from Rio Conference. In the end, on a time axis, the period before 2000 is marked by two main concepts: *CSR* and *Sustainable development.* Corporate sustainability is considered as part of sustainable development, while corporate social performance is included in CSR (Loew et al. 2004).

Between 2000 and 2010, *sustainability* and *corporate social responsibility* merge together towards certain common goals of social and environmental nature and stakeholder orientation.

Figure 4.2 presents the main stages in the development of sustainability and corporate social responsibility between 1990 and 2010 that have been previously described.

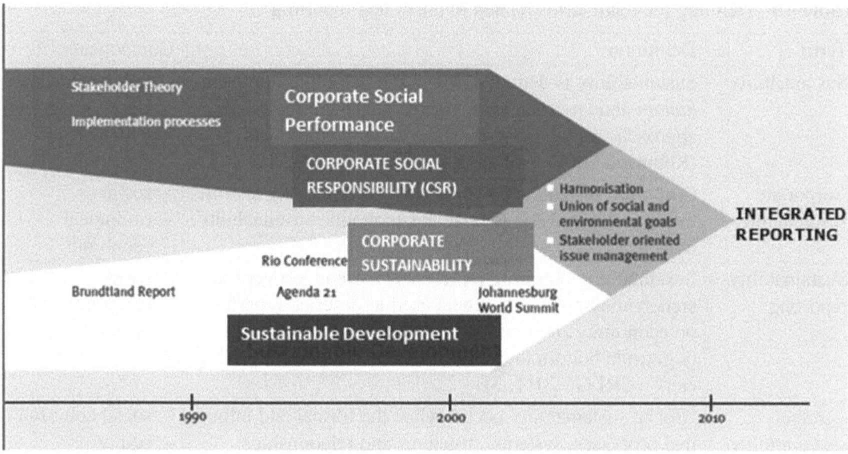

Fig. 4.2 CSR and Sustainability debate between 1990 and 2010. Source: Adapted after Loew et al. 2004: 9

According to Loew et al. (2004), corporate sustainability and corporate social responsibility show the contribution of the company to sustainable development, CSR being included in corporate sustainability.

United Nations Principles declaration (Lemke and Lins 2002) underlines that corporate social responsibility stands for 'responsibility' (or accountability), and the word 'social' is more used for sustainability.

Further on, we are going to expand the terms of sustainability and CSR as *terms* and *reporting*, to present their components, and find common elements.

Analysing the definitions expressed below (Table 4.1), we can deduce that the common elements of sustainability and corporate social responsibility are the social and environmental dimensions. In addition to these, each concept develops divergently, in different directions. For sustainability, the social and environmental pillars are completed by the economic opportunities and the obligations to present and future generations. Sustainability also develops on ESG considerations that assume the corporate governance element besides the social and environmental ones. CSR or corporate social responsibility is by definition a trio of social, environmental, and stakeholder concerns. It focuses mainly on human capital and human rights, in contrast to sustainability that makes reference to social cohesion and equity. Finally, corporate social responsibility argues for social and environmental *performance*, while sustainability is concerned with social, environmental, and economic *impacts* (Fig. 4.3).

We will continue with a discussion on sustainability and corporate social responsibility reporting as elements of an *integrated report*.

The *integrated report* contains, by definition, both sustainability and CSR information (Aceituno et al. 2012; IIRC, 2012). In the international literature on non-financial disclosure, it is often presented as the unique, single report (Lozano

Table 4.1 Defining the main terms related to integrated reporting

Term	Definition	Components
Sustainability	Sustainability is defined through the two pillar approach- *human* and *environmental* concerns or the three pillars approach- *social, economic,* and *ecological* aspects (Robinson 2004)	– social (human) – environmental (ecological) – economic
Corporate sustainability	Corporate sustainability illustrates the "company's efforts to achieve social, ecological and economic sustainability objectives" (Loew et al. 2004: 13)	– social – ecological – economic
Sustainability reporting	Sustainability reporting represents "a broad term considered synonymous with others used to describe reporting on economic, environmental, and social impacts (e.g. triple bottom line, corporate responsibility reporting, etc.)" (GRI G3 2011: 3)	– social – environmental – economic
– *social sustainability*	"Social sustainability occurs when the formal and informal processes, systems, structures and relationships actively support the capacity of current and future generations to create healthy communities." (McKenzie 2004: 18)	– social cohesion – equity
– *environmental sustainability*	Environmental sustainability is connected to "knowledge of ecosystems and resources. . .ability to initiate, advocate and absorb radical shifts in desired lifestyles, values and technology" and ultimately the acknowledgment of environmental problems (McKenzie 2004: 10)	– ecosystems and resources – lifestyles, values and technology – environmental problems
– *economic sustainability*	Economic sustainability means the "obligation to preserve the present-day economic opportunities (such as productive capacity) for the future, not necessarily to increase them" in order to attain the "intergenerational equity" (Anand and Sen 2000)	– economic opportunities – present and future generations
Sustainable development	Sustainable development means to fulfill "the needs of present without compromising the ability of future generations to meet their own needs" (Baker 2006: 20)	– present and future generations needs
Social responsibility	Social responsibility relates to the "social conscience" of the business that goes beyond profit maximization (Friedman 2007: 1) and contributed to the development of the well-known notion of *corporate social responsibility*- or social responsibility of corporations. The term is also linked to *socially responsible investments* that assumes the mixture of social objectives with the financial ones (Haigh and Hazelton 2004)	– social conscience
Corporate social responsibility	CSR is, "a concept whereby companies integrate *social and environmental* concerns in their business Operations and in their interaction with their *stakeholders* on a voluntary basis" (European Commission, 2002: 3)	– social – environmental – stakeholders
Corporate social responsibility reporting (CSR reporting)	Through corporate social responsibility reporting companies disclose their social responsible investments (Anand and Sen 2000) informing the users on "social and environmental performance, including the human rights dimension" (European Commission 2002: 5)	– social responsible investments – social and environmental performance – human rights dimension

(continued)

Table 4.1 (continued)

Term	Definition	Components
Corporate social performance	Academics and scholars define corporate social performance as a function of social responsibility principles, social responsiveness, and social policies (Wartick and Cochran 1985), representing in a broader meaning the social obligation of any business (Davis 1973)	– social responsibility – social responsiveness – social policies – social obligation

Source: Author's design

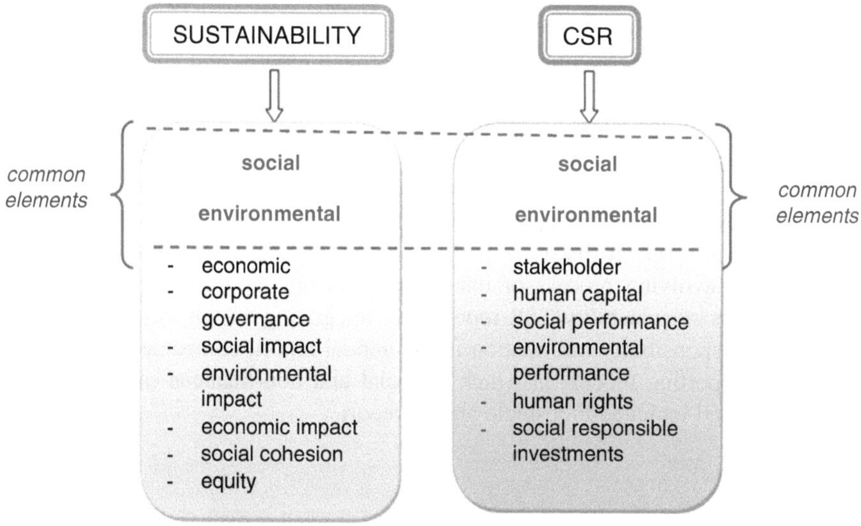

Fig. 4.3 Comparison between Sustainability and CSR from a terminology point of view. Source: Author's design

and Huisingh 2011; Eccles et al. 2010) that presents the common elements of CSR and sustainability: environmental information (Xuan et al. 2014; De Villiers and Van Staden 2011; Fifka and Drabble 2012) and social information (Fifka and Drabble 2012), or simply *corporate social responsibility aspects* (Dhaliwal et al. 2012; Andrew et al. 2012), respectively *sustainability issues* (De Villiers and Van Staden 2011; Kirklin et al. 2013; Fifka and Drabble 2012).

Aceituno et al. (2012) underline the need for corporations to include sustainability information in a single report (Lozano and Huisingh 2011, Eccles et al. 2010) and integrate corporate social responsibility in the same unique report. Based on the assumption that both type of corporate social responsibility information and sustainability information are included in the integrated report, besides the financial information, the paper developed by Aceituno et al. (2012) studies the influence of the legal system on the integrated reports. The authors sustain that corporate social responsibility reporting and practice is more common to stakeholder oriented countries (also known as civil- law countries). In addition, they argue that sustainability reporting is less characteristic to countries with a common- law origin in which shareholder value is a priority, while civil-law countries will generate more sustainability reports.

The authors identify a need for connecting the information in corporate reports, and discuss the implication of a country's legal system on integrated reporting (the perspective of institutional theory- IT). The study has two dimensions: the first one involves the analysis on civil versus common law legal systems based on normative isomorphism, and the second investigates strong versus weak regulation enforcements from a coercive isomorphism perspective. The methodology involves a logit analysis on 750 corporations from 20 countries during 2008–2010. Results demonstrate the impact of the political factor on integrated reports, and note that civil law countries favour the diffusion and adoption of integrated reporting, while strong legal enforcement countries generate higher reporting transparency and positively influence integrated reporting.

The IIRC defines integrated reporting as a mixture of environmental, social, and corporate governance information (IIRC 2012).

The current research discusses both sustainability and CSR reporting in order to show the evolving process of integrated reporting- as these reports emerged from previous sustainability/CSR reports. By integrating sustainability and corporate social responsibility information in the annual report, companies can develop successful reporting models in which financial and non-financial information are interrelated and included in a single annual report.

4.2 Sustainability: Definition and Pillars

"The concept of sustainability involves operating in a way that takes full account of an organization's impacts on the planet, its people, and the future" (ICAEW 2004: 4). Mentioned for the first time in the *Brundtland Report Commission* "Our Common Future" *in 1987* (Kuhlman and Farrington 2010; Tovey 2009), the notion of sustainability refined the view that economic development has to adjust its progression to the limited resources of our planet. In Brundtland report, sustainable development is defined "the needs of present without compromising the ability of future generations to meet their own needs" (Baker 2006: 20). Still, Gibson et al. (2013: 48) find the Brundtland agenda as "contradictory" and believed that the economic growth (as current need) will not eliminate poverty, inequity, or resource scarcity (future needs).

In the contemporary literature there is a lack of consensus about the sustainability definition. Robinson (2004) suggests that the definition varies and has identified two approaches:

(a) Dualistic approach: Emphasizing the relationship between humanity and nature
(b) Three pillar approach: Emphasizing the economic, social and ecological dimensions

The three pillar approach is the most common attempt to break down the concept of sustainability. Robinson and Tinker (1998) consider that the three components, economic, environmental and social have direct effect on each other and can never be used in isolation.

Blume (2009) outlines the sustainability pillars and explains the alternatives: economic-environmental elements incline to the usage of the term *viable* as a capacity to solve problems according to on-hand tools and given the circumstances. For social-environmental purposes, the academic employs the concept of *equitable*: "fair to all parties as dictated by reason and conscience" (Princeton University 2006). The convention between economic and social aspects leads to a *bearable* situation. According to Blume (2009), sustainability is complete only by integrating social, economic and environmental aspects.

Other two issues on the list of sustainability pillars, in the form of political and cultural factors are added by Gibson (2006). The author is against stratifying sustainability between these sectors, and proposes instead a series of principles. We agree that the pillars cannot be dissipated, and they must go together, as sustainability means dedication for people, planet and profit altogether. Tovey (2009) argues that the degree of importance of the three elements from the pillar depends on what we choose to sustain and ultimately the author underlines the importance of the environment. We consider that the three pillars have the same importance therefore, should be equally treated by corporations.

However, critics of the sustainability approach (Marshall and Toffel 2005) advocate the difficulty of applying this concept in practice, especially when involved in the decision- making process. In addition, the triple bottom line does not cover the ethical perspective, which should be at the core of business' activities.

Companies rely heavily on their stakeholders, who use reporting information for decision-making purposes. Transparency becomes a sensitive subject as an on-going element for building credibility and trust. In addition, firms need to be aware of what is important for stakeholders and how should they disclose sustainability and corporate social responsibility information in their reports. Nowadays, is getting more and more difficult for large corporations to fulfil all their obligations from sustainability and CSR reporting perspective. The balance between users and producers of non-financial information should be the best solution to this dilemma. We kindly recommend other firms to adopt sustainability and CSR reporting practices, in order to incorporate the benefits and advantages from this type of reporting.

G3 Guidelines define sustainability reporting as "a broad term considered synonymous with others used to describe reporting on economic, environmental, and social impacts" (GRI G3 2011: 3) often indicated as 'ESG reporting' or 'reporting on ESG information'. ESG stands for environmental (input and output in terms of energy, water, emissions, waste, climate change etc.), social (leadership, employees, customers, communities etc.) and governance (transparency, independence, compensation, shareholder rights etc.). The ESG factors: environmental, social, and corporate governance are used to measure sustainability and fundament the perspective of socially responsible investments. Sustainability reporting is the practice of measuring, disclosing, and being accountable to internal and external stakeholders for organizational performance towards the goal of sustainable development. When computing the overall corporate performance, one must take note of the 3 sustainability pillars— economic/social/environmental- and the ESG factors—environmental/ social/corporate governance (Fig. 4.4):

The users of the sustainability reports include suppliers, customers, governments, communities, or other stakeholders who may enforce or stimulate the production and quality of sustainability reports by means of various policy instruments, CSR (reporting) organizations and educational institutions. KPMG (2008) grouped the users of sustainability reports into the following categories: business, consultancy, civil society, research academic, investment/rating agencies, individuals, public agency. This distribution is different from the one in The International Integrated Reporting Framework, where the main groups of stakeholders are defined as "employees, customers, suppliers, business partners, local communities, legislators, regulators and policy-makers"(IIRC 2013: 7). Nevertheless, investors are interested in policies adopted by corporations (GRI or CSR) as well as sustainability level-stakeholder engagement, reporting principles, performance, innovation, objectives and strategies.

4.3 Empirical Research on the Connection Between Sustainability and CSR Disclosure and Other Possible Determinants

4.3.1 Purpose of the Research

The current section of the report seeks to demonstrate existing links between the two dimensions of integrated reporting- sustainability and CSR information-, and organizational business size—*sales* and *number of employees*. Table 4.2 shows the sales and employees figures for the sample companies (data extracted from the annual reports), while adding the disclosure index for sustainability and corporate social responsibility information (from the case studies described in the previous section). The organizational data for revenues is expressed in mil. EUR.

We developed two hypotheses:

H_{01}: The organizational size of the company impacts the disclosure of sustainability information and CSR information in annual reports.
H_{02}: The level of sustainability information disclosure is influenced by the degree of CSR information disclosure and vice-versa.

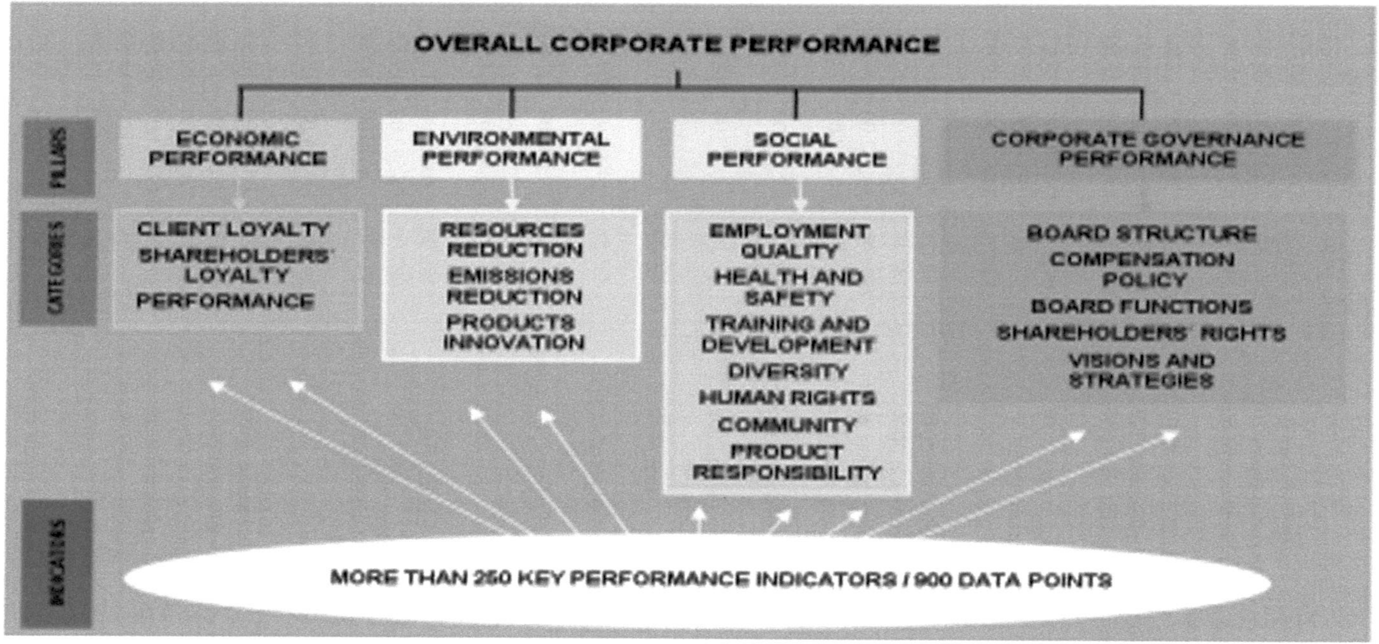

Fig. 4.4 Overall performance of company. Source: Kocmanová and Šimberová 2013: 546

Table 4.2 Additional data on our sample of integrated reports

Company	Document title	Sales-in mil. EUR-	No. of employees	DI sustainability[a]	DI csr[b]
AEP	2010 annual report	12,236.02	3268.00	89.75	0.75
Altron	Integrated report 2010	2532.90	12,311.00	125.97	1
Amlinplc	Amlinplc annual report 2009	2238.98	1161.00	112.96	0.5
Amlinplc	Annual report 2010	1317.30	1391.00	112.96	0.5
Anglo Platinum	Annual report 2009	3455.92	50,681.00	121.40	0.5
Anglo Platinum	Integrated annual report 2009	4335.56	48,654.00	121.40	0.5
BASF	Annual report 2010	50,693.00	104,779.00	125.45	1
BASF	Integrated report	63,873.00	109,140.00	125.45	1
Capita Group	Annual report and accounts 2009	3514.28	34,625.00	117.67	0.75
Capita Group	Annual report 2010	3440.75	36,800.00	117.67	0.75
Eskom	Annual report 2009	4782.16	37,857.00	120.89	1
Eskom	Integrated report 2010	7594.40	39,222.00	120.83	1
Gold Fields	Integrated annual report 2009	1458.47	51,122.00	120.59	0.75
Gold Fields	Integrated annual report 2010	2076.14	47,268.00	120.59	0.75
Great Portland Estates	Annual report 2009	135.04	77.00	118.91	0.25
Great Portland Estates	Annual report 2010	116.55	66.00	116.51	0.25
Implats	Integrated annual report 2010	2885.58	4000.00	109.74	0.75
Logicaplc	Annual report and accounts 2009	201.16	39,501.00	121.90	0.25
Logicaplc	Annual report and accounts 2010	197.95	38,963.00	121.90	0.25
Massmart	2010 annual report	5380.94	30,649.00	113.22	0.5
Metso Corporation	Annual report 2009	1687.00	28,593.00	119.77	1
National Grid	Annual report and accounts 2009	2873.36	6650.00	118.39	0.5
National Grid	Annual report and accounts 2010	3297.50	6311.00	118.39	0.5
Natura	Natura report 2010	4119.74	5135.00	120.45	0.75
Novo Nordisk	Annual report 2009	6860.00	29,329.00	117.66	1
Novo Nordisk	Annual report 2010	8161.00	30,483.00	117.66	1
Philips	Annual report 2010	25,419.00	119,001.00	123.08	1
Potash Group	Financial review 2009	2768.18	5000.00	114.58	0.75
RB Platinum	Integrated annual report 2010	238.91	3207.00	113.68	0.75
Standard Bank	Integrated report 2010	10,028.76	21,872.00	123.41	0.75
UTC	Annual report 2010	40,567.53	208,200.00	107.31	0.25

(continued)

Table 4.2 (continued)

Company	Document title	Sales-in mil. EUR-	No. of employees	DI sustainability[a]	DI csr[b]
Vodacom	Integrated report 2010	3553.73	7643.00	123.61	0.75
Wolesely	Annual report and accounts 2010	15,848.88	17.00	116.63	0.5
Xstrata plc	Annual report 2009	15,823.75	37,741.00	121.51	0.5
Xstrata plc	Annual report 2010	15,314.80	38,449.00	121.51	0.5

Source: Author's design

[a]This represents the final score obtained when computing the *Deloitte Sustainability Reporting Scorecard* for each company (in absolute value)—can be found in the previous section of the current report

[b]The *disclosure index for corporate social responsibility* has been computed using the disclosure index in the CSR section of this chapter

4.3.2 Methodology

The methodology involves SPSS data processing, by establishment of the correlation level between our four elements: sustainability and CSR disclosure indexes, sales, and number of employees.

We computed three statistical coefficients, namely: *Pearson, Kandel, and Spearmen* correlation coefficients. The results are discussed from the perspective of 5% and 10% statistical significance correlation for the sample of 35 integrated reports.

Financial and profitability ratios have been used by academics and scholars to show the effect they have on corporate social responsibility or sustainability. Aras et al. (2010) aims to determine the linkage between corporate social responsibility and sales, revenues, equity, etc. The findings indicate a correlation between the firm size (measured as total number of employees and sales) and CSR. Morhardt (2010) uses the Pacific Indicator for measuring the CSR and sustainability reporting on 25 industrial segments and then investigates if there is a connection between this indicator and the companies' sales. They find evidence for strong correlations in America, while Asia and Europe did not result in relevant influences. Gallo and Christensen (2011) proves that sustainability reporting is highly linked to organizational size and ownership. Balluchi and Furlotti (2013) investigate the importance attributed by Italian companies to CSR and environmental information according to certain factors such as the juridical form of the firm, district area, number of employees, and activity sector. The results show that the firm size positively influences the focus on CSR and environmental issues. Overall, we can state that according to previous studies in the literature review, corporate social responsibility and sustainability are impacted by the business size of the company. Finally, we fundament our choice of *sales* and *number of employees* as determinants for the disclosure of sustainability and non-financial information as they are the most used in literature for investigating correlations and modelling different economic phenomenon.

4.3.3 Results

This section presents the output results obtained through SPSS regression analysis. Table 4.3 shows the Pearson Correlation Coefficients determined for our sample of 35 integrated reports.

This particular coefficient does not illustrate any significant influence between the degree of sustainability and CSR information versus organizational data (sales and number of employees). Although all the values indicate positive correlation between the variables, the significance of 1% is met only in case of sales and number of employees.

On the other hand, checking Kendal and Spearman coefficients (Table 4.4), the disclosure index for sustainability is positively correlated to the number of employees (0.390 and 0.5 correlations at a 0.01 significance level), which means that the higher the employment rate in a company, the more likely for its reports to produce sustainability information. Further on, CSR disclosure has a correlation of 0.277 and 0.36 against sales (at 0.05 significance level), suggesting that organizations with consistent revenues will positively impact the presentation of corporate social responsibility information in annual reports. This proves that the first hypotheses are true.

On the whole, we can conclude that sustainability and corporate social responsibility information do not influence each other (denying the last hypothesis), preposition that is sustained by the fact that they are to be added to the common function of *integrated reporting*. Summing up all our findings, integrated reporting is positively correlated with organizational or business size of a company (measured by its sales and employees).

Table 4.3 Pearson Co-efficient

Correlations					
		Sales	No_of_empl	DI_sustainab	DI_csr
Sales	Pearson correlation	1	0.740**	0.109	0.243
	Sig. (2-tailed)		0.000	0.531	0.160
	N	35	35	35	35
No_of_empl	Pearson correlation	0.740**	1	0.126	0.071
	Sig. (2-tailed)	0.000		0.470	0.687
	N	35	35	35	35
DI_sustainab	Pearson correlation	0.109	0.126	1	0.194
	Sig. (2-tailed)	0.531	0.470		0.263
	N	35	35	35	35
DI_csr	Pearson correlation	0.243	0.071	0.194	1
	Sig. (2-tailed)	0.160	0.687	0.263	
	N	35	35	35	35

Source: SPSS processing
**Correlation is significant at the 0.01 level (2-tailed)

Table 4.4 Kendal and spearman coefficients

Correlations						
			Sales	No_of_empl	DI_sustainab	DI_csr
Kendall's tau_b	Sales	Correlation coefficient	1000	0.297*	0.129	0.277*
		Sig. (2-tailed)		0.,012	0.280	0.034
		N	35	35	35	35
	No_of_empl	Correlation coefficient	0.297*	1000	0.390**	0.180
		Sig. (2-tailed)	0.012		0.001	0.168
		N	35	35	35	35
	DI_sustainab	Correlation coefficient	0.129	0.390**	1000	0.185
		Sig. (2-tailed)	0.280	0.001		0.159
		N	35	35	35	35
	DI_csr	Correlation coefficient	0.277*	0.180	0.185	1000
		Sig. (2-tailed)	0.034	0.168	0.159	
		N	35	35	35	35
Spearman's rho	Sales	Correlation coefficient	1000	0.406*	0.185	0.360*
		Sig. (2-tailed)		0.016	0.286	0.033
		N	35	35	35	35
	No_of_empl	Correlation coefficient	0.406*	1000	0.539**	0.208
		Sig. (2-tailed)	0.016		0.001	0.230
		N	35	35	35	35
	DI_sustainab	Correlation coefficient	0.185	0.539**	1000	0.257
		Sig. (2-tailed)	0.286	0.001		0.137
		N	35	35	35	35
	DI_csr	Correlation coefficient	0.360*	0.208	0.257	1000
		Sig. (2-tailed)	0.033	0.230	0.137	
		N	35	35	35	35

Source: SPSS processing
*Correlation is significant at the 0.05 level (2-tailed)
**Correlation is significant at the 0.01 level (2-tailed)

This research is not meant to observe corporations that adopt sustainability reporting and corporate social responsibility reporting for enhancing sustainability and CSR practice. We cannot guarantee that a certain company complies with the information presented in the integrated report, or is possible that they do not include all the mentioned elements regarding sustainability/CSR. However, we can only rely on those facts published in annual reports, as evidence for our case studies.

Therefore, we investigate the information presented in integrated reports in search for elements that prove the integration of sustainability reporting—and CSR reports.

Our sample is subject to geographical bias, as 75% of the companies are headquartered in South Africa and United Kingdom. This homogenous distribution is explained by the limited access to reports in English, state implication or pressure regarding sustainability/CSR information and its integration within the annual report, the industry factor (e.g. oil and gas companies are more willing to demonstrate their sustainability/CSR commitment with respect to environment), and ultimately corporations' attitude towards stakeholders and the importance and relevance attributed to non-financial information in connection with the financial aspects.

4.4 Conclusions

The current research is also meant to encourage companies that have already adopted integrated reports, or are in the process of adoption, to use the Deloitte Sustainability Scorecard for receiving feedback on their improvements on sustainability reporting practices. The Deloitte Sustainability Scorecard is meant to highlight the transparency of annual reports, by underlining the importance of presenting information that is relevant for stakeholders, in order to build credibility and trust. The structure of the scorecard concentrates on stakeholder related disclosure, as target audience, along with their needs and expectations. The scorecard provides both the producer and the user of reports the opportunity to establish the level of sustainability information disclosed by the company, also setting a feedback for firms that want to find how they can improve sustainability reporting. We consider the issue of the German state, which appears to be the first in line on sustainability scorecard. The companies from this state register the higher values for sustainability information disclosure, in average of 90.48%. The controversy of having Germany on the first place, instead of, for instance South Africa, that has strict regulation for integrated reporting, can be connected to the German Sustainability Code (GSC) established in 2011, that concentrates on transparency and comparability as well as serving the interests of the stakeholders. Furthermore, the German focus on sustainability practices has its roots some years earlier, in 2009 and 2010, when companies register high degree of disclosure for sustainability information. From an industry perspective, the circumstance of having chemical companies with maximum degree of sustainability reporting, can only mean that this sector is accountable for the eventual negative effects that it can produce upon its stakeholders, and by providing transparent information in their reports, they try to reduce the impact and to bring explanations upon sustainability issues.

The main beneficiaries of CSR reporting practices are the communities- at global, regional and local levels. Organizations that practice CSR are implied in community programs and contribute to the well-being of society, promoting a safe and clean environment for future generations. Therefore, we notice the pressure on world scale to incorporate best corporate social responsibility practices. From a CSR point of view, integrated reports concentrate on environmental related information. We can add that organizations can still improve their CSR reporting. However, the results of our study show that corporate social responsibility information is generally presented in high proportions. In addition, the figures demonstrate medium and high levels for CSR disclosure. We believe that by disclosing corporate social responsibility information, corporations can properly respond to a more than obvious stakeholder pressure for CSR practice.

The empirical research on sustainability and corporate social responsibility versus the business size of the organizations (expressed in sales and number of employees) illustrates positive correlations. Therefore, the higher the organizational size of a company, the greater the probability for a corporation to register good scores for sustainability and CSR disclosure in an annual report.

References

Aceituno, F., Ariza, R., & Sanchez, G. (2012). Is integrated reporting determined by a country's legal system? *Journal of Cleaner Production, 44*, 44–45.

Anand, S., & Sen, A. (2000). Human development and economic sustainability. *World Development, 28*(10), 2029–2049.

Andrew, N., Wickham, M., O'Donohue, W., & Danzinger, F. (2012). Presenting a core-periphery model of voluntary CSR disclosure in Australian annual reports. *Corporate Ownership & Control Journal, 9*(2), 438–448.

Aras, G., Aybars, A., & Kutlu, O. (2010). Managing corporate performance: Investigating the relationship between corporate social responsibility and financial performance in emerging markets. *International Journal of Productivity and Performance Management, 59*(3), 229–254.

Baker, S. (2006). *Sustainable development*. New York: Routledge.

Balluchi, F., & Furlotti, K. (2013). Small and medium enterprises and environmental issues: Empirical evidences in Italian context. *European Scientific Journal, 9*, 19.

Blume, D. (2009). *Port sustainability*. Balancing Competing Priorities, AAPA (American Association of Port Authorities) (2009) Executive Management Conference. Available at http://aapa.files.cms plus.com/SeminarPresentations/2009Seminars/09ExecMan/BlumeDana.pdf

Davis, K. (1973). The case for and against business assumption of social responsibilities. *Academy of Management Journal, 16*, 312–322.

De Villiers, C., & Van Staden, C. (2011). Shareholder requirements for compulsory environmental information in annual reports and on websites. *Australian Accounting Review, 21*(4), 317–326.

Dhaliwal, D. S., Radhakrishnan, S., Tsang, A., & Yang, Y. G. (2012). Nonfinancial disclosure and analyst forecast accuracy: International evidence on corporate social responsibility disclosure. *The Accounting Review, 87*(3), 723–759.

Donaldson, T., & Preston, E. L. (1995). The stakeholder theory of the corporation: Concepts, evidence, and implications. *The Academy of Management Review, 20*(1), 65–91.

Eccles, R., Cheng, B., & Saltzman, D. (2010). *The landscape of integrated reporting reflections and next steps*. Boston, MA: Harvard Business School.

European Commission. (2002). *Corporate social responsibility: A business contribution to sustainable development*. http://trade.ec.europa.eu/doclib/docs/2006/february/tradoc_127374.pdf

Fifka, M. S., & Drabble, M. (2012). Focus and standardization of sustainability reporting – A comparative study of the United Kingdom and Finland. *Business Strategy and the Environment, 21*(7), 455–474.

Foote, J., Gaffney, N., & Evans, R. J. (2010). Corporate social responsibility: Implications for performance. *Total Quality Management, 21*(8), 799–812.

Freeman, R. E. (1984). *A stakeholder approach. Strategic management*. Boston, MA: Pitman Publishing.

Friedman, M. (2007). The social responsibility of business is to increase its profits. In *Corporate ethics and corporate governance* (pp. 173–178). Berlin: Springer.

Gallo, P. J., & Christensen, L. J. (2011). Firm size matters: An empirical investigation of organizational size and ownership on sustainability-related behaviours. *Business & Society, 50*(2), 315–349. Available at http://bas.sagepub.com/content/early/2011/03/17/0007650311398784.abstract

Gibson, R. B. (2006). Sustainability assessment: Basic components of a practical approach. *Impact Assessment and Project Appraisal, 24*(3), 170–182.

Gibson, B., Hassan, S., & Tansey, J. (2013). *Sustainability assessment: Criteria and processes*. London: Routledge.

GRI G3. (2011). *Sustainability reporting guidelines*. Global Reporting Initiative, from www.globalreporting.org

Haigh, M., & Hazelton, J. (2004). Financial markets: A tool for social responsibility? *Journal of Business Ethics, 52*(1), 59–71.

Hill, C. W. L., & Jones, T. M. (1992). Stakeholder agency theory. *Journal of Management Studies, 29*(2), 131–154.

ICAEW. (2004). *Information for better markets sustainability: The role of accountants*. An initiative from the Institute of Chartered Accountants in England & Wales. Available at https://www.google.ro/url?sa=t&rct=j&q=&esrc=s&source=web&cd=1&ved=0CCEQFjAA&url=https%3A%2F%2Fwww.icaew.com%2F~%2Fmedia%2Fcorporate%2Ffiles%2Ftechnical%2Ffinancial%2520reporting%2Finformation%2520for%2520better%2520markets%2Fifbm%2Fsustainability%2520execsum.ashx&ei=QtZVVc7yKoHqgwSUk4DgDg&usg=AFQjCNEiCKYL53CZTwyIqcRUBkno1-6qfg&sig2=N-013I7StZnqqiivpAt2LQ&bvm=bv

IIRC. (2012). *Consultation draft of the international framework*. Available at http://integratedreporting.org/wp-content/uploads/2013/03/Consultation-Draft-of-the-InternationalIRFramework.pdf

IIRC. (2013). *The international IR framework*. Available at http://integratedreporting.org/wp-content/uploads/2015/03/13-12-08-THE-INTERNATIONAL-IR-FRAMEWORK-2-1.pdf

Jones, T. M. (1995). Instrumental stakeholder theory: A synthesis of ethics and economics. *The Academy of Management Review, 20*(2), 404–437.

Kirklin, J. K., Naftel, D. C., Kormos, R. L., Stevenson, L. W., Pagani, F. D., Miller, M. A., & Young, J. B. (2013). Fifth INTERMACS annual report: Risk factor analysis from more than 6000 mechanical circulatory support patients. *The Journal of Heart and Lung Transplantation, 32*(2), 141–156.

Kocmanová, A., & Šimberová, I. (2013). Multidimensional modelling of social performance indicators in processing industry companies in the Czech Republic. *International Journal of Mathematical Models and Methods in Applied Sciences, 7*(5), 500–507.

KPMG. (2008). *International survey of corporate responsibility reporting 2008*. Amsterdam: KPMG International.

Kuhlman, T., & Farrington, J. (2010). What is sustainability? *Sustainability, 2*, 3436–3448.

Lemke, T. P., & Lins, G. T. (2002). *Regulation of investment advisers (securities law series)*. Seattle, WA: Amazon Publisher.

Loew, T., Ankele, K., Braun, S., & Clausen, J. (2004). *Significance of the CSR debate for sustainability and the requirements for companies*. Future E.V. an Institute for Ecological Economy Research GMBH (IÖW).

Lozano, R., & Huisingh, D. (2011). Inter-linking issues and dimensions in sustainability reporting. *Journal of Cleaner Production, 19*, 99–107.

Marshall, J. D., & Toffel, M. W. (2005). Framing the elusive concept of sustainability: A sustainability hierarchy. *Environmental Science & Technology, 39*(3), 673–682.
McKenzie, S. (2004). *Social sustainability: Towards some definitions* (Hawke Research Institute Working Paper).
Mitchell, R. K., Agle, B. R., & Wood, J. D. (1997). Toward a theory of stakeholder identification and salience: Defining the principle of who and what really counts. *The Academy of Management Review, 22*(4), 853–886.
Morhardt, J. E. (2010). Corporate social responsibility and sustainability reporting on the internet. *Business Strategy and the Environment, 19*(7), 436–452.
Princeton University. (2006). *WordNet search – 3.0*. Princeton University Cognitive Science Laboratory. Retrieved June 15, 2009, from WordNet.
Robinson, J. (2004). Squaring the circle? Some thoughts on the idea of sustainable development. *Ecological Economics, 48*, 369–384.
Robinson, J., & Tinker, J. (1998). Reconciling ecological, economic and social imperatives. In J. Schnurr & S. Holtz (Eds.), *The cornerstone of development: Integrating environmental, social and economic policies*. Ottawa: International Development Research Centre.
Ruff, B. M., Muralidhar, K., & Paul, K. (1998). The development of a systematic, aggregate measure of corporate social performance. *Journal of Management, 24*(1), 119–133.
Tovey, H. (2009). Sustainability: A platform for debate. *Sustainability, 1*(1), 14–18.
Wartick, S. L., & Cochran, P. L. (1985). The evolution of the corporate social performance model. *Academy of Management Review, 10*, 758–769.
Xuan, Z. H. A. O., Lieke, Z. H. A. N. G., & Yuwei, S. O. N. G. (2014). Environmental information disclosure of listed company study on the cost of debt capital empirical data: Based on thermal power industry. *Canadian Social Science, 10*(6), 90–96.

Chapter 5
Small and Medium-Sized Organisations: Why and How Do They Implement Integrated Reporting?

Mara Del Baldo

5.1 Introduction

The topic of IR (integrated reporting) has been attracting a growing interest of scholars, managers, professionals and organisations belonging to different countries around the world (Massaro et al. 2016). IR has been defined as "a process founded on integrated thinking that results in a periodic integrated report by an organisation about value creation over time and related communications regarding aspects of value creation" (IIRC 2013: 7). An integrated report is "a concise communication about how an organisation's strategy, governance, performance and prospects, in the context of its external environment, lead to the creation of value over the short, medium and long term" (IIRC 2013: 33). Evolving business models are in fact more inclusive of non-financial information, encompassing environmental, social and governance (ESG) issues (Dumay et al. 2015).

As an emerging phenomenon, in recent years there is a huge amount of the studies focused on IR. In this regard De Villiers et al. (2014) proposed a comprehensive agenda for future research, pointing out several theoretical and empirical challenges because of the different ways in which IR is understood and enacted within institutions and companies. They stressed several areas of inquiry including the concepts of materiality and integrated thinking, the "reconciliation" between sustainability reporting standards and IR principles, the incorporation of compliance methodologies into performance and assurance frameworks (KPMG 2013a) and the changing of auditing standards by regulatory bodies.

Indeed, IR is currently in an experimental phase, and empirical approaches lack homogeneity (KPMG 2013b; Vaz et al. 2016; Dumay et al. 2016). Similarly, the theoretical perspectives are not always in agreement. IR has in fact been considered

M. Del Baldo (✉)
Department of Economics, Society and Politics, University of Urbino Carlo Bo, Urbino, Italy
e-mail: mara.delbaldo@uniurb.it

S. O. Idowu, M. Del Baldo (eds.), *Integrated Reporting*, CSR, Sustainability, Ethics & Governance, https://doi.org/10.1007/978-3-030-01719-4_5

both as the evolution of the annual report (Tweedie and Martinov-Bennie 2015) and the evolution of the sustainability report (Kolk and van Tulder 2010; Seuring and Mueller 2008; GRI 2013). Some contributions are of a highly critical nature, especially with regard to the integrated report model proposed by the IIRC (Brown and Dillard 2014; Flower 2015; Thomson 2015), while others argue that integrated and sustainability reporting can coexist (Adams 2015). Overcoming the controversial positions, some authors (Fasan 2013; Silvestri et al. 2017) consider IR as a tool that has the potential to overcome the limitations of both the annual report (complexity, short-termism, shortage of non-financial information) and the sustainability report (low reliability and trust from investors, disconnection from financial performance).

Some studies have reviewed the emerging literature on IR, by highlighting antecedents and consequences of IR, such as country and firm characteristics correlated to early adopters of IR (Frías-Aceituno et al. 2013a; Perego et al. 2016; Dumay et al. 2017), or a relation between the adoption of IR and quality of management (Churet and Eccles 2014). According to Dumay et al.'s structured literature review (Dumay et al. 2016), the relevant studies in IR can be analysed according to jurisdiction, organisational focus, country of research, research methods, IR frameworks and models, and the inclusion of academics, practitioners and consultants.

Previous research has also focused on the outcomes and benefits, pointing out that IR provides forward-looking information (Adams and Simnett 2011), demonstrates how organisations create and sustain value (Hampton 2012; Watson 2013), reduces reputational risk, enables companies to make better financial and non-financial decisions (Hampton 2012), breaks down operational and reporting silos in organisations, leads to improved systems and processes (Roberts 2011), improves resource allocation decision-making (Frías-Aceituno et al. 2013b) and supports challenges of integrated performance measurement systems (Giovannoni and Maraghini 2013).

Studies have also addressed the analysis of theories (mostly the institutional theory, the stakeholder theory and the legitimacy theory) to which the IR process relates (Ioana and Adriana 2014), as well as the theoretical and practical challenges arising from a non-unique interpretation of the integrated report (De Villiers et al. 2014; Camodeca and Almici 2017, p. 152).

In particular, both institutional and stakeholder theories provide a rationale for the IR reporting framework (Larrinaga-Gonzàlez 2007; Jensen and Berg 2012; Vaz et al. 2016). Institutional pressure exerts a positive effect on IR (De Villiers et al. 2014). Using institutional theory (DiMaggio and Powell 1983) Jensen and Berg (2012) analysed the external determinants (country-level drivers, i.e., political and legal systems, economic development and cultural characteristics) and internal determinants (company-level, such as size, profitability or industry) that explain the choice to implement IR. Stakeholder theory (Freeman 1984) has been used to analyse value creation at a firm level resulting from the collective efforts described in IR (Haller and Van Staden 2014). Moreover, the legitimacy theory represents a useful view for understanding the nature and impact of IR and the potential of this tool to manage organisational legitimisation and drives change in the extent of disclosure of non-financial capitals (Solomon and Maroun 2012; Setia et al. 2015). In addition, Perego et al. (2016) conducted a literature review of 17 studies by distinguishing the papers that analyse key institutional drivers or firm-level determinants of IR (antecedents) from those focused on the effects generated by its diffusion (consequences).

Recently, Velte and Stawinoga, after having analysed the current state of empirical research on IR, have developed a research agenda based on the level of analysis, the main theories used by researchers, research methodologies, the main input factors and the output achieved. According to their findings, IR research is mainly focused on the market and organisational levels of analysis, with few studies at the group or individual levels of investigation. As for the research methodology in use, surveys, interviews and archival studies are the most common methods (Velte and Stawinoga 2016).

Despite the fact that—as aforementioned—over the last decade IR has progressively gained prominence in the academic, professional and standard setters world, the problems and the opportunities of IR for SMEs (small and medium-sized enterprises) have been partially and only rarely addressed. Namely, studies on implementations of IR in this context are still rare (Del Baldo 2017b). This occurs despite SMEs' diffusion and their worldwide relevance within the socio-economic context (Smallbone and Wyer 2000; Dobbs and Hamilton 2007; Amorós et al. 2013; Eurostat 2015). For example, we see the almost total lack of studies aimed at analysing the motivations at the base of IR implementation in SMEs, the choice being voluntary rather than mandatory. Furthermore, there is a shortage of empirical research aimed at assessing the process of IR implementation.

Literature has only recently dealt with the issues of IR that focus on the analysis of the IR framework and its implementation among large organisations (Wild and Van Staden 2013; Lai et al. 2013; Incollingo 2014; Salvioni and Bosetti 2014; Ruiz-Lozano and Tirado-Valencia 2016; Lai et al. 2017; Silvestri et al. 2017; Macias and Farfan-Lievano 2017), while empirical studies on the actual degree to which the integrated report has been adopted by SMEs within the different countries are very rare, especially with regard to unlisted SMEs. This practice is in fact just beginning in Europe. It is still an uncommon practice, especially in Italy where only a few listed companies have decided to face the challenges that the integrated report implies (Camodeca and Almici 2017), and SMEs that are currently issuing an integrated report are quite rare. To be exact, SMEs seems to be completely neglected from the IR discourse.

Drawing from these premises the present work aims to fill this gap. Accordingly, the purpose of this chapter is to examine the reasons, the main steps, as well as the criticalities and the perceived benefits of a medium-sized Italian company (Dellas Spa) in implementing the IR process and releasing the integrated report (Dellas 2016, 2017). Given that the Italian scenario is mainly formed by SMEs, we focused on a specific medium-sized company that stands out in Italy for having adopted the integrated disclosure process much earlier than most other domestic large and international companies. Namely, Dellas has been selected because it represents the first best practice relative to an Italian SME (Eccles and Krzus 2015; Del Baldo 2017a) that released the integrated report.

The empirical research is based on the case study methodology and aims to answer the following research questions: Why did Dellas move to IR? How did Dellas perform the IR journey? What were Dellas' main steps towards the integrated report? In order to answer the above questions, an interpretativist research approach has been performed, since Dellas was included among the pioneers of Italian companies adhering to the IIRC/NIBR (Italian Network Business Reporting), a specific multi-stakeholders working group created in 2013 (<IR > and SMEs Working Group of NIBR) with the aim to develop and diffuse specific guidelines for the implementation of IR in SMEs at the national and international level (Del Baldo and Girella 2017; NIBR 2018 upcoming). Moreover, site visits and interviews were carried out, and company materials were analysed.

This chapter is structured as follows. Section 5.2 provides a brief theoretical literature review about IR in SMEs, and Sect. 5.3 describes the research approach. Section 5.4 deals with the case study, and Sect. 5.5 contains some concluding remarks.

5.2 Theoretical Background

The IIRC framework states that "an integrated report and other communications resulting from the IR will be of benefit to all stakeholders interested in an organisation's ability to create value over time" (IIRC 2013, p. 8). In addition, the framework states that: "an integrated report should be prepared primarily for providers of financial capital in order to support their financial capital allocation assessments" (IIRC 2013, p. 8). Therefore, it has been argued that it is mainly targeted at large, publicly listed companies, excluding small to medium-sized organisations and those in the government and non-government sectors (Burritt 2012; Milne and Grey 2013; Dumay et al. 2016).

Among the literature on IR, only a few pioneering contributions have been addressed to SMEs (James 2013a, 2013b; Del Baldo 2015, 2017a, 2017b; Camodeca and Almici 2017; Demartini 2017). A call for more research at the firm level has been requested (Higgins et al. 2014; Stubbs and Higgins 2014; Kolk 2010; Havlováa 2015) in order to engage with actors responsible for reporting, gain more understanding of the issues associated with implementing reporting innovations (Parker 2005; Eccles and Krzus 2015) and explore whether IR is stimulating innovative disclosure mechanisms (Higgins et al. 2014). However, the research agenda (De Villiers et al. 2014) does not explicitly mention the need to inquire the opportunities and difficulties faced by SMEs in operationalising the IR framework.

The meta-analysis of Dumay et al. (2016)—aimed to critically evaluate past and future IR research—found that most published IR research presents normative arguments for IR, while there are minimal contributions examining IR practice. Therefore, authors point out the need to move beyond the first stage of research (focused on creating a discourse and triggering engagement) and the second stage aimed to clearly demonstrate the benefit of IR[1], to subsequent research stages, in order to bridge the gap between practice and academic research (see Dumay et al. 2016: 13). Drawing from the results of their analysis one can note that based on the criterion "organisational focus" (consisting of six attributes: publicly listed, private-SMEs, private-others, public sector, not-for-profit and general) out of 56 published papers, none addressed private-SMEs. Thus it is clear that research on IR in SMEs is lacking. What are the possible reasons for this gap?

On the one hand, we can claim that IR is currently in an experimental phase for all companies, independent from their size (Vaz et al. 2016: Dumay et al. 2016). However, previous empirical studies do not agree that legal, political, economic and cultural characteristics (at the country level) and size, profitability and industry (at the company level) explain the decision to present an IR. Based on the stakeholder theory, the empirical analysis carried out by Frías-Aceituno et al. (2013b) concluded that size and gender diversity of the board are positively correlated with IR and that large companies with high growth opportunities produce the most IRs. In addition, Frías-Aceituno et al. (2013a, 2014) found that profitable companies are more likely to produce an IR. In contrast, García-Sánchez et al. (2013) reported that larger, more profitable companies with lower growth opportunities are more likely to present IRs. Focusing on the internal determinant, they found a significant association between IR and the assured sustainability report, company size and industry supplemental reports. Finally, considering both the country and company level of analysis, the empirical study carried out by Vaz et al.'s (2016) shows no influence of company-level characteristics (size, industry, dependence on capital markets and the existence of an assured sustainability report) on the decision to prepare and release an IR.

On the other hand, the gap could be due to the absence of guidelines "able to speak to" SMEs, which are easily adoptable by their consultants (particularly by chartered accountants), who still encounter difficulties in operationalising the IR principles (such as materiality, integrated thinking and connectivity). SMEs' specificities are mainly related to the different governance and organisational structures, the strategic approach (the logic of the strategic management process is usually incremental-based and process-driven and does not follow an object-driven logic, nor is it formalised) and the limited provision of managerial and financial resources, which mirrors the different culture of management control and reporting (Marchini 2005; Gnan et al. 2015). These features drive the hypothesis that, in the context of SMEs, the IR approach needs to be consistent with the entrepreneur's attitudes and purposes, the managerial culture, as well as the regulatory framework and the company's local operating environment. This hypothesis is partially confirmed by the fact that IIRC and other international organisations (such as IFAC and GRI) are interested in releasing specific frameworks/guidelines addressed to SMEs, which

[1] "The research that we can classify as second stage identifies that the application and impact of <IR> are fragmented and inconclusive about <IR>'s benefits" (Dumay et al. 2016: 178).

still encounter difficulties in operationalising the sustainability reporting and IR principles.

In fact, in the last few years, the relevance of IR for SMEs has been internationally acknowledged by the Chartered Institute of Management Accountants (CIMA–Integrated Reporting for SMEs–Helping Business Grow–Case studies, August 2015) within the SME & Entrepreneurship Taskforce of B20/G20. Policy makers have also expressed an interest in supporting IR diffusion within SMEs through the World SMEs Forum (WSF) which has been set up to concretise recommendations and assist SMEs to overcome hurdles in the process of IR. In such a context, the IIRC has become the promoter of SMEs' need to develop a more holistic and integrated approach to accountability which emphasises the way value is created. Accordingly, a new specific section has been included in the Example Database of the IIRC entitled "<IR > Reporters", incorporating the integrated reports of best practices released by SMEs. Furthermore, within the IIRC/NIBR-Italian Network Business Reporting, a specific working group was set up with the aim to release specific guidelines for the implementation of IR in SMEs at the national and international level (NIBR 2018 upcoming).

Namely, in Italy some pioneering initiatives of soft regulation (Morth 2004, 2006) have been diffused (Commissione Bilancio Integrato ODCEC Milano 2016). These initiatives stress the benefits deriving from the adoption of IR: the internal benefits; the external market benefits; benefits from managing regulatory risk (Eccles and Saltzman 2011). The first type of benefits regards lower reputational risk, increased involvement of stakeholders and better use of internal resources, while external market benefits regard the fact that stakeholders may be more and better informed about the financial and non-financial performance of a company (Arguelles et al. 2015). Finally, the third class of benefits regards advantages that firms can have regarding regulators, such as the possibility of being involved as the main actor in developing frameworks and standards (Eccles and Saltzman 2011). Despite such considerations, the low rate of adoption could be due to the costs and resources needed to produce IR and to communicate this to external users (Prado-Lorenzo and Garcìa-Sànchez 2010).

Based on the above discourse, the inquiry of the case of the first SME adopters that have successfully experienced the IR journey appears to be relevant.

5.3 The Research Approach

The research has been conducted by focusing on Dellas S.p.a, a medium-sized Italian company that has been selected because it is one of the first adopters of the integrated report among the Italian medium-sized companies. Dellas has been involved, in addition to two other Italian SMEs (Del Baldo 2018), in the Italian Network Business Reporting in the drafting process of, a guidance for IR in SMEs. The analysis examines Dellas' transition from the financial and corporate social responsibility report to the integrated report, aiming to answer the following research questions: Why did Dellas move to IR? How did Dellas perform the IR journey? Which steps did drive Dellas towards the integrated report? In order to reply to these questions, the empirical study has been developed using a qualitative approach and a case-study method (Eisenhardt and Graebner 2007; Patton 2002; James 2013b; Doni and Gasperini 2014; Veltri and Silvestri 2015; Camodeca and Almici 2017; Macias and Farfan-Lievano 2017; Cerbone and Maroun 2017). The case-study approach allows us to undertake an in-depth investigation of a phenomenon in its real-life context and answer "how" or "why" questions about contemporary events (Yin 2014; Qu and Dumay 2011). It also offers the possibility to understand management accounting in practice (Scapens 1990; Ryan et al. 1992; Dumay et al. 2016).

Within the research field on IR, attention has been paid to the need to investigate, through the use of an in-depth qualitative analysis, the internal mechanisms employed by early adopters of IR in order to explore how they are managing their reporting process (Stubbs and Higgins 2014; Havlováa 2015). Accordingly, the study of this significant case (Yin 2003; Naumes and Naumes 2006) follows the action research approach (Benbasat et al. 1987) and is considered particularly useful for the improvement of knowledge in the field. Through this approach the action researcher does not stand as an independent observer but becomes a participant, a part of the NIBR working group formed by scholars and practitioners, such as IIRC representatives, chartered accountants, banks officers, consultants, managers and entrepreneurs directly involved in issuing the integrated report and drafting the guide for the implementation of IR in SMEs.

The analytical approach based on narrative analysis and qualitative data collection draws on an interpretivist approach (Crane 1999; Currie et al. 2009; Brown and Dillard 2014), which is considered appropriate for studying evolving organisational practices (Higgins et al. 2014) and new reporting practices as per the objective of this explorative research (Stubbs and Higgins 2014). This approach was used searching for an explanation of at least three issues: the reason for Dellas' move to integrated reporting; the main steps of this transition; the current state of progress. Accordingly, the analysis (developed across a multi-year period, beginning in 2013 and continuing today) benefited from 4 in-depth semi-structured interviews that focused on the above mentioned main features with the Dellas CFO (Chief Financial Officer) who represents the key informant person since he was involved in both the preparation of IR for a firm that pioneered this form of reporting and the NIBR working group. In detail, the interviews aimed to investigate the following main aspects: Dellas' route towards the integrated report; the conditions under which Dellas' engagement with IR was enacted and evolved throughout the years; the main recommendations Dellas has complied with in the preparation of its integrated report; Dellas' compliance with the IR Framework's guiding principles and content elements; the drivers and challenges that emerged from this process; the difficulties encountered in producing the integrated report and adopting the IR framework and the way they faced them; and the benefits and future perspectives. All interviews were recorded, transcribed and validated.

In addition, we performed a participant observation—typical of the action research approach (Adams and McNicholas 2007)—and we benefited from informal discussions and conversations during focus groups, meetings, round tables and technical workshops. The interviews were supplemented with the analysis of Dellas' reports (from 2011to 2016), reports issued by the NIBR working groups and data drawn from the company web sites. Triangulating data sources enhance trustworthiness (Lincoln and Guba 1985) and contextualise the data (Yin 2014; Van Bommel 2014).

On this basis, the next subsections introduce the company, then describe and discuss the main findings drawn from the analysis.

5.4 The Case Study

5.4.1 Dellas Spa: Company Profile

Dellas is an Italian industrial group that was founded in 1973. Its headquarters are in Lugo Grezzana (Verona). With over 40 years of activity, Dellas has gained a top position as an international market leader in the manufacturing and sales of a wide range of diamond tools for marble, granite and agglomerate. The group is highly internationalised and has foreign subsidiaries and branches in several European and extra EU countries. Dellas is owned by the Ferrari family and is currently led by its second generation. The founder is the Honorary President, who holds 19% of the shares; 17% is held by his wife and the two children, who are respectively the current CEO and Vice-President and Marketing, Communications and HR Manager of the company, each owning 18.5% of the shares. The remaining (27%) is held by three other individual members.

Since its beginnings, Dellas has been strongly inclined towards research and development, dynamism and the innovation of processes and products. Its excellence comes from human resources and distinctive entrepreneurial and managerial capabilities. The enterprise started its life in the valley of Valpantena, that like the nearby Valpolicella has always been known for the local skills in the cutting of marbles and granites from this area, from Italy as a whole and internationally. In 2016 Dellas achieved a consolidated turnover of approximately 18 million euro and an invested capital of 34 million euro, counting 124 employees.

5.4.2 The Why of IR: A Will to Disclose Dellas' Value Creation Model

Through a process of evolution of the corporate reporting, which started in 2011, Dellas released its first integrated report in 2012. Drawing from the international tendencies in business reporting following a series of subsequent steps, in 2014, Dellas applied the principles of the International <IR> Framework (December 2013) in order to disclose its value creation model, being aware that efficient financial-economic communication promotes the setting up of a continuous and well-informed relationship with its stakeholders and reinforces the company's strategic and operational credibility.

> The main motive for implementing the IR was (and is) value creation, the will to understand, assess and communicate how Dellas creates value. Integrated thinking has raised awareness of the fact that the company can build its future on solid foundations only if it produces positive economic results and, in the meantime, value for the community (social, environmental and ethics value) (Dellas CEO, March 7th, 2016).

The motivations for starting the IR journey are tied both to external and internal factors. Among the first ones Dellas considers external pressure, pressures exerted from within the company structure and reasons connected to new communications technology. Among the second ones are the family owners' set of values that forged the company's cultures since its foundation: respect, responsibility, honesty, willingness to help (Table 5.1). The Charter of values and behaviours (Dellas' code of ethics) strived for by the group's sales staff is the basis for extending the company's values to all its employees and collaborators. These values created the first pillar toward responsibility, transparency and attention to intangibles. As one can read on the Dellas company report "*Many small and medium-sized businesses, despite having a significant social impact themselves, have, to date, failed to suitably provide adequate information on their social responsibility work, either because they think that such information remains a matter solely for large businesses or because they deem social responsibility to be a matter of wholly secondary importance. There are in reality many tangible reasons why businesses should subscribe to ethical values and apply them to their company operations*" (Dellas Annual Integrated Report 2016, Dellas 2017: 33).

As one can draw from the integrated annual report "environmental, social and governance performance indicators directly contribute to a company's financial performance" (Dellas 2017: 33).

Table 5.1 Company values

Respect	We are convinced that the prosperity of a company springs from the human and professional growth of its employees and its customers
Responsibility	The commitment, professionalism and loyalty provide the basis for the establishing of relationships founded on trust and the sense of responsibility, with benefits for all persons involved and the community in which we operate
Honesty	The experience we have gained over 40 years of working in this sector of industry means we are able to guide the customer towards transparent purchases
Willingness to help	We listen, we appraise and we decide together with our customers on what is most appropriate to them for the achieving the best possible performance

Source: our adaptation from Dellas Integrated Annual Reports (Dellas 2016, 2017)

Dellas, aware of the importance of monitoring and managing non-financial drivers, has thus approached IR in order to fully understand and communicate, how and to what extent these drivers were related to the ability of the company to create economic and financial value, together with a wider shared social value. Dellas' integrated report has become the main communication tool for the management and employees, and all other stakeholders. The company wanted to illustrate to its stakeholders their capacity to sustainably increase the long-term value through its core-drivers that have an impact on the income statement (pre-financial), namely: customer satisfaction in terms of flexibility and delivery times; good relationships with key suppliers according to their reliability and accuracy of payments; very low staff turnover due to a good level of training, organisational climate and occupational safety.

5.4.3 The How of IR: A Step by Step Process

Dellas had decided to go beyond the financial report, initially taking on sustainability reporting and subsequently implementing integrated reporting (Table 5.2).

> However, initially, in 2011 and 2012, the <IR> Framework was still its current version, and its principles were unknown to us. By the 2013 Annual report, we started to implement Integrated Reporting. Starting with the 2013 Annual Report, we have provided the integration of the economic and financial, social and environmental profiles and specific sections of the report, according to the IIRC Framework, defining the key performance indicators of the areas of governance and social and environmental impact and opening channels of dialogue with stakeholders. In the 2014 report, we expanded the reporting model, adopting the GRI Sustainability Reporting Guidelines, the G4 version, which enabled us to better define the material issue and KPIs to measure and communicate and to better manage stakeholder engagement for the materiality analysis. We have also improved the implementation of the IIRC Framework, implementing additional standards, and we have divided the KPIs in relation to capital described in the IIRC Framework (Dellas CEO, January 13, 2017).

Through the 2011 report, the company produced its first structure notification of Corporate Social Responsibility according to the first European Commission guidelines.[2] The document went well beyond the minimum content regarding the economic, capital and financial information required by civil regulations and national accounting standards to embrace and also illustrate the social and environmental profiles of the company.

Since 2012, Dellas has put more effort into expanding the scope of the account rendering of the company. A picture of the Dellas Group was given in its entirety, involving all the main economic, financial, social and environmental aspects of the

[2]The main documents concerning the guidelines published by the European Commission, are the following: European Commission, "Guide to Communicating about CSR", 2009 European Commission, "Awareness-Raising Questionnaire", 2005; European Commission, "CSR in Small Businesses: Good Practice Examples", 2007; European Commission, "Opportunity and Responsibility", 2007.

Table 5.2 The evolution of Dellas' corporate reporting

Report 2011	• First structured notification of the CSR according the first European Commission guidelines • Breakdown of business model, governance, risks and opportunities • Determination of the areas of social impact
Report 2012	• Adoption of the «Principles for the drafting of the financial statement company» in 2001 from GBS • Expansion of account rendering scope for the whole group • Introduction of the economic-financial, social and environmental framework of operations, following the draft framework published by the International Integrated Reporting Council
Report 2013	• Following the framework of the International Integrated Reporting Council, having carried out the integration between the economic-financiall, social and environmental profiles of the company management, entering specific connections between the various sections of the report • Performance indicators of the areas of social impact • Listening and dialogue channels with stakeholders, and improvement objectives
Report 2014	• Adoption of account rendering standard GRI-G4 • Stakeholder engagement for materiality analysis • Communication to partners of the company approach to sustainability by the creation of the ethical code • Breakdown of KPIs with reference to company capitals as specified in the IIRC Framework
Report 2015	• Improvements to the connectivity of information within the report and between the report And the website in response to the corresponding guidelines of the IIRC framework • Having reorganised and added more detail to the section relating to the business model Most connected with value creation • Online survey of annual report information and content • Inclusion of the accounting tables for the consolidated financial statements
Report 2016	• Analysis of intangibles for the better representation of the human, intellectual and rational capital on the basis of the WICI intangible reporting framework • The section regarding the business model and the value Creatin process through an explanation of the principal impacts • Review of the consolidated financial statements

Source: our elaboration from 2016 Integrated Annual Report Dellas Spa (2017: 19)

operations of all entities over which Della Spa exerts an influence. The principles for the drafting of the financial statement and social report were adopted as established in 2001 by the GBS standard.[3]

Since the year 2013, with the publication of the annual report, taking into account the <IR> International Framework, the document has included: indications of the main integration items for the social and environmental management profiles, through inclusion of specific cross-references and connections between the various

[3]The GBS is the Italian Study Group set up to establish the GBS Social report drafting standard established in 1998 by entities, associations, universities, companies and certification bodies.

sections of the report; and information relative to its internal and the external environment, the governance and business model, risks and opportunities, strategy and allocation of resources, performance and prospects.

In 2014 Dellas adopted the GRI International Standard 4.0 (GRI 2013) with reference to the ESG (environmental, societal and governance) aspects relating to areas that do not strictly regard accounting matters: such a standard has enabled Dellas to identify and report specific indicators relating to the social and environmental performance of the company governance.

In 2015 the company introduced the following new elements: indications of notifications on the uses of the various areas of company capital, as in the <IR> International Framework; and reorganisation of the business model, more closely connected to the concept of value creation.

Finally, through the 2016 integrated annual report (Dellas 2017), Dellas has continued the communication plans, already begun through the use of the GRI 'core' option with reference to international account rendering of ESG aspects within the <IR> International framework. Furthermore, to better represent the human, intellectual and relational capital, the company has further examined the WICI suggestions regarding the Intangible Reporting Framework (WICI 2014).[4]

Moreover, taking into account the EU Directive 2014/95 on Communication of non-financial information (assimilated by the Italian Legislative Decree 254/2016),[5] the Dellas Group has chosen to make a voluntary non-financial information statement. However, Dellas has waived the supervisory activity.

The part that regards "Company Social Responsibility" has been made possible, as in previous financial years, thanks to the gradual and targeted involvement of some of the most managerial figures of the company (the inclusiveness principle), which included programmed encounters and focus groups designed to identify the most significant accounting information (relevance principle) regarding the actions carried out by Dellas during the year 2015 (completeness principle). In the account rendering of the contents of the balance sheet, it has been Dellas intention to provide an impartial picture of the performance, indicating both the positive and negative aspects (principle of balance).

The next Dellas report shall be drawn up to improve, increasingly and year by year, the annual information available to ensure the publication of an integrated report that complies with the best international standards, that is fit to truly meet the

[4]WICI Intangibles Reporting Framework—September 2016—www.wici-global.com. The World Intellectual Capital/Assets Initiative has developed a framework to improve the representation of intangibles, in a context of value creation, in direct relation to the international <IR> Framework. www.wici-global.com

[5]The EU Directive 2014/95, assimilated by Legislative Decree 254/2016, establishes new minimum reporting standards on environmental and social matters in relation to the management of personnel, with respect to human rights and the combating of active and passive corruption. The directive aims to introduce and reinforce good conduct and has the objective to increase transparency in the communication of non-financial information and increasing investor trust as well as that of the stakeholders in general.

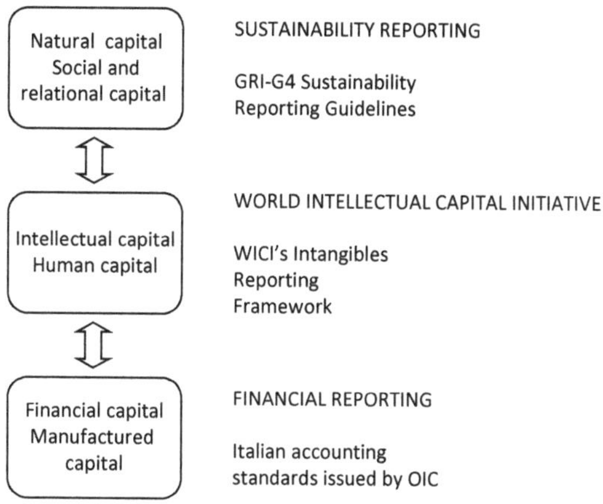

Fig. 5.1 The evolution of Dellas' corporate reporting. Source: our adaptation from Dellas' Integrated Annual Report, 2016 (Dellas 2017: 19–20). Italian National Association of Chartered Accountants. www.fondazioneoic.eu

information needs of all the stakeholders and provide an exhaustive and integrated picture of the impact of the company's activities on its own socio-economic impact.

Indeed, Dellas' latest integrated report is the result of an evolutionary process, passing through different reporting models since 2011 (Fig. 5.1).

Under the process view, Dellas approached the journey towards IR following a planned route, starting with the definition of its foundations in terms of strategy, business model, team and action plan.

First, in order to illustrate, in the integrated report, the way in which the company works and its processes, by ensuring a real and complete narration of the company history of value creation, Dellas considered it necessary to directly involve different areas of the company. For this purpose, a team formed by the Chief Executive Officer (CEO) and 5 leaders has been identified: Chief Sales Officer (CSO), Chief Financial Officer (CFO), Chief Operations Officer (COO), Chief Human Resources Officer (CHRO) and R&D Manager.

Secondly, the IR working group realised that more than a year was necessary to introduce integrated decision-making, in terms of resources and timing. Accordingly, a roadmap of 3/4 years, named "Corporate Reporting Evolution", was planned and scheduled (Fig. 5.2). Moreover, Dellas understood that in order to achieve a good level of reporting it was necessary to provide the team with the support of an external technical partner and the public financial support obtained from the Veneto Region.

Finally, the group made it clear that the results of the reporting could be useful for each business function (for better focusing on their KPI's) and for the investors (to allow for a better evaluation of the company's strategies and the value of the key processes of each business area).

The experience continued with the process of value creation (Table 5.3).

> The main question to be answered with an Integrated Report was: "How does Dellas create value?". We have tried to answer this question using a "waterfall plot" useful to summarise, integrate and monitor different aspects of our value creation, from the main objective to the KPI's (Dellas CEO, March 3, 2017).

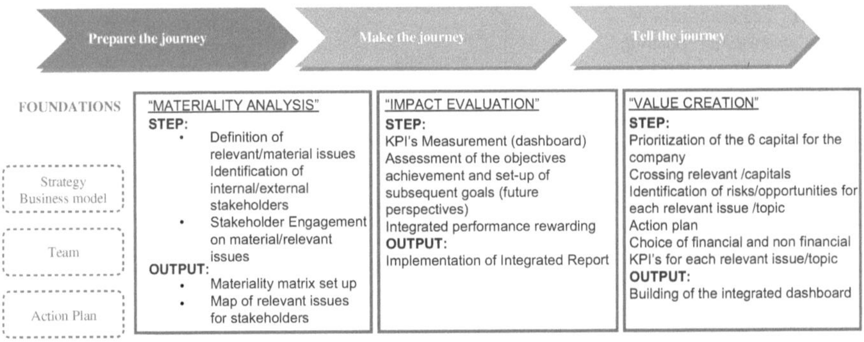

Fig. 5.2 The Roadmap. Source: our adaptation from NIBR working group "Guidelines for IR in SMEs" (2018) draft

Table 5.3 Dellas value creation

Steps	Output
Prioritisation of 6 capitals for the company Intersection between relevant topics and capitals Individuation risks and opportunities for each relevant topics Action plan Choice of financial and non-financial KPI's for each relevant topic	Construction of the integrated dashboard

Source: our elaboration from interviews addressed to Dellas CEO

Each capital was attributed to different categories of internal and external stakeholders. Each category of stakeholders has been involved in specific initiatives and with variable frequency from quarterly to annually. Specific KPIs were selected to monitor the performance achieved, strengthen the capitals involved and the stakeholder interactions and implement specific improvements. Starting with defining the company's relevant activities, Dellas continued with the identification of the internal/external stakeholders and the process of stakeholder engagement on material aspects.

In July 2013 the study of materiality and the building of the materiality matrix was carried out to investigate: which themes have an impact on the economic, social and environmental performance of the group and to what extent and which themes influence the decisions of the stakeholders and their opinions on the activities of the group. The analysis of materiality is annually updated through a process which encompasses two main operational steps: the analysis of relevancy (to define the most important themes for Dellas' business) and the assignment of a priority (low, medium, high) to the relevant themes according to the impact on economic/social/environmental performances (drawn from the analysis of previous annual company reports, internal policies and interviews that addressed top managers) and the stakeholders influence on the activities of the group.

Table 5.4 Dellas Impact Evaluation

Steps	Output
KPI's measurement (dashboard) Assessment of the objectives achieved and setting-up of next targets (future perspectives) Integrated performance rewarding	Integrated report releasing

Source: our elaboration from interviews addressed to Dellas CEO

The last step—"Impact evaluation"—was performed in order to implement continuous improvements through the identification and monitoring of specific KPIs for each category of stakeholders (Table 5.4).

Dellas introduced integrated thinking through the IR process. What began as a reporting of ESG aspects highlighted within the sustainability report, became a more integrated section in the annual report.

5.4.4 *The Benefits*

According to the CEO: "*Through the improvement of the integration of social and environmental aspects in the development of strategic objectives and the overall value of a company, the foundations were laid for learning about and better understanding the link between environmental and social performance and the success of the business. The IR approach has allowed the company to understand the connections between financial and non-financial dimensions and how sustainability can influence profitability and vice versa. In other words, it allowed us to improve the global strategy to generate and communicate the company's value*" (Dellas CEO, January 13, 2017).

In order to carry out a structured stakeholder engagement, Dellas conducted a series of interviews with various categories of corporate stakeholders in 2014 to obtain feedback on the effective communication of Dellas' 2013 integrated report. Drawing from the interviews the following benefits have been pointed out: consolidation of mutual trust between the company and its stakeholders; growth of Dellas' reputation; increase in transparency through a reporting tool able to fully understand Dellas' competitive advantage. Among the external benefits, it should be mentioned that the Integrated Report has become the preferred means of external communication with stakeholders.

> Our suppliers have received the integrated report with interest, acknowledging that Dellas is one of the few Italian SMEs to provide structured communication; Particular appreciation has been expressed for the visibility given to the various projects of R&D, developed to improve the product. Large interest has been shown by employees and collaborators, who personally contribute every day, in order to achieve these high standards of quality (Dellas CEO, March 2017).

Dellas' annual report has been inserted in the IIRC database and contains emerging best practices of integrated reports.

> Being reported as a best practice is a source of pride for Dellas (Dellas CEO, March 7, 2017).

5.4.5 *Difficulties and Future Goals*

> The major difficulty in the application of integrated reporting principles is that the <IR> Framework is principle based. We must first understand the principles proposed and then find the most suitable ways to express them within the company. In our case, to reach a good level of reporting was crucial, early on, to receive external support with a training nature that involved a group of 5 key-leaders; managers put the "principle-based notions" of ' <IR> Framework into place (Dellas CEO, October 14, 2016).

Dellas performed a process of continuous improvement for the reporting model; however, the principle of connectivity was one of the most difficult to apply in a complete manner. "*We are currently engaged in improving connectivity, as the IIRC Framework suggests, to more effectively measure the factors that affect the organisation's ability to create value over time. We are also improving the performance of the business model, which is based on different forms of capital. The organisation's output and its activities are reflected in the outcome, in terms of the effects on different capital and the categories of stakeholders involved*" (Dellas CEO, March 7, 2016).

> It is understanding and translating in numbers how the different forms of capital, defined as input factors, through business activities, are converted into outputs (products, services, and waste) and how these outputs are substantiated in outcomes on different capital and the categories of stakeholders. The goal is to explain the process of translating logical flows into real financials (Dellas CEO, October 14, 2017).

> Historically, the founder followed all business activities closely and steadily spread out, however informally, the company's values and the importance of considering how significant the pre-financial drivers are, with their own way of doing things and directing the company daily. Following the change, it was important that the new corporate managers and stakeholders, particularly shareholders, recognised the central value of a sound management of intangible drivers to generate sustainable economic value, and thus integrated reporting was a lever that made it spread in a more formal manner, and the culture of the economic, social and environmental global value was fully accepted, through new management tools and encoded and shared KPI (Dellas CEO, January 13, 2017).

One of Dellas' long-term goals is integrated performance rewarding, namely the incorporation of the IR KPIs in Dellas' Corporate Performance Management. Finally, a further goal that is a critical aspect is relative to the assurance. The report of the Board of Statutory Auditors and the Accounts Auditor only relates to financial statements and the annual business report. As we can read in the annual report "*Organisation's policy and current practice with regard to seeking external assurance for the report on Corporate Social Responsibility was not subject to external assurance*" (Dellas Integrated Annual Report 2017, 76: G4-33).

> The assurance is currently a sore point; every year we wait to tackle this issue while deciding. For Dellas, it would represent a value, which increases the value of the report. The assurance is not expected, however, it is voluntary. There is currently no third-party auditor. At this moment, we are not able to identify who (and how) will get it done (Dellas CEO, March 7, 2016).

5.5 Discussion

Drawing from the case, we can argue the following insights.

First, from Dellas' point of view, moving to an integrated annual report expressed the transition from a partial approach of corporate disclosure to a holistic type of vision. Currently, Dellas is releasing an integrated report as envisaged by the IR framework even if they have not implemented the guidelines in an uncritical manner as Dellas focused its attention on the achievement of sustainable conditions, seen as the reconciliation of all the company's stakeholders' expectations. Therefore, Dellas does not identify the shareholders and the lenders as the privileged recipients of the integrated report, but all stakeholders (employees, customers, suppliers, public administration, etc.) potentially interested in learning about corporate performance in the medium and long term and understanding the impact the business has on the social and environmental dimension. With reference to the literature framework, relative to the main issues raised about the framework (Higgins et al. 2014; Brown and Dillard 2014; Flower 2015; Thomson 2015; Ruiz-Lozano and Tirado-Valencia 2016), the case points out that releasing the integrated report has encouraged the adoption of more responsible behaviour over time, giving prominence to corporate responsibility, transparency and sustainability principles. In this regard, we can note that corporate sustainability is one of the main objectives of Dellas' report alongside the value relevant information. In addition, Dellas' transition towards the integrated report have been characterised by three main stages: the listing and corporate social responsibility reporting (2011–2012); the integration between financial and non-financial information (2013–2015); the integration of the multidimensional perspectives (economic, social, environmental, governance and intangibles) report according to the IR and the WICI framework (2016). Therefore, IR represents, in this case, both an evolution of the annual report and the sustainability report and allows them to overcome their respective limitations (Fasan 2013; Silvestri et al. 2017).

Secondly, we acknowledge that the implementation of IR required three fundamental aspects: 1) the "re-definition" and critical revision and assessment of the business model and strategies; 2) the need to create a team that involves the main business functions and 3) the need to set the targets, time and costs for the IR implementation, given that "IR is a journey" that involves an evolution of the organisational structure, as well as the approach. The roadmap to get to the IR progressively requires at least a period of 2/3 years and is structured around 3 basic steps: a) Materiality Analysis: definition of the areas where the organisation can create more value and, at the same time, potentially risk the greatest value loss; b)

Value Creation: definition of the way in which the organisation uses its capital (input) for the value creation (outcomes) towards the stakeholders and; c) Impact Evaluation: measurement of relevant indicators and definition of the objectives to be achieved in order to ensure value creation. Dellas' successful experience thus points out that even an SME can succeed in setting up a specific team (an IR working group), although formed by a few members. Such a team (still existent) has driven the evolution of the reporting model through the years and has made it possible to complete the process and release the integrated report. Moreover, with the presence of skilled internal/external resources, the availability of a suitable informative and technological system emerges as a key factor in supporting the journey.

Thirdly, Dellas is aware that the internal benefits deriving from IR can be obtained only with the maximum involvement of the organisational structure, taking into account the available resources. Accordingly, both the strong and authentic commitment of the board (Adams and McNicholas 2007; Cormier et al. 2004; Maharaj and Herremans 2008) and the role and tasks of the working group's head are crucial. The team leader performs multiple roles:

- ensures participants' cooperation and access to all sources of information;
- guarantees reliability and quality of the data in a transparent way;
- allocates roles, responsibility and times to all participants;
- establishes the meetings schedule and the reporting date and the date to release the reporting;
- identifies the key issues to work on;
- agrees on an action plan in which all activities (and who are responsible for them) are scheduled;
- identifies the role and relationship with both internal and external stakeholders and the most significant aspects that affect the firm's ability to create value by assessing different priorities;
- defines the business model and how it responds to the strategic objectives of the organisation, through the use of available capital;
- checks if the report preparation follows the guiding principle and content elements provided by the IR Framework;
- identifies KPIs and KRIs;
- ensures the conciseness and truthfulness of the narrative;
- ensures the continuity of the reporting process, which must be improved continuously.

Thirdly, the importance of participating in the NIBR working group emerges. This experience has enhanced the awareness of IR advantages and limitations and has encouraged a cultural and organisational change. Currently, Dellas is going through the development stage of the IR process and is committed to improving the quality of the report.

The perceived internal and external benefits of the IR journey are significant, and this awareness is the basis of the choice of IR adoption. From the internal point of view, it has helped to strengthen integration within the organisation and nurture a holistic vision of the value process creation, positively affecting the decision-making process and favouring sharing, cohesion and transparency. The internal benefits include an assessment of intangibles (Lev 2004; Goodridge et al. 2014; WICI 2014; Zambon and Guenther 2011) that would otherwise remain hidden (i.e., specific technical and organisational skills, product quality, excellent relations with clients, employees, suppliers and the local community) and strengthening of the cohesion of the company culture. Notably, the following feedback has emerged from the interviews: the IR journey contributes to the improvement of the internal decision-making process; it raises awareness about medium and long-term risks; it develops the capacity for strategic sharing with collaborators (internal engagement); it produces a growth in the visibility of the management team and entrepreneurial leadership. In other words, the IR process contributed (and contributes) to transforming corporate processes (Phillips et al. 2011), breaking down operational and reporting silos, leading to improved systems and processes (Roberts 2011) and improving resource allocation decision-making (Frías-Aceituno et al. 2013b). From the external point of view, the IR represents a tool capable of increasing reputation, strengthening relationships with stakeholders and improving stakeholder engagement (Eccles and Armbrester 2011; Eccles and Krzus 2010; Hampton 2012; Watson 2013).

These considerations allow us to answer the first research question (RQ1), related to the why of IR. IR is perceived as the bearer of internal and external benefits and the tool that allows you to give more visibility to the value creation process. This feature constitutes the main engine of choice, especially in companies such as Dellas, which have a high level of intangibles.

However, it represented a complex and challenging choice. Many barriers emerged during the IR journey, due to the difficulties in operationalising some principles (i.e., connectivity and integrated thinking) tied to both internal and external factors, relative to the "high level of technicalities" of the IR framework (Ruiz-Lozano and Tirado-Valencia 2016) that should be necessarily simplified with regard to SMEs.

Therefore, trying to answer the second research question (RQ2), related to the how of IR implementation, drawing from Dellas' experience, we can hypothesise that IR represents a possible goal for an SME, if conceived as a process to perform in a medium to long-time perspective, in accordance with the complex of contextual variables (Perego et al. 2016) and communication strategy used to disclose the value of the company. In this regard, findings highlight scholars and practitioners, in accordance with managers and entrepreneurs, should work together to ameliorate the IR framework and render it more suitable for SMEs, as in Dellas' current experience.

5.6 Conclusions

The paper addresses the issue of IR implementation and diffusion within SMEs. Among the emerging literature focused on IR, both theoretical arguments and empirical research are almost entirely absent concerning SMEs. In order to deepen and enrich this "niche" but relevant and still under-investigated topic, this work attempts to provide insights and discuss the reasons that can lead an SME to adopt IR and the possible way they reach this goal. Accordingly, we performed a case analysis, focusing on an Italian SME that is part of the pivotal companies included among a working group aimed to produce a guide capable of helping small and mid-sized entities to adopt an integrated report. The case study we examined in this chapter identifies a best practice in IR within the Italian context. The paper has attempted to investigate the reasons that led Dellas to move from the corporate social responsibility report and sustainability report to the integrated report; the main steps of Dellas' IR process, its criticalities and benefits and the current state of progress of Dellas' integrated reporting process. Dellas' choice in drawing up the integrated report is an answer to a cultural incentive in terms of sustainability and social responsibility that the company has decided to implement, according to a continuous improvement approach.

Firstly, among the leading motivations to start the IR process in SMEs the will to reinforce stakeholders relationships (particularly with customers, employees and local community) emerges, to illustrate the companies' capacity to increase its value in the long-term on the basis of its core-drivers (i.e., the R&D capabilities) and to monitor and manage non-financial key-value drivers (RQ1). Drawing from the empirical analysis, a relevant challenge in the initial phase of IR implementation that influences the decision to adopt the IR and thus affect "the why" question is attributable to the weakness of the culture in disclosing non-financial information and the difficulty in fully understand the IR framework guidelines, which are difficult to "digested" for SMEs. SMEs are affected by a weak culture of accountability, and there is a gap in their diffusion of management and performance control systems (Silvi and Bartolini 2011; Marchini 2005). This evidence, relative to the SME's typical attributes, tied to the "company characteristics", appears in contrast to the findings of previous studies (Vaz et al. 2016), which demonstrate that size is not significant. Nevertheless, it aligns with research results, which remark on the influence of structural weaknesses tied to the specificities of SMEs (Frías-Aceituno et al. 2013a; García-Sánchez et al. 2013). In this regard the case-study emphasises the importance to have a simplified and handy guidance for implementing the IR process, releasing the integrated report and operationalising the necessary steps and the main basic concepts and principles. The lack of "*ad hoc rules*" for SMEs inhibit a more widespread adoption of IR.

Moreover, findings provide a reply to the "how" question (RQ2) by emphasising that the implementation of IR principles, such as the materiality, connectivity and integrated thinking principle requires a gradual process, consistent with a low level of formalisation adopted by SMEs in their strategic management process. These fundamental concepts should be developed through different time frames, because it may take more time before innovative disclosure mechanisms in SMEs emerge (Stubbs and Higgins 2014). SMEs adopting IR moves away from the outside-in approach, such as the GRI, to an inside-out (strategy led), or they adopt a twin approach (Higgins et al. 2014). The empirical analysis suggests that IR should be conceived as a transitional—rather than radical—initiative that carries out a "transformation" from other forms of reporting (i.e., sustainability reporting) or financial reporting and contributes it into developing the entrepreneurial and managerial culture of SMEs (Adams and Frost 2008; Adams and McNicholas 2007).

Secondly, findings point out the positive impact and the importance of two "enabling conditions": (1) the direct involvement of the top management and entrepreneurial team and the involvement of external consultants (i.e., professional accountants) in the IR process and (2) the adoption of a "step by step" approach whose roadmap is spanned throughout several years. An essential challenge for SMEs in implementing IR is in fact represented by the evolution of both the entrepreneurial and management control culture. Basic steps of the process include the involvement of the entrepreneurial team, the internal staff engagement process, aimed to gain a common understanding of the reporting process and data/performance indicators (Adams and McNicholas 2007) and the creation of cross-functional teams that act as key mechanism for implementing the IR process (Stubbs and Higgins 2014). In addition, the case study points out the importance of both the support of adequate information and technology systems, and specific software solutions and the setting up of partnerships with external consultants. In particular, the key role of practitioners able to explain the advantages and the potential downsides arising from the adoption of IR to SMEs (Parker 2005) emerges. Taking into consideration specific elements of SMEs, such as the family governance system and the entrepreneurial culture, practitioners serve as an integral part of the management team (Greenwood et al. 2002; IFAC 2008a, 2008b; Kolk 2010; Owen 2013; Havlováa 2015) and play a key role in driving SMEs toward incremental changes that are oriented to prepare and realise the IR journey.

Third, findings provide the basis for suggesting the effectiveness of a path (a "journey" to be prepared, managed and explained) through different steps, which are necessary to drive the evolution of the entrepreneurial and organisational culture of a medium-sized company towards IR.

Accordingly, within the theoretical and practical implication of the work, we can mention its contribution to drive the attention of scholars and practitioners to the diffusion of IR among SMEs, focusing on a research field which has not yet been adequately investigated and is currently underestimated. The analysis of the successful early adopter experiences helps to support managers, entrepreneurs and practitioners to carefully evaluate the implications of IR implementation by emphasising criticalities and benefits deriving from the IR process in medium-sized companies.

Finally, we are aware that the work is not free of limitations, which could be partly emended through future research steps. The main limitation is due to the fact that the case analysis is related to a single case study. However, SMEs that are adopting IR are currently "white flies" in the world and rare in Italy. Therefore, further research steps could be performed only when a considerable number of integrated reports issued by SMEs are available, in order to extend the comparative analysis including a significant number of companies. Moreover, the study needs to be improved by analysing the IR process in selected companies, using a medium or long-term period of time in order to assess the steps of the process. As such, despite the limitations of the empirical analysis, the reflections that have emerged are intended to influence change in the IR framework by suggesting a contingent-based approach for the improvement of IR in SMEs, considering specific factors affecting medium-sized companies and their stakeholders' expectations and needs.

References

Adams, C. A. (2015). The international integrated reporting council: A call to action. *Critical Perspetctives on Accounting, 27*, 23–28.

Adams, C. A., & Frost, G. R. (2008). Integrating sustainability reporting into management practices. *Accounting Forum, 32*(4), 288–302.

Adams, C. A., & McNicholas, P. (2007). Making a difference; sustainability reporting, accountability and organisational change. *Accounting, Auditing & Accountability Journal, 20*(3), 382–402.

Adams, S., & Simnett, R. (2011). Integrated reporting: An opportunity for Australia's not-for profit sector. *Australian Accounting Review, 21*(3), 292–301.

Amorós, J. E., Bosma, N. S., & Levie, J. (2013). Ten years of global entrepreneurship monitor: Accomplishments and prospects. *International Journal of Entrepreneurial Venturing, 5*(2), 120–152.

Arguelles, M. P. M., Balatbat, M., & Green, W. (2015). Is there an early-mover market value effect for signalling adoption of integrated reporting (Working paper). University of New South Wales.

Benbasat, I., Goldstein, D. K., & Mead, M. (1987). The case research strategy in studies of information-systems. *MIS Quarterly, 11*, 369–386.

Brown, J., & Dillard, J. (2014). Integrated reporting: On the need for broadening out and opening up. *Accounting, Auditing & Accountability Journal, 27*(7), 1120–1156.

Burritt, R. L. (2012). Environmental performance accountability: Planet, people, profits. *Accounting, Auditing & Accountability Journal, 25*(2), 370–405.

Camodeca, R., & Almici, A. (2017). From the corporate social responsibility reporting to the integrated reporting: The case of Sabaf S.p.a. *Problems and Perspectives in Management, 15* (1–1), 150–157. https://doi.org/10.21511/ppm.15(1-1).2017.01.

Cerbone, D., & Maroun, W. (2017). Exploring the challenges of preparing an integrated report. *Meditari Accountancy Research, 25*(4), 481–504. https://doi.org/10.1108/MEDAR-10-2016-0085.

Churet, C., & Eccles, R. G. (2014). Integrated reporting, quality of management, and financial performance. *Journal of Applied Corporate Finance, 26*, 56–64. https://doi.org/10.1111/jacf.12054.

CIMA. (2015). *Integrated reporting for SME – Helping business grow*. Case studies. Chartered Institute of Management Accountants.

Commissione Bilancio Integrato ODCEC Milano. (2016). (a cura di). *Il bilancio integrato per le PMI*. I Quaderni della Scuola di Alta Formazione, No. 69, Dicembre 2016, Scuola Alta Formazione Luigi Martino, Fondazione dei Dottori Commercialisti di Milano. Ordine dei Dottori Commercialisti ed Esperti Contabili: Milano.

Cormier, D., Gordon, I., & Magnan, M. (2004). Corporate environmental disclosure: Contrasting management's perceptions with reality. *Journal of Business Ethics, 49*(2), 143–165.

Crane, A. (1999). Are you ethical? Please tick yes or no. On researching ethics in business organizations. *Journal of Business Ethics, 20*(2), 237–248.

Currie, G., Humpreys, M., Waring, J., & Rowley, E. (2009). Narratives of professional regulation and patient safety: The case of medical devices in anaesthetics. *Health, Risk & Society, 11*(2), 117–135.

De Villiers, C., Rinaldi, L., & Unerman, J. (2014). Integrated reporting: Insights, gaps and an agenda for future research. *Accounting, Auditing & Accountability Journal, 27*(7), 1042–1067.

Del Baldo, M. (2015). Is it time for integrated reporting in small and medium-sized enterprises? Reflections on an Italian experience. In S. O. Idowu, C. S. Frederiksen, A. Y. Mermod, & M. E. J. Nielsen (Eds.), *Corporate social responsibility and governance. Theory and practice* (pp. 183–209). Heidelberg: Springer.

Del Baldo, M. (2017a, April 11). The "why" and the "how" of the IR journey in medium-sized firms. Insights and implications from empirical cases. In Proceedings SIW – Sidrea International Workshop, Integrated Reporting and Non-financial Information Assurance, *Integrated reporting and non-financial information assurance. SMEs vs large firms*, University of Roma Tre-Universitá Politecnica delle Marche (pp. 98–116). Rome: A.S.P.I. Editore.

Del Baldo, M. (2017b). The implementation of integrating reporting <IR> in SMEs. Insights from a pioneering experience in Italy. *Meditari Accountancy Research, 25*(4), 505–532. https://doi.org/10.1108/MEDAR-11-2016-0094.

Del Baldo, M. (2018). Integrating reporting in SMEs. Issues and perspectives from Italy. In S. O. Idowu, C. Sitnikov, D. Simion, & C. Bocean (Eds.), *Current issues in corporate social responsibility. An international consideration, chapter 13* (pp. 195–215). Heidelberg: Springer.

Del Baldo, M., & Girella, L. (2017, September 21–22). Soft regulating integrated reporting for SMEs: The case of Italy, 13th Interdisciplinary workshop on intangibles and intellectual Capital, EURAM Conference, Ancona, Italy.

Dellas. (2016). *Annual report 2015*. Lugo di Grezzana (VR).

Dellas. (2017). *Dellas annual integrated report 2016*. Lugo di Grezzana (VR).

Demartini, C. (2017). Integrated reporting. In C. Demartini & S. Trucco (Eds.), *Integrated reporting and audit quality. Contributions to management science, chapter 2* (pp. 9–35). Cham: Springer. https://doi.org/10.1007/978-3-319-48826-4_2.

DiMaggio, P. J., & Powell, W. W. (1983). The iron cage revisited: Institutional isomorphism and collective rationality in organizational fields. *American Sociological Review, 48*(2), 147–160.

Dobbs, M., & Hamilton, R. T. (2007). Small business growth: Recent evidence and new directions. *International Journal of Entrepreneurial Behaviour and Research, 13*(5), 296–322.

Doni, F., & Gasperini, A. (2014). The internal integrated reporting and value creation: A case study approach. In: *Knowledge and management models for sustainable growth*. E-book of proceedings of 9th IKFAD forum (International Forum on Knowledge Asset Dynamics), pp. 2666–2704.

Dumay, J., Frost, G., & Beck, C. (2015). Material legitimacy: Blending organisational and stakeholder concerns through non-financial information disclosures. *Journal of Accounting & Organizational Change, 11*(1), 2–23.

Dumay, J., Bernardi, C., Guthrie, J., & Demartini, P. (2016). Integrated reporting: A structured literature review. *Accounting Forum, 40*, 166–185. https://doi.org/10.1016/j.accfor.2016.06.001.

Dumay, J., Bernardi, C., Guthrie, J., & La Torre, M. (2017). Barriers to implementing the international integrated reporting framework: A contemporary academic perspective. *Meditari Accountancy Research, 25*(4), 461–480. https://doi.org/10.1108/MEDAR-05-2017-0150.

Eccles, R., & Armbrester, K. (2011). Integrated reporting in the cloud. *IESE Insight, 8*(1), 13–20.

Eccles, R., & Krzus, M. (2010, March). Integrated reporting for a sustainable strategy. *Financial Executive*, *26*(2): 29–32.

Eccles, R. G., & Saltzman, D. (2011). Achieving sustainability trough integrating reporting. *Stanford Social Innovation Review, 9*(3), 57–61.

Eccles, R. G., Kruz, M. P., & with Ribot S. (2015). *The integrated reporting movement – Meaning, momentum, motives and materiality*. New York: Wiley.

Eisenhardt, K. M., & Graebner, M. E. (2007). Theory building from cases: Opportunities and challenges. *Academy of Management Journal, 50*(1), 25–32.

Eurostat. (2015). *Statistics on small and medium-sized enterprises Jump to: Navigation, search dependent and independent SMEs and large enterprises*. Accessed July 7, 2017, from http://ec.europa.eu/eurostat/statistics-explained/index.php/Statistics_on_small_and_medium-sized_enterprises

Fasan, M. (2013). Annual reports, sustainability reports and integrated reports: Trends in corporate disclosure. In C. Busco (Ed.), *Integrated reporting* (pp. 41–58). Heidelberg: Springer.

Flower, J. (2015). The international integrated reporting council: A story of failure. *Critical Perspectives in Accounting, 17*, 1–17.

Freeman, R. E. (1984). *Strategic management: A stakeholder approach*. Boston, MA: Pitman Publishing.

Frías-Aceituno, J. V., Rodrìguez-Arìza, L., & Garcìa-Sànchez, I. M. (2013a). Is integrated reporting determined by a country's legal system? An exploratory study. *Journal of Cleaner Production, 44*, 1–210.

Frías-Aceituno, J. V., Rodriguez-Ariza, L., & Garcìa-Sànchez, I. M. (2013b). The role of the board in the dissemination of integrated corporate social reporting. *Corporate Social Responsibility and Environmental Management, 20*, 219–233.

Frìas-Aceituno, J. V., Rodrìguez-Ariza, L., & Garcìa-Sànchez, I. M. (2014). Explanatory factors of integrated sustainability and financial reporting. *Business Strategy and the Environment, 23*(1), 56–72.

García-Sánchez, I. M., Rodríguez-Ariza, L., & Frías-Aceituno, J. V. (2013). The cultural system and integrated reporting. *International Business Review, 22*, 828–838.

Giovannoni, E., & Maraghini, M. P. (2013). The challenges of integrated performance measurement systems: Integrating mechanisms for integrated measures. *Accounting, Auditing & Accountability Journal, 26*(6), 978–1008.

Gnan, L., Montemerlo, D., & Huse, M. (2015). Governance systems in family SMEs: The substitution effects between family councils and corporate governance mechanisms. *Journal of Small Business Management, 53*(2), 355–381.

Goodridge, P., Haskel, J., & Wallis, G. (2014). *Estimating UK*. Investment in Intangible Assets and Intellectual Property Rights. Accessed September 8, 2017, from https://www.gov.uk/government/publications/estimating-uk-investment-in-intangible-assets-andip-rights.html

Greenwood, R., Suddaby, R., & Hinings, C. (2002). Theorizing change: The role of professional associations in the transformation in institutionalized fields. *Academy of Management Journal, 45*(1), 58–80.

GRI – Global Reporting Initiative. (2013). *GRI International Standard 4.0*. G4 – Reporting Principles and Standard Disclosures. www.globalreporting.org/reporting/g4/Pages/default.aspx

Haller, A., & Van Staden, C. (2014). The value added statement – An appropriate instrument for integrated reporting. *Accounting, Auditing and Accountability Journal, 27*(7), 1190–1216.

Hampton, R. (2012, May). Brace yourself: More regulatory changes. *Accountancy SA, 14*: 22–23.

Havlováa, K. (2015). What integrated reporting changed: The case study of early adopters. *Procedia Economics and Finance, 34*, 231–237.

Higgins, C., Stubbs, W., & Love, T. (2014). Walking the talk(s): Organisational narratives of integrated reporting. *Accounting, Auditing & Accountability Journal, 27*(7), 1090–1119.

IFAC. (2008a). *International good practice guidance principles for effective business reporting process*. New York: International Federation of Accountants (IFAC).

IFAC. (2008b). *The crucial roles of professional accountants in business in mid-sized enterprises*. New York: International Federation of Accountants (IFAC).

IIRC. (2013). *The international <IRS> framework*. London: The International Integrated Reporting Council.

Incollingo, A. (2014). *Le prime esperienze di bilancio integrato*. Torino: Giappichelli.

Ioana, D., & Adriana, T. T. (2014). Research agenda on integrating reporting: New emergent theory and practice. *Procedia Economics and Finance, 15*, 221–227.

James, M. L. (2013a). Sustainability and integrated reporting. Opportunities for small and midsize entities. Paper presented at the 2013 Allied Academies international conference, Las Vegas, Nevada. *Entrepreneurial Executive, 18*, 17–28.
James, M. L. (2013b). Sustainability and integrated reporting: A case exploring issues, benefits and challenges. *Journal of the International Academy for Case Studies, 19*(8), 95–103.
Jensen, J. C., & Berg, N. (2012). Determinants of traditional sustainability reporting versus integrated reporting. An institutionalist approach. *Business Strategy and the Environment, 21* (5), 299–316.
Kolk, A. (2010). Trajectories of sustainability reporting by MNCs. *Journal of World Business, 45* (4), 367–374.
Kolk, A., & van Tulder, R. (2010). International business, corporate social responsibility and sustainable development. *International Business Review, 19*, 119–125.
KPMG. (2013a). *Integrated reporting*. Performance Insight Through Better Business Reporting, Issue 2, KPMG. Accessed March 20, 2016, from www.kpmg.com/AU/en/IssuesAndInsights/Articles Publications/Better-Business-Reporting/Documents/integrated-reporting-issue-2.pdf
KPMG. (2013b). *KPMG international survey of corporate responsibility reporting*. Accessed September 28, 2016, from www.kpmg.com/Global/en/IssuesAndInsights/ArticlesPublications/corporateresponsibility/Documents/corporate-responsibility-reporting-survey-2013-execsummary.pdf
Lai, A., Melloni, G., & Stacchezzini, R. (2013, September 19–21). *Disclosing business model in the "Integrated report": Evidencefrom European early adopters*. Paper presented at Convegno Bicentenario Aidea, Lecce.
Lai, A., Melloni, G., & Stacchezzini, R. (2017). What does materiality mean to integrated reporting preparers? An empirical exploration. *Meditari Accountancy Research, 25*(4), 533–552. https://doi.org/10.1108/MEDAR-02-2017-0113.
Larrinaga-Gonzàlez, C. (2007). Sustainability reporting: Insights from neo-institutional theory. In J. Unerman, J. Bebbington, & B. O'Dwyer (Eds.), *Sustainability accounting and accountability* (pp. 150–167). London: Routledge.
Lev, B. (2004). Sharpening the intangibles edge. *Harvard Business Review, 82*(6), 190–116.
Lincoln, Y. S., & Guba, E. G. (1985). *Naturalistic inquiry*. Newbury Park, CA: Sage.
Macias, H. A., & Farfan-Lievano, A. (2017). Integrated reporting as a strategy for firm growth: Multiple case study in Colombia. *Meditari Accountancy Research, 2*(4), 605–628. https://doi.org/10.1108/MEDAR-11-2016-0099.
Maharaj, R., & Herremans, I. M. (2008). Shell Canada: Over a decade of sustainable development reporting experience. *Corporate Governance: International Journal of Business in Society, 8* (3), 235–247.
Marchini, I. (2005). *Il governo della piccola impresa. La gestione strategica* (Vol. II, 2a ed.). Genova: Aspi/Ins-Edit.
Massaro, M., Dumay, J., & Guthrie, J. (2016). On the shoulders of giants: Undertaking a structured literature review. *Accounting, Auditing & Accountability Journal, 29*, 767–801.
Milne, M. J., & Grey, R. (2013). W(h)iter ecology? The trible bottom line, the global reporting initiative, and corporate sustainability reporting. *Journal of Business Ethics, 118*, 13–29. https://doi.org/10.1007/s10551-021-1543-8.
Morth, U. (Ed.). (2004). *Soft law in governance and regulation*. Cheltenham: Edward Elgar.
Morth, U. (2006). Soft regulation and global democracy. In M. L. Djelic & K. Sahlin-Andersson (Eds.), *Transnational governance: Institutional dynamics of regulation* (pp. 119–136). Cambridge: Cambridge University Press.
Naumes, W., & Naumes, M. J. (2006). *The art and craft of case writing* (2nd ed.). London: ME SHARPE Inc.
NIBR. (2018 forthcoming). *Italian network on business reporting*. Linee Guida per la redazione di un Report Integrato per le PMI *Guidelines for IR in SMEs*, a cura dei Gruppi di Lavoro "PMI" del NIBR – Network Italiano Business Reporting.
Owen, G. (2013). Integrated reporting: A review of developments and their implications for the accounting curriculum. *Accounting Education, 22*(4), 340–356.

Parker, L. D. (2005). Social and environmental accountability research: A view from the commentary box. *Accounting, Auditing & Accountability Journal, 18*(6), 842–860.
Patton, M. Q. (2002). *Qualitative research & evaluation methods* (3rd ed.). Thousand Oaks, CA: Sage.
Perego, P., Kennedy, S., & Whiteman, G. (2016). A lot of icing but little cake? Taking integrated reporting forward. *Journal of Cleaner Production, 136*(Part A), 53–64. https://doi.org/10.1016/j.jclepro.2016.01.106.
Phillips, D., Watson, L., & Willis, M. (2011). Benefits of comprehensive integrated reporting. *Financial Executive, 27*(2), 26–30.
Prado-Lorenzo, J. M., & Garcìa-Sànchez, I. M. (2010). The role of the board of directors in disseminating relevant information on greenhouse gases. *Journal of Business Ethics, 97*, 391–424.
Qu, S., & Dumay, J. (2011). The qualitative research interview. *Qualitative Research in Accounting & Management, 8*(3), 238–264.
Roberts, L. (2011, December). Integrated reporting: The status quo. *Accountancy SA*, 22–23.
Ruiz-Lozano, M., & Tirado-Valencia, P. (2016). Do industrial companies respond to the guiding principles of the integrated reporting framework? A preliminary study on the first companies joined to the initiative. *Revista de Contabilidad, 19*(2), 252–260.
Ryan, B., Scapens, R., & Theobal, M. (1992). *Research method and methodology in finance and accounting*. London: Academic.
Salvioni, D., & Bosetti, L. (2014). *Stakeholder engagement and integrated reporting: Evidence from the adoptionof the IIRC Framework.* Paper presented at International Multidisciplinary Academic Conference, Institute of Strategic and International Studies, Paris.
Scapens, R. W. (1990). Researching management accounting practice: The role of case study methods. *The British Accounting Review, 22*(3), 259–281.
Setia, N., Abhayawansa, S., Joshi, M., & Huynh, A. V. (2015). Integrated reporting in South Africa: Some initial evidence. *Sustainability Accounting, Management and Policy Journal, 6*(3). https://doi.org/10.1108/SAMPJ-03-2014-0018.
Seuring, S., & Mueller, M. (2008). Core issues in sustainable supply chain management – A Delphi study. *Business Strategy and the Environment, 17*, 455–466.
Silvestri, A., Veltri, S., Venturelli, A., & Petruzzelli, S. (2017). A research template to evaluate the degree of accountability of integrated reporting: A case study. *Meditari Accountancy Research, 25*(4), 675–704. https://doi.org/10.1108/MEDAR-11-2016-0098.
Silvi, R., & Bartolini, M. (2011). The role of performance indicators in management commentary. *Financial Reporting, 3*, 61–79.
Smallbone, D., & Wyer, P. (2000). Growth and development in the small firm. In S. Carter & D. James-Evans (Eds.), *Enterprise and small business*. Harlow: Prentice Hall.
Solomon, J. F., & Maroun, W. (2012). *Integrated reporting: The influence of King III on social, ethical and environmental reporting* [WWW Document]. http://www.accaglobal.com/content/dam/acca/global/PDF-technical/integrated-reporting/tech-tp-iirsa.pdf
Stubbs, W., & Higgins, C. (2014). Integrated reporting and internal mechanisms of change. *Accounting, Auditing & Accountability Journal, 27*(7), 1068–1089.
Thomson, I. (2015). But does sustainability need capitalism or an integrated report – A commentary on "the international integrated reporting council: A story of failure" by flower, J. *Critical Perspectives on Accounting, 27*, 18–22.
Tweedie, D., & Martinov-Bennie, N. (2015). Entitlements and time: Integrated reporting's double-edged agenda. *Social and Environmental Accountability Journal, 35*(1), 49–61. https://doi.org/10.1080/0969160X.2015.1007466.
Van Bommel, K. (2014). Towards a legitimate compromise? An exploration of integrated reporting in the Netherlands. *Accounting, Auditing & Accountability Journal, 27*(7), 1157–1189.
Vaz, N., Fernandez-Feijoo, B., & Ruiz, S. (2016). Integrated reporting: An international overview. *Business Ethics: A European Review, 25*(4), 577–591.

Velte, P., & Stawinoga, M. (2016). Integrated reporting: The current state of empirical research, limitations and future research implications. *Journal of Management Control*, 1–46. https://doi.org/10.1007/s00187-016-0235-4.

Veltri, S., & Silvestri, A. (2015). The free state university integrated reporting: A critical consideration. *Journal of Intellectual Capital, 16*(2), 443–462.

Watson, A. (2013, December–January). Early days yet for integrated reporting, but excellent examples are emerging, *Accountancy SA*, 14–17.

WICI –The World Intellectual Capital Initiative. (2014). Comment Letter on the ESMA Consultation Paper on Alternative Performance Measures – final 21 May 2014 "ESMA –European Securities and Market Authority" – "Guidelines on Alternative Performance Measures (APMs)" issued on February 13, 2014 European Securities and Markets Authority (ESMA) Paris, France www.esma.europa.eu.

WICI–The World Intellectual Capital Initiative. (2016). *WICI intangible reporting framework. 1.0 September 2016, stakeholders.* www.wici-global.com/wirf/WICI_Intangibles_Reporting_Framework_v1.0.pdf

Wild, S., & Van Staden, C. (2013). *Integrated reporting: Initial analysis of early reporters – An institutional theory approach*, 7th Asia Pacific Interdisciplinary Research in Accounting Conference, Kobe, Japan.

Yin, R. K. (2003). *Case study research.* Thousand Oaks, CA: Sage.

Yin, R. K. (2014). *Case study research: Design and methods.* Los Angeles: SAGE.

Zambon, S., & Guenther, T. W. (2011). Special issue: Measuring and managing intangibles for accounting use. *Journal of Management Control, 22*(3), 237–239.

Part II
Critical Perspectives on Integrated Reporting in Theory and Practice

Chapter 6
From Sustainability to Integrated Reporting: How the IIRC Framework Affected Disclosures by a Financial Institution in Australia

Federica Casonato, Federica Farneti, and John Dumay

6.1 Introduction

This chapter presents a case study on the sustainability reporting and <IR> practices of CBD, a financial institution, and its journey of disclosure. The material presented explores CBD's need to establish trust and legitimacy after a scandal forced them to embark on an economic, social and governance (ESG) reporting program and traces their subsequent <IR> journey. In this setting, <IR> has its origins in corporate social responsibility and sustainability reporting initiatives (Dumay et al. 2016).

Previous researchers have considered the evolution of the ESG reporting process at CBD but have not specifically focused on the years during which CBD participated in the Pilot Program or after it adopted the IIRC's Guidelines (IIRC 2011), the <IR> Prototype (IIRC 2012), and the <IR> Framework (IIRC 2013), or their most recent reporting years up to 2016. Existing literature examines <IR> in its early stages and the limitations and shortcomings of the IIRC's framework and practices (de Villiers et al. 2014). Therefore, studying companies in the process of implementing <IR> is valuable for identifying improvements to both the framework and how it is practised.

Identifying improvements to both the <IR> Framework and ESG reporting practice is the motivation for this research. In doing so, this study picks up from where prior research by Beck et al. (2017) left off through an analysis of CBD's integrated

F. Casonato (✉)
Ernst & Young, Milan, Italy

F. Farneti
Department of Sociology and Business Law, University of Bologna, Bologna, Italy
e-mail: federica.farneti@unibo.it

J. Dumay
Macquarie University, Sydney, NSW, Australia
e-mail: john.dumay@mq.edu.au

S. O. Idowu, M. Del Baldo (eds.), *Integrated Reporting*, CSR, Sustainability, Ethics & Governance, https://doi.org/10.1007/978-3-030-01719-4_6

Table 6.1 List of CBD source documents and reference codes

Document name	Reference code
2004 Corporate Social Responsibility Report	CBD01
2011 Annual Review	CBD02
2005 Corporate Social Responsibility Report	CBD03
2009 Corporate Responsibility Review	CBD04
2010 Annual Review	CBD05
2012 Annual Review	CBD06
2013 Annual Review	CBD07
2014 Annual Review	CBD08
2014 Dig Deeper	CBD09
2015 Annual Review	CBD10
2016 Annual Review	CBD11

reports from 2010 to 2016. We analyse how CBD aligned its reporting to the IIRC's principles during this period, considering the fact that CBD started its transition to <IR> before in 2010.

Beck et al. (2017) explore the reasons why CBD began their journey toward <IR> and examine how their reports developed over the years. Their findings indicate that CBD's reporting practices mainly focused on investors and other stakeholders to repair lost legitimacy after a scandal in 2004. It was not until 2010 that CBD seems to have regained trust and re-established legitimacy with its shareholders and stakeholders. However, despite deploying the <IR> framework, CBD, once again, became embroiled in further major scandals. It appears that implementing reporting practices only affected the public face of CBD; the organisational culture that allowed these scandals to happen did not change in line with the changes to reporting. Moreover, scandals are endemic to Australia's financial industry, which further adds to the validity of our findings.

The data used to develop this chapter was sourced through a review of publicly available documents published by the IIRC and CBD and builds on research by Dumay et al. (2015) and Beck et al. (2017). CBD is a pseudonym used to preserve anonymity in the previous studies and is continued here. Hence, traditional in-text citations are used to refer to the source documents produced by the IIRC, while the following table lists the source documents produced by CBD (Table 6.1).

6.2 The Journey Towards <IR>

<IR> is the result of a process of evolution of different ESG reporting practices that have been adopted to evaluate the activities of a company with the aim of creating an integrated report that groups together information about management, sustainability, governance, and finance (IIRC 2011, p. 7). The main objective of the <IR> is to

consider financial sustainability and stability for the benefit of a broad group of stakeholders to disclose how an organisation creates value:

> An integrated report benefits all stakeholders interested in an organization's ability to create value over time, including employees, customers, suppliers, business partners, local communities, legislators, regulators and policy-makers. (IIRC 2013, p. 4).

CBD joined the IIRC's Pilot Program when it began in 2011. This meant CBD participated in the process of improving <IR> until the release of The International <IR> Framework in December 2013 (IIRC 2013). The development of the <IR> Framework (IIRC 2013) was preceded by a discussion paper (IIRC 2011), an argument for the creating the Framework (IIRC 2012), and the <IR> Prototype (IIRC 2012), which were initially used by CBD to develop their first few integrated reports.

The final <IR> Framework (IIRC 2013) has increased the amount of research on <IR>, generating a number of studies on various aspects of creating integrated reports and its effects on the companies that adopt an <IR> approach (Dumay et al. 2016). One critique of the Framework is that the IIRC is mainly interested in increasing the quality of the information provided to shareholders about a company's value, not its sustainability. <IR> is therefore seen as focusing mainly on creating financial value for shareholders, while neglecting social or environmental issues (Flower 2015). de Villiers et al. (2014) highlight that <IR> is only at the beginning of its journey; and that it is highly interconnected with management and, consequently, the contents of the report may be biased. Higgins et al. (2014) looked at the holistic approach taken by the first companies to adopt <IR>, which did not revolutionise the way companies disclosed information but did advance research into sustainability reporting by integrating financial and non-financial information. On the other hand, Flower (2015) harshly criticises <IR> for its focus on shareholders and its lack of attention to sustainability from a socio-environmental point of view. Flower argues that, this is what will potentially cause <IR> to fail and that, for these reasons, it is necessary to change <IR> and find alternative ways to think about sustainability.

When companies use the Framework, they need to think about the effect it has on their activities. Because a company's reports can affect the external environment, which can, in turn, potentially increase or decrease a company's value, the effects these reports provoke deserve consideration. However, when using an investor's perspective as advocated by the <IR> Framework, companies might overlook the fundamental impacts they have on other groups and focus too much on short-term economic indicators rather than indicators significant to the long term. Therefore, there is evidence to support the argument that the current <IR> Framework does not place enough emphasis on social and environmental issues, and that failing to include human and natural capitals as inputs and outputs in a business model (IIRC 2013) might be detrimental to value creation over the long term. Instead, Adams (2015) counters Flower's argument and issues a call for action in support of <IR> because of the potential benefits she believes <IR> will bring to companies.

<IR> may have internal benefits brought about by the concept of integrated thinking. The IIRC highlights that one of its objectives is to "support integrated

thinking, decision-making and actions that focus on the creation of value over the short, medium and long term" (IIRC 2013, p. 2). The IIRC (2013, p. 33) defines integrated thinking as:

> The active consideration by an organization of the relationships between its various operating and functional units, and the capital that the organization uses or affects. Integrated thinking leads to integrated decision making and actions that consider the creation of value over the short, medium and long term.

As a concept, integrated thinking is widely accepted by managers and report preparers, but the issue is then translating this concept into practice. The main problem is that integrated thinking causes more confusion than clarity because the "IIRC has not fully defined and articulated the concept of integrated thinking, and there is no shared consensus among practitioners" as to what it means (Feng et al. 2017, p. 330). Similarly, Dumay and Dai (2017) identify that for integrated thinking to work as anticipated by the IIRC some of the existing organisational culture must be replaced. However, strong organisational cultures are not readily or easily replaced, especially when they are associated with an organisation's past success. Therefore, while the IIRC's rhetoric towards integrated thinking is appealing, research on integrated thinking finds it not quite so easy to implement.

<IR> should support a change in the way companies look at their business and make them think about the impact of their company's activities. However, this requires an understanding of the underlying theoretical arguments behind the IIRC's claims. In this context, it is useful consider <IR's> potential as a tool for justifying and legitimising a company's actions from a moral and institutional perspective, rather than a tool promoting integrated thinking and value creation. Arguably, the IIRC wants companies to communicate to external stakeholders to justify management's actions, not just to explain value-creation, and this may be one of the rationales for a particular organisation to adopt <IR>. Suchman's (1995) take on institutional legitimacy helps explain why companies interact with stakeholders in environments where institutions construct and interpret the environment they are part of. This view is opposed to strategic legitimisation, whereby managers have the ability to modify reporting according to their own needs, the company's needs, or to justify tangible outputs (Suchman 1995, p. 576). However, one problem when analysing reporting concepts, such as <IR>, from a legitimacy perspective, is that most researchers concentrate on either institutional or strategic legitimacy and rarely combine the two.

CBD is an Australian financial institution operating worldwide. It is a leader in ESG reporting practices within Australia, having won several national corporate reporting awards and a silver medal from the Dow Jones Sustainability Index (CBD02 2011, p. 12). However, all was not always well at CBD. In January 2004, CBD suffered losses of AUD$360 million following a currency trading scandal, which eventually saw several employees jailed because of their involvement. A subsequent report by PWC recommended four changes to be addressed: the Board of Directors, senior management, risk and control frameworks, and corporate culture (Beck et al. 2017, p. 196).

In most organisations <IR> is the result of a journey, and it is not possible to predict its final output (Stubbs and Higgins 2014; Dumay and Dai 2017). Therefore, CBD did not adopt <IR> because it was a new and interesting accounting

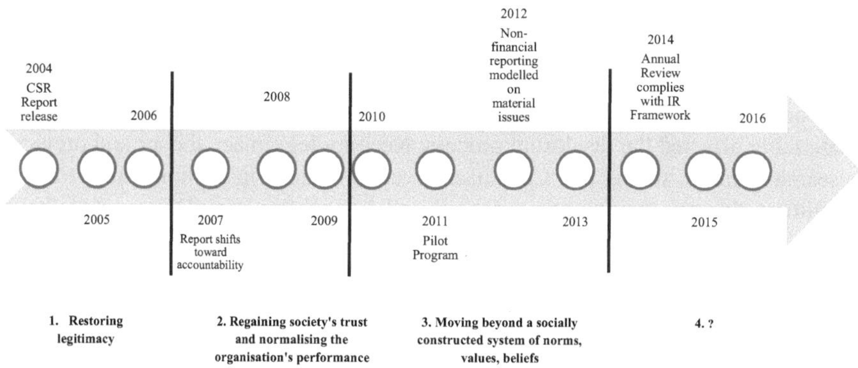

Fig. 6.1 Timeline of reporting and related motivation (2013). Source: Dumay et al. (2015); Beck et al. (2017)

technology that was intended to radically change its reports. As such, when analysing how a company comes to use <IR>, it is important to understand the journey of how it arrived at using <IR>. This journey begins in 2004, with CBD's first Corporate Responsibility Report after admitting fault in the currency trading scandal. Dowling and Pfeffer (1975) four-stage framework for restoring legitimacy has been used to understand the evolution of the reports produced by CBD and how the <IR> Framework affected their disclosure process (see Fig. 6.1).

Figure 6.1 represents the timeline of the journey undertaken by CBD. It has three phases: restoring legitimacy, regaining society's trust and normalising organisational performance, and moving beyond a socially constructed system of norms, values, and beliefs. The fourth stage, missing in the timeline, concerns the last years of reporting following the introduction of the Framework, which is addressed later in this chapter.

6.3 CBD and Its <IR> Practices

One study that combines institutional and strategic legitimacy is that by Dumay et al. (2015) who investigated how companies develop <IR> and what they include in their reports—one of which was CBD. Here, Dumay et al. (2015, p. 2) add to legitimacy theory, introducing the term "material legitimacy". They define it as "the form of legitimacy that enables organisations to blend what is important to the organisation (strategic legitimacy) with the primary concerns of its major stakeholders (institutional legitimacy)". In integrated reports, this is often called the materiality matrix: a matrix that reconciles the concerns of stakeholders with the concerns of the company. However, sometimes not all material issues are disclosed. As Dumay et al. (2015, p. 7) point out, "an issue can be highly important to stakeholders and the company, but that item might not be publicly disclosed due to the confidentiality of the parties involved or commercial sensitivity". Therefore, regardless of the promises to build transparency through reporting, inevitably any report, including an integrated report, will only tell part of the story.

To analyse CBD's reports, observations were made about: how the Framework was used; the type of information that was disclosed; and how the concept of materiality affected the disclosure process. Keeping legitimacy theory and previous research in mind, stating that CBD has strived to achieve institutional and strategic legitimacy through their corporate reports and Annual Reviews (Dumay et al. 2015; Beck et al. 2017). Thus, this analysis observes where the company stands in its reporting journey. Four phases were identified. The first two phases are briefly described as background; phases III and IV are discussed in-depth given that <IR> was developed during these phases.

6.4 Phases I and II

The need to change the culture of an organisation deeply embroiled in a scandal that was blamed on its culture moved CBD to try to repair trust and legitimacy by exposing what went wrong. In phase one, CBD needed to restore legitimacy amid scandal and was eager to find a way to re-establish itself as one of the best and most reliable financial institutions in Australia. Thus, in 2004, CBD commenced production of its first CSR Report, which focused on the company's mistakes (insider trading and miscommunication), and its lack of control. Admitting fault, CBD began its reporting journey to restore its legitimacy. Dumay et al. (2015, p. 22) liken the first 2 years of this path (2004–2006) to a phase of legitimacy restoration (Fig. 6.1). Producing a CSR report allowed the firm to begin the legitimacy restoration process of discovery, explanation, penance, and rehabilitation outlined by Suchman (1995).

Analysing the 2004 CSR report (CBD001), the company fully recognised its mistakes and reacted to them. They realised the need to regain trust among customers, employees, and shareholders:

> We introduced our new Corporate Principles to ensure we create and deliver consistently superior value to our customers, shareholders, employees and the communities in which we operate. (CBD01 2004, p. 3).

Numbering 40 pages, the first report is far from concise. It was structured according to the various GRI guidelines and contained a long discussion on sustainability and the practices adopted by the company. Over 50 non-financial metrics set out in the GRI were presented to a broad group of stakeholders to disseminate non-financial information about the firm's activities. CBD was working to acquire institutional legitimacy and re-establish its reputation:

> Culture change is now a high priority. In 2004, we developed a new set of Corporate Principles and behaviors as a first step in changing our culture. We are in the process of embedding these Principles into the way we do business every day (CBD03 2005, p. 3).

CBD had begun to highlight the need to change its culture to repair trust and its legitimacy.

CBD also identified their core operating principles, with a focus on how to incorporate them into everyday performance. Their choice of words confirms that CBD was building a path towards corporate responsibility to re-establish trust as well as its position in the market and society:

> Several years ago we began a journey to embed CSR into the way we do business. CSR is a key element of our strategy of growing long-term shareholder value. We are still in the early stages of this journey. We recognise this is a journey we cannot make on our own, and it will involve engagement and collaboration with our customers and other key stakeholders along the way. (CBD03 2005, p. 6)

Throughout this first phase (2004–2006), the company restored its legitimacy in order to operate in society by setting new objectives and striving to be more in line with their stakeholders' expectations.

The second phase of CBD's journey toward legitimacy started in 2007 (see Fig. 6.1). This phase involved restoring society's trust and normalising organisational performance (2007–2009) (Dumay et al. 2015). They focused on how they achieved the performance targets set during previous years, and their reports were more concise. The reports highlight the progress of activities undertaken in previous years and how or why their objectives were achieved or otherwise. GRI guidelines were still used, but the number of metrics was reduced to only those of greatest significance to CBD's activities. Even though CBD was able to regain public credibility and justify its activities, it still acted to strengthen its image as a sustainable company, as evidenced by the Chairman's message:

> It is also clear that the public perception of [financial institutions], including CBD, is not good enough. To build long-term sustainable businesses into the future, I believe our industry and CBD must act to improve its reputation. (CBD04 2009, p. 3)

The analysis in Phase II is limited, as CBD was focused on verifying whether the objectives and milestones set in previous years had been met. This 2-year period connects CSR's reporting Phase I (2004–2006) with the introduction of <IR> at the beginning of Phase III in 2010.

6.5 Phase III

CBD entered the third phase of its legitimacy process in 2010 grounded in a well-structured system of regulations, values, and beliefs (Fig. 6.1). This phase lasted until 2013 when CBD re-established its position in society and became capable of justifying its actions through a new holistic strategy aimed at bolstering shareholder returns through sustainable business performance (CBD05 2010, p. 2). In 2010, CBD's Annual Review was assured by Environmental Resources Management Pty Limited (ERM) for the first time, further increasing its validity. Approval from this well-recognised assurance company bolstered the reliability of the information contained in the report. As ERM reports,

> CBD employed a more comprehensive and formal materiality determination process than in previous years. Material CR performance issues were identified based on stakeholder engagement and business risks highlighted by the Executive. These issues formed the basis of the CR content of the 2010 Annual Review, which provides a balanced representation of material CR performance issues. (CBD05 2010, p. 41)

Notably, CBD referred to their 2010 Review as an integrated report in interviews, despite the report preceding the IIRC Framework. With this report, CBD had moved toward a single document capable of satisfying both stakeholder and investor needs. Considerable changes were observed. First, CBD introduced the concept of materiality when drafting the review. All the information disclosed was required to convey the financial value of activities (Dumay et al. 2015). Stakeholders lost the central role they had played in the CSR Report, although it is important to note stakeholders still found mention when their participation in CBD activities resulted in an increase in firm value.

CBD eventually achieved both strategic and institutional legitimacy because it complied with external requirements and regulations that were created by institutions and organisations (pragmatic legitimacy)—even those that were voluntary in nature. At the same time, it could justify and legitimise its actions as part of a strategy that considered both the internal and external environment (moral and cognitive legitimacy) (Suchman 1995). The greater the disclosure of information relevant to stakeholders, the more institutional legitimacy is achieved. By contrast, moral legitimacy is achieved by providing information relevant to the organisation. This leads to an important observation: only issues relevant to *both* stakeholders and the firm were reported in the Annual Review. That said, issues relevant to stakeholders might have been excluded from their reports because they were deemed to be of insufficient merit to financial value creation, or were deemed too sensitive for disclosure (Dumay et al. 2015).

From 2011 onwards, the reporting focus shifted toward future and long-term objectives, with the company aiming to implement a strategy capable of yielding short-term profitability without losing sight of long-term goals. In addition, the CEO and management attributed their business success to a broader number of stakeholders and the involvement of all actors in the company's activities:

> The long-term success of our business depends on us doing the right thing by many—our shareholders, our customers, our community and our people. This is CBD's second integrated report combining both our business and corporate responsibility outcomes for the year (CBD02 2011, p. 3)

Unlike its predecessors, the 2011 report saw the first integration of financial and non-financial information. Holistic approaches require a higher level of internal integration and communication along with a common strategy across all areas of a business. CBD claimed to have fully integrated CSR into its strategy and spread the concept of social responsibility throughout the entire organisation from the bottom up:

> We have internal mechanisms at every level of the organisation to shape and manage our approach to important CSR issues. (CBD02 2011, p. 7)

Integrating financial and non-financial information also required CBD to change the way it included CSR within its governance framework (CBD02 2011, p. 8). CBD had originally focused on CSR, and accordingly on stakeholders. The company had instituted a dedicated CSR governance framework (CBD01 2004, p. 6) that comprised a CSR Global Council reporting directly to the Group Executive Committee and the Board of Directors (CBD01 2004, p. 6) and operating business areas. Reviewing the way its governance structure was now presented, CBD did not make clear whether a complex CSR governance framework was still in place (CBD02 2011, p. 8). There was a much clearer description of how the company had integrated CR at each company level, but this does not imply that CBD had bestowed CR functions on different departments, or that sustainability was part of the company's culture. On the contrary, this could be symptomatic of better legitimisation of its activities. Moreover, segregation between the information regarding shareholders and the information relating to CSR persisted. Again, this could be interpreted as a lack of sustainability integration within the company culture or attributed to a misalignment between the objectives of the firm and its shareholders.

During the second year of the Pilot Program, compliance with the IIRC's guidelines and <IR> became more visible. The 2012 Annual Review was more investor-oriented and had a different structure than previous reviews and areas previously treated as marginal, such as risk management, were now a focus. There were extensive discussions on sustainability and several pages dedicated to CSR. Compared to the previous year, CBD also provided a more extensive discussion on how it had integrated CR into its corporate governance framework: CSR was now embedded into the Review, and the structure of the report shifted the primary focus to performance, strategy, and governance.

> As part of our approach to integrated reporting, we have incorporated our CR strategy and performance throughout this review. (CBD06 2012, p. 2).

A relevant section identifies the company's objectives and briefly describes how these objectives were expected to affect various participants in the company's success. CBD justified their objectives by aligning them with those of stakeholders, thus legitimising its activities:

> We have internal mechanisms at every level of the organisation to shape and manage our approach and ensure we are focused on the issues that matter to our stakeholders. (CBD06 2012, p. 29).

In line with the previous discussion on legitimacy, CBD was able to identify how to align itself with its investors' interests. Interestingly, CBD used the IIRC's guidelines to create the report but used GRI's metrics as its indicators, from which it can be deduced that the IIRC does not suggest how to measure and communicate performance.

In 2013, CBD still used the <IR> Prototype (IIRC 2012) and therefore the changes were not determined by IIRC requirements. From the outset, the Chairman's message highlights the financial issues in 2013. Also, there was no mention of sustainability in the discussion on sustainable plans and actions. This Annual Review is characterised by a more cohesive discourse:

> As you read through the detailed performance commentary in this report, it is important to keep in mind the broader social and economic operating environment, as these themes affect both our performance and the strategic decisions we make for our future planning. (CBD07 2013, p. 4)

Again the sections of the report focus on strategy and company performance with a special section on stakeholder engagement, not CR as was previously the case. What differs is that CBD identifies its greatest strengths and how to manage risk. In a dedicated section, CBD explains its integrated strategy and how that strategy aligns with stakeholders.

The 2013 Review was designed to inform investors, but there is still evidence of engagement with different stakeholders. The report is peppered with many stories about the experiences of customers and employees that have been able to start their own business or grow professionally. The most successful stories bolster the value of those businesses. (CBD07 2013, pp. 13, 16, 22, 26). However, the audience is different to 2012, with charts and numbers designed to communicate value to shareholders.

In Phase III, non-financial indicators were represented graphically. While there was a willingness to pursue the company's objectives by looking at impacts on the different actors involved, without a doubt, the main aim was to show an increase in economic value. CBD had started pay more heed to its investors' interests in order to strengthen its financial stability and sustainability. Even if CBD did not fall into the "shop-list trap" (Stubbs and Higgins 2014), where companies merely comply with each guideline in a schedule, it was trying to standardise its reporting by following the process of "accountingisation" prescribed by the <IR> (Dumay et al. 2015). Moreover, although CBD understood the effects of its business, it failed to identify any outcomes resulting from its activities that might have a negative impact on its brand image. This is indicative of a subjective approach and an inability to improve disclosure that contain both positive and negative outcomes. As highlighted in previous studies, this happens because financial capital is often considered more relevant than other capitals (de Villiers et al. 2014).

6.6 Phase IV

Following the third phase, the IIRC completed the final version of the <IR> Framework (IIRC 2013) in December 2013, and CBD adopted it to produce their 2014 Annual Review. The IIRC centred development of their Framework on materiality and engagement with shareholders, which naturally determined a stronger focus on financial value for CBD despite limited structural changes.

Analysis of a supporting document, the Dig Deeper report (CBD09 2014), revealed interesting information about CBD's real intentions:

> We have evolved our approach to reporting, moving away from the issues and stakeholder-based approach (customer, people, environment, community and supply chain) that we have used over the past five years, towards a format with a stronger focus on materiality and greater alignment to our Wealth of Opportunity story and how it can help to deliver sustainable, satisfactory shareholder returns. (CBD08 2014, p. 1)

The Chairman's message in the 2015 Annual Review confirms this objective:

> Our vision is to be the most respected [financial institution] in Australia and New Zealand [...]. We will be focused on improved and consistent returns for our shareholders and deliver a great customer experience. (CBD10 2015, p. 6)

CBD was striving to involve suppliers and interactions with them were considered more valuable than compliance with environmental regulations that did not really concern the bank. Only information that was relevant and material was selected for inclusion in the 2015 Annual Review. For example, environmental indicators were secondary in nature and were removed in favour of information about the supply chain. Flower (2015) argues that one of the limitations of <IR> is its vague interpretation of value, and how it limits the extent to which a company can generate value for others while generating value for itself in the process. In line with this critique, the 2015 Review provided no clear interpretation of the concept of value, but rather a simplified explanation of how CBD intended to increase financial value for itself:

> Our goal is to deliver sustainable and satisfactory returns to our investors. We believe that what is good for our customers is also good for business. (CBD10 2015, p. 2)

During Phase IV, there was no longer a need to integrate sustainability into their business strategy. However, what does appear relevant is that, through an integrated strategy which should hold benefit for a broad group of stakeholders, the board showed how it had achieved objectives for shareholders. In line with Flower's (2015) idea of the transformation of reporting, CBD had abandoned CSR reporting in favour of a corporate reporting practice largely aligned to the IIRC's concept of interpreting value creation through the eyes of financial capital providers. As advocated by the IIRC, CBD had shifted its reporting to a standardised output that focused on a specific group of stakeholders—its shareholders. This heralds the beginning of the fourth stage in the evolution of CBD's reporting (Fig. 6.2) and can be interpreted as an alignment with institutional legitimacy (Suchman 1995).

Despite its focus on financial measures, CBD still presented activities concerning other stakeholders. These mostly concentrated on the most important socially responsible initiatives for businesses to participate in, such as microfinancing and pension schemes for employees and customers (CBD08 2014, p. 24). All these programs were both financially-oriented and highly optimal for illustrating the alignments between CBD and other stakeholders. Several renewable energy initiatives had been supported by loans with lower rates to bolster environmental sustainability. CBD had also increased their involvement in the Aboriginal community through education and professional programs to improve the company's potential for recruiting talent as well as the native population's banking knowledge.

However, it is not clear how much these initiatives contributed to overall company profitability. As the aim was to communicate mainly with investors, it would

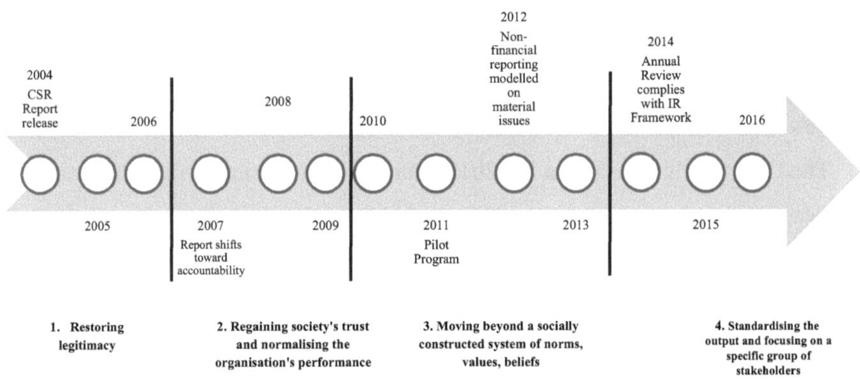

Fig. 6.2 Timeline of reporting and related motivation (2016). Source: Dumay et al. (2015); Beck et al. (2017)

have been more relevant to understand whether and how these programs were making a difference to CBD. Undoubtedly, these types of activities are helpful for communities, but they do not represent the most relevant part of a financial institution's business. Nothing was said about how the company selects its investment choices or invests its customers' money. Therefore, CBD did not provide any information about whether they had included CR in their value creation activities.

In general, the information disclosed mainly related to best practices and initiatives with positive impacts. In a few cases, activities and events with a negative effect on non-financial capitals were reported, for example, insider-trading events and other non-ethical practices that resulted in layoffs and firings. In 2015, several of CBD's employees were terminated after being found guilty of forging customer signatures and advising people to make hazardous investments and risky loans. CBD paid millions of dollars in compensation to the customers that had received bad advice. However, the Annual Review contains scant information on the scandal; a text box at the bottom of the report (CBD10 2015, p. 10) contains a brief description of the issue and the resolution. As theorised by Suchman (1995), a delegitimised organisation must immediately address such issues, admit their failures, and highlight any disruptions to demonstrate their reconnection with moral and cognitive beliefs. Should a company fail to recognise an issue, it risks losing its legitimacy to operate and will incur all the reciprocal consequences including loss of trust, loss of customers, major stakeholder issues, and, ultimately, negative financial results (Lawrence and Weber 2014).

Another example occurred in 2016 when one of CBD's financial advisors was found guilty of engaging in misleading and deceptive behaviour to favour particular clients. He was able to retrieve sensitive information about a superannuation fund by pretending to be a member and then use that information to improperly transfer funds to his clients. The 2016 annual review does not include any information about this scandal, reinforcing the findings of this study: that CBD cherry-picks what it includes and what it omits from its reports. Failing to mention this event is a clear indication of bias by CBD's management.

The fact that CBD has faced repeated scandals implies a lack of sustainability integration within the corporation and a lack of control. It is also evident that CBD has tried to dissociate itself from its employees' behaviour. Therefore, <IR> has not been able to modify the way CBD discloses information and how it conducts its business.

The argument above is further confirmed by the change in management that occurred in 2016. However, despite the appointment of a new Chairman, there has been no noticeable change in the objectives of the company:

> We are strongly motivated to be the most respected [financial institution] in Australia and New Zealand. To realize that vision, we have to make the [financial institution] simpler, with a stronger customer focus. (CBD11 2016, p. 7).

The fact that CBD has repeatedly stressed this idea over the years suggests that it has not been able to achieve its goal of becoming Australia's most respected financial institution. Implicitly, this indicates that there has been no significant change in the organisation's culture over the last 10 years and that the culture in place when the first scandals occurred during Phase I has remained substantially the same. Previous studies have demonstrated that CSR is something to be achieved every day, through business practices, and something that cannot be achieved by <IR> alone (Dumay and Dai 2017).

To conclude, in 2014 CBD entered into Phase IV, signaling its definitive acceptance of, and compliance with, a new set of accounting rules that shifted its focus toward shareholders and aimed to standardise its output (Fig. 6.2). The key insight is a focus on investors rather than any awareness of the impact of environmental or social issues. This finding is in contrast to Adams (2015) who, when analysing companies in the process of developing <IR>, found that the use of <IR> indicated a greater focus on sustainability development issues. Our findings are also in contrast to Montecalvo et al. (2018), where the use of <IR> had the potential to enhance sustainability reporting for state-owned enterprises. At CBD, the Annual Review is no longer a CSR reporting tool but a corporate one, as has emerged from other studies. Respect for stakeholders is necessary to guarantee financial sustainability, and it is subordinate to the creation of value. CBD has moved from sustainability reporting to corporate reporting.

6.7 Conclusion

The fact that CBD is an avid reporter and supporter of the <IR> Framework helps us understand the impact of <IR>. According to Dumay et al. (2015), CBD began using the term integrated reporting in 2010, well before the formation of the IIRC. However, their purposes were similar: to integrate all reporting into one simple, more concise document, including their sustainability reporting. Exactly where the term 'integrated report' originated from is not clear. But, it was around this time that the Eccles and Krzus (2010) book *One Report: Integrated Reporting for a*

Sustainable Strategy was released in the US. CBD's approach to reporting then was closer to *One Report*, than what we now know as <IR>. However, CBD did join the IIRC's Pilot Program and subsequently used their Framework to develop their integrated reports.

One of the key features of <IR> is its investor focus and its desire to build "financial stability and sustainability" (IIRC 2013, p. 3). This feature is not a stakeholder view of reporting. Consequently, one of the dangers of adopting <IR> is that an organisation might narrow and concentrate their reporting focus on shareholders as the most powerful stakeholders and relegate the other stakeholders to secondary consideration. Albeit, this shift is unlikely. As Dumay and Dai (2017) report, organisational cultures do not change quickly, especially when the culture is strong, so it is difficult for <IR>, or any other form of reporting, to induce any major impact.

The CBD case is interesting from an organisational culture perspective because the culture at CBD was one of the primary reasons for its original foray into CSR reporting in 2004. As CBD outlines in 2004, "We are also changing our leadership teams, our structures, our culture and our strategies. This report is a tangible demonstration that we will continue to strengthen our business practices and lift the level of transparency and accountability of the [CBD] on issues that are important to our stakeholders" (CBD01 2004). Thus, the need to change the culture of an organisation deeply embroiled in a scandal blamed on its culture moved CBD to try to repair trust and legitimacy by exposing what went wrong, showing that the guilty parties had been removed, and not allowing panic to set in with investors and the government (see Suchman 1995, p. 597). By moving to repair their legitimacy, CBD was imploring change, while at the same time carrying on business as usual. Yet, if there is one thing that is difficult to change in an organisation, it is culture (Merchant and Van der Stede 2007; Dumay and Dai 2017).

'Business as usual' is the result of entrenched systems and processes. These are an organisation's key resources, and they are sticky, especially in the short term (Teece et al. 1997; Dumay and Guthrie 2012). Culture is no different because people cannot change themselves or the way they behave overnight. So, even if a firm did need to fire undesirable people and acquire new people with a different mindset, change would only be effected in limited form; organisations cannot throw out the old and take in the new very easily. People who form part of the undesirable culture can remain in the organisation in the short to medium term, and those entering the organisation may even find they need to fit in with the existing culture, rather than introducing their own value sets to change the culture of existing employees (Dumay and Dai 2017).

So, if resources and cultures are sticky, what chance does any form of reporting have to change a culture? In this case, very little. As evidenced in the CBD case, it took several years (2004–2011) for CBD to declare that "trust" had been repaired (Dumay et al. 2015, p. 11). When it did, CBD declared their customers and their community came first, and this principle underpinned the returns they generated for shareholders. But, how long was this to last? If CBD's organisational culture did, in fact, put their customers and community first, it is unlikely that further scandals would have occurred because their culture would have been strong enough guide to individual behaviour. However, in 2015, another scandal rocked CBD when it identified that many customers were at risk due to inappropriate information given by their financial planners over the last 5 years—well after implementing integrated reporting, after adopting the <IR> Framework, and after joining the Pilot Program. That means that, since at least 2010, some CBD employees in investment advisory roles had not been putting their customers and the community first, but rather were prioritising their own wealth creation. Oddly, this was happening at the same time CBD was declaring that trust had been repaired. Were they remiss to conclude, "we continue to pursue a strategy that is founded on the value of advice in assisting more Australians to have trust and confidence in their financial futures" in their 2015 Annual Review? (CBD10 2015, p. 10) The trust CBD claims to have restored in 2010 appears not have penetrated the employees in their financial advisory business.

It is true that no organisation is immune to the behaviour of one or two rogue employees, and it is often difficult to make behavioural constraints foolproof, especially when an organisation is dealing with disloyal, deceitful employees (Merchant and Van der Stede 2007, p. 77). However, while CBD argues its problems were contained to a handful of individuals, its wealth advice process is an integral part of CBD's systems and day-to-day business, and, thus, are part of the culture CBD identified as the problem. These problems are very much cultural and cannot be changed by simply removing a few employees. By CBD's own admission, the changes needed demanded: "several initiatives to substantially improve outcomes for our financial advice customers" (CBD10 2015, p. 10).

Acknowledgment Federica Casonato, would like to express the deepest appreciation to Macquarie University that accepted her as a visiting student to conduct her research in Sydney, Australia and to all the staff members who contributed directly or indirectly to this book chapter. She would like to acknowledge the assistance of Professor John Dumay who offered her valuable suggestions for this study and provided her with relevant advice. She wishes to express her gratitude to her supervisor at the University of Bologna, Professor Federica Farneti, who supported her during all relevant stages of work and gave her the opportunity to get involved in the research.

References

Adams, C. A. (2015). The international integrated reporting council: A call to action. *Critical Perspectives on Accounting, 27*(1), 23–28.

Beck, C., Dumay, J., & Frost, G. (2017). In pursuit of a 'Single Source of Truth': From threatened legitimacy to integrated reporting. *Journal of Business Ethics, 141*(1), 191–205.

CBD01. (2004). *Corporate social responsibility report*. Melbourne. Available at https://www.CBD.com.au/content/dam/CBDrwd/documents/reports/corporate/2004-corporate-responsibility-report.pdf

CBD02. (2011). *Annual review*. Melbourne. Available at http://cr.CBD.com.au/download-centre

CBD03. (2005). *Corporate social responsibility report*. Melbourne. Available at https://www.CBD.com.au/content/dam/CBDrwd/documents/reports/corporate/2005-corporate-responsibility-report.pdf

CBD04. (2009). *Annual corporate responsibility review*. Melbourne. Available at https://www.CBD.com.au/content/dam/CBDrwd/documents/reports/corporate/2009-corporate-responsibilityreview.Pdf
CBD05. (2010). *Annual review*. Melbourne. Available at http://cr.CBD.com.au/download-centre
CBD06. (2012). *Annual review*. Melbourne. Available at http://cr.CBD.com.au/download-centre
CBD07. (2013). *Annual review*. Melbourne. Available at http://cr.CBD.com.au/download-centre
CBD08. (2014). *Annual review*. Melbourne. Available at http://cr.CBD.com.au/download-centre
CBD09. (2014). *Dig deeper*. Melbourne. Available at https://www.CBD.com.au/content/dam/CBDrwd/documents/reports/corporate/dig-deeper-2014.pdf
CBD10. (2015). *Integrated reporting*. Available at http://cr.CBD.com.au/how-we-work/integrated-reporting
CBD11. (2016). *Annual review*. Melbourne. Available at http://cr.CBD.com.au/how-we-work/integrated-reporting
de Villiers, C., Rinaldi, L., & Unerman, J. (2014). Integrated reporting: Insights, gaps and an agenda for future research. *Accounting, Auditing & Accountability Journal, 27*(7), 1042–1067.
Dowling, J., & Pfeffer, J. (1975). Organisational legitimacy: Social values and organizational behaviour. *Pacific Sociological Review, 18*(1), 122–136.
Dumay, J., & Dai, T. (2017). Integrated thinking as a cultural control? *Meditari Accountancy Research, 25*(4), 574–604.
Dumay, J., & Guthrie, J. (2012). IC and strategy as practice. *International Journal of Knowledge and Systems Science, 3*(4), 28–37.
Dumay, J., Frost, G., & Beck, C. (2015). Material legitimacy: Blending organisational and stakeholder concerns through non-financial information disclosures. *Journal of Accounting & Organizational Change, 11*(1), 2–23.
Dumay, J., Bernardi, C., Guthrie, J., & Demartini, P. (2016). Integrated reporting: A structured literature review. *Accounting Forum, 40*(2016), 166–185.
Eccles, R., & Krzus, M. (2010). *One report: Integrated reporting for a sustainable strategy*. Hoboken, NJ: Wiley & Sons.
Feng, T., Cummings, L., & Tweedie, D. (2017). Exploring integrated thinking in integrated reporting – An exploratory study in Australia. *Journal of Intellectual Capital, 18*(2), 330–353.
Flower, J. (2015). The international integrated reporting council: A story of failure. *Critical Perspectives on Accounting, 27*(1), 1–17.
Higgins, C., Stubbs, W., & Love, T. (2014). Walking the talk(s): Organisational narratives of integrated reporting. *Accounting, Auditing & Accountability Journal, 27*(7), 1090–1119.
International Integrated Reporting Council (IIRC). (2011). *Towards integrated reporting: Communicating value in the 21st century*. London: International Integrated Reporting Council.
International Integrated Reporting Council (IIRC). (2012). *Integrated reporting prototype framework*. London: International Integrated Reporting Council.
International Integrated Reporting Council (IIRC). (2013). *The International <IR> Framework*. London: International Integrated Reporting Council.
Lawrence, A., & Weber, J. (2014). *Chapter 3. In business and society: Stakeholders, ethics, public policy* (14th ed.). New York: McGraw-Hill Higher Education.
Merchant, K. A., & Van der Stede, W. (2007). *Chapter 3: Action, personnel and cultural controls* (pp. 76–94). London: Management Control Systems.
Montecalvo, M., Farneti, F., & de Villiers, C. (2018). The potential of integrated reporting to enhance sustainability reporting in the public sector. *Public Money & Management, 38*(5), 365–374.
Stubbs, W., & Higgins, C. (2014). Integrated reporting and internal mechanisms of change. *Accounting, Auditing & Accountability Journal, 27*(7), 1068–1089.
Suchman, M. C. (1995). Managing legitimacy: Strategic and institutional approaches. *Academy of Management Review, 20*(3), 571–610.
Teece, D. J., Pisano, G., & Shuen, A. (1997). Dynamic capabilities and strategic management. *Strategic Management Journal, 18*(7), 509–533.

Chapter 7
The Impact of Triple Bottom Dispersal of Actions on Integrated Reporting: A Critical Perspective

Adriana Schiopoiu Burlea

7.1 Introduction

The aim is to apply the integrated reporting to explain how organizations use the triple bottom dispersal of actions in order to reduce the amount of pressure caused by the increasing number of relevant information that needs to be disclosed.

Integrated reporting (IR) is considered in literature (Busco et al. 2013; Del Baldo 2015; Eccles and Krzus 2010) as a practice that brings many benefits to organizations.

We used one of the three concerns of the critical researcher identified by Alvesson and Deetz (2000), critique produced through deconstruction and we made the difference by combining insight and our critique perspective. Thus, our critical approach of IR is based on the relationship between IR and triple bottom dispersal actions. It consists of three dispersal dimensions: *dispersal of responsibilities, dispersal of information, and dispersal of knowledge*.

The triple bottom dispersal of actions is a theory explaining the source of the problems and the crisis at an individual, organizational, national and international level (Burlea Schiopoiu 2010; Burlea Schiopoiu 2015; Burlea Schiopoiu and Idowu 2016; Rost and Osterloh 2010). At the beginning, Burlea Schiopoiu elaborated this theory, with the aim to explain one of the causes leading to crises and she started from the assumption that crises, regardless of their nature, have always been and will continue to be present in every field of activity and in every sphere of the society. Experts and scientists have reached to the conclusion that crises cannot be entirely avoided because a crisis, in its structure, is a cyclic phenomenon that affects different sectors of activities in different moments. Therefore, it can be noted an increasing interest to prevent and diminish the effects of crises. However, it is expected that no main results to be ever obtained as long as the main cause that generates a crisis is

A. Schiopoiu Burlea (✉)
University of Craiova, Craiova, Romania

S. O. Idowu, M. Del Baldo (eds.), *Integrated Reporting*, CSR, Sustainability, Ethics & Governance, https://doi.org/10.1007/978-3-030-01719-4_7

neglected. She reached to the conclusion that the general cause of crisis, regardless of the sphere in which it occurs and the moment of its occurrence, is represented by the triple bottom dispersal of actions.

The structure of the papers provides some background information on the triple bottom dispersal of actions as core element of Integrated Reporting. This is will be followed by a critical analysis of the relationship between principles of Integrated Reporting and Triple bottom dispersal of actions which provides the theoretical dimension for the seven principles. The lessons from the international organizations as promoters of triple bottom dispersal of actions in integrated reporting approach are discussed. The paper concludes with a set of reflections on how IR diminishes the triple bottom dispersal of actions.

7.2 It Is the Triple Bottom Dispersal of Actions a Core Element of Integrated Reporting?

Integrating reporting highlights the importance of the measurement of information-responsibilities-knowledge by increasing the value-added. These reports underline the reactive role of accounting and provide the key stakeholders (i.e. shareholders, investors, employees, customers, communities) with valid and reliable data about the company.

Integrating reporting is a complex approach focused on value creation which is founded on the analysis of the following capitals: information (financial, manufactured)—responsibilities (human, social and relationship)—knowledge (intellectual, natural).

The sustainability report is focused in protecting the investment of shareholders and the company's reputation, for this reason the sustainability report doesn't use risk management as a strategic tool (Hess 2014) and the integrated reporting has to play the role of the corporate reporting norm, through voluntary mechanisms.

7.3 Why Dispersal of Information?

The dispersal of information is described by a large amount of information that is presented in a chaotic and repetitive way (Burlea Schiopoiu 2010). IR aims to bring together financial and non-financial information, but organizations are disclosing only information that does not threaten its reputation and legitimacy in the market (KPMG International 2013). The dispersal of information (DI) is recognized by Michael Meehan, *Chief Executive of GRI*, who stated that 'The powerful work in unlocking the value of sustainability data has just begun.' (GRI 2016, p. 5). However the actions of international organizations are proving that it will be difficult to reduce the negative impact that DI has on the value of IR (Tschopp and Huefner 2015). The information asymmetry is a source of DI (Burlea Schiopoiu and Remme 2017) and IR has the potential to reduce it (Frias-Aceituno et al. 2014).

Companies appear to use the voluntary disclosure as a mean to justify for the lack of meaningful insights or for transmitting incomplete and selective information, but also as a motivation for transforming IR into a corporate reporting norm through voluntary mechanisms.

Corporate legitimacy is supported by a high level of un-sustainability (Milne and Gray 2013) and DI led, paradoxically, to the improvement of disclosure and accountability of the company. How dispersal of information play a positive role?

The answer is: by omitting the relevant information and the company fails to provide the relevant information to which shareholders make reference in assessing the company's value.

The depth of analysis of the IR indicators depends on the degree of the DI, because the higher the DI, the less meaningful is the analysis and, therefore, decisions made show a high level of risk and uncertainty. In this context, the disclosure report is mainly interpreted as a mean that the organization is using to advertise itself, rather than a critical instrument to account the organization's performance. The quality of reporting will increase not by disclosing meaningless information. On contrary, it is expected to improve by the means of disclosing their actions which drive the social and environmental impact.

7.4 Why Dispersal of Responsibilities?

IR involves many actors (i.e. directors, accountants, consultants, and auditors) and the responsibilities are dispersed between them, because every actor starts from the premise that is not his/her responsibility to give some information and, for this reason, the IR lack of information. Lao Tzu, captured very well the essence of dispersal of responsibly: 'If you look to others for fulfillment, you will never be truly fulfilled.' (Sasson 2012, p. 4) The IIRC uses the voluntary (self-regulation) approaches to implementing IR as an excuse of the company to emphasis on the value for shareholders and to diminish the value for society (Flower 2015). As a result, the company could present some unsustainable practices as being sustainable and, hence, to transfer the responsibilities form the company to society (Thomson 2015). Thus, the dispersal of responsibilities (DR) translates the risks from a person to another and the consequences of their actions, respectively lack of actions will be translated to the organization and will damage its reputation.

7.5 Why Dispersal of Knowledge?

IR reflects a comprehensive framework, a tool that can be used as means to gain a deeper level of knowledge and understanding of the shareholders' always higher expectations and of stakeholders' increasing requirements regarding the organizational

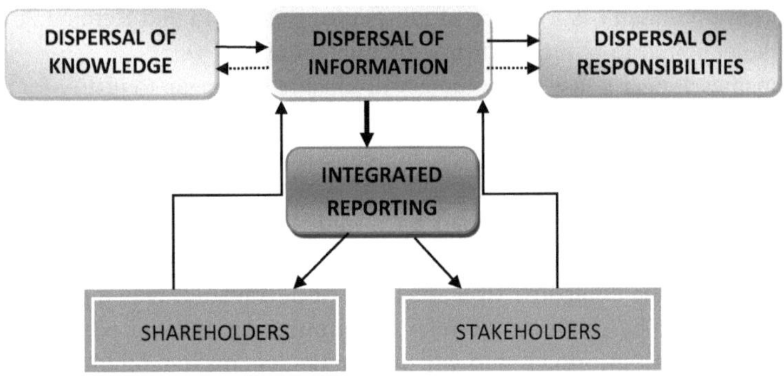

Fig. 7.1 The relationship between triple bottom dispersal of actions and Integrated Reporting. Source: Author contribution

performance outcomes. In the "always-on" economy, when companies are faced with increased scrutiny from a wide range of stakeholders, organizations actively seek new avenues to foster growth and generate value. While some scholars consider standardization as a main cause contributing to the lack of relevant and insightful knowledge encompassed by the IR (King and Lenox 2000), the dispersal of knowledge (DK) is increasingly recognized as being one of the most important aspects that hinders the attractiveness of any company in the eyes of the beholders. For global organizations, few challenges are more daunting than providing inadequate content to its stakeholders and shareholders and creating a false impression that they truly cover expectations raised by modern society. The DK is related to the process of reducing the business risk and IR is used as an imperfect public disclosure tool that proves to be unable to fully capture the relationship established between the sustainable and un-sustainable items (see Fig. 7.1).

7.6 The Relationship Between Principles of Integrated Reporting and Triple Bottom Dispersal of Actions

The IR principles can be used by companies as a source of dispersal of responsibilities (DR) and can use the market competition and the pressure of stakeholders as a motivation for promoting DR. To offer a better understanding of the relationship between IR principles and DR, we will start from Cooper and Morgan (2013, p. 431) who stated that IR standards create a false impression that they cover the entire set of requirements of the stakeholders.

The first principle—*Strategic approach and Future orientation*—underlines the importance of risks and opportunities to the development of the company. The DK is strongly related with this principle because it is very difficult for any company to identify the significant risks and good opportunities in relationship with its strategic objectives and business model.

The triple bottom dispersal, based on *strategic approach and future orientation of the companies*, is used as initiative to foster the unethical behavior and a recent example is scandal of Volkswagen AG (VW) and previous corporate scandals, such as Enron (Dembinski et al. 2006) or AIG (Ferrell and Ferrell 2010) have already shown that poor governance or unethical behaviors can destroy reputations, lead to significant financial losses and legal prosecutions or even to catastrophic consequences resulting in corporate shutdowns.

The second principle—*Connectivity of information*—is a reflection of the interdependence between the inoffensive information disclosed by a company to their stakeholders and the effort of the company to promote this information as an achievement. The consistency of the information must be based on quality and not on quantitative indicators that, most often, lack sustainability. Wong and Millington stated that 'Access to private disclosures is estimated as the percentage of the stakeholder's time spent on using private information (Private), relative to information from corporate reports and infomediaries. Time is limited for both providers and users of information. Users being rational will only spend time on information which contributes to decision making.' (Wong and Millington 2014, p. 871). The DI is related to this principle because the more information is dispersed and is presented in an ambiguous way, the more the role of IR will be less important to the decision making process.

The third principle—*Stakeholders relationship*—is a key principle and it is a building block of DR because it creates the impression that stakeholders have the opportunity to take a look inside the company and to evaluate the degree of its responsibility in meeting their needs and expectations. The stakeholders' engagement represents a part of IR which has to be transparent and to disclose the real factors which affects the capital and the activities of the company. To prove its efficiency, it is highly important that the communication process to be reinforced by corporate actions and not only backed by quantitative indicators (i.e. the needs of the stakeholders meet the priorities of the company? and the communication channel is open for every group of stakeholders or for the key stakeholders only?).

The fourth principle—*Materiality*—is related to entire triple bottom dispersal because, in order to demonstrate its sustainability, the company must use valid information, to take the responsibilities of the process of value creation and to share the resulted knowledge with its stakeholders.

The fifth principle—*Conciseness*—is also a source of DI because the cross-references are used with the purpose to increase the number of pages of the report and not for adding value to the company's mission, values, strategy and objectives. The DI discloses the less relevant information and reduces the accuracy of the IR, because it better reflects the accounting and governance in relationship with finance capital, but does not serve the needs and or does not meet the expectancies of the other stakeholders who want to evaluate the organization's degree of accountability and sustainability (i.e. civil society groups).

Ruiz-Lozano and Tirado-Valencia have dedicated their work to the study of the level of importance that industrial companies have assigned to IR principles in conducting their business and they reached the conclusion that 'Only the principle of "Conciseness" does not seem to be correlated significantly with any other principle.' (Ruiz-Lozano and Tirado-Valencia 2016, p. 257).

The sixth principle—*Reliability and completeness*—refers to the different aspects of the company and, by omitting the information; it is source of triple bottom dispersal (DI, DR and DK). Both the internal and the external verification and validation of the report are performed taking into account the information and not the facts (i.e. the negative aspects are presented as problems in process of being solved or already solved).

In Volkswagen Sustainability Report 2014 (Volkswagen 2014, p. 144) the indicators concerning the 'Consumption and emission data' are presented, while at the page 145, the responsibilities of the company's Board of Managing Directors (in the section 'Management's Responsibility') and of the practitioner (in the section 'Practitioner's Responsibility') are mentioned. These responsibilities are not clearly delimited at an internal level. In consequence, one of recommendation presented at the page 147 is related to the sixth principle as follows: 'For a materiality analysis according to the G4 standard it is necessary to further improve the analysis of consequences for the defined topics in order to enable a stronger focus on the material aspects'.

In many ethical scandals, the DK and the ignorance are both a consequence of dispersal of the ethical responsibilities. IR encompasses all sorts of economic, social, financial and various others indicators, but it also provides a measure of the risks (i.e. Volkswagen 2014 on 'Strategy' section presents the information about risk management), however no disclosure is made about the potential threats that could alter the company's image as a good social contributor. This scandal came as enormous unexpected news to most shareholders and stakeholders of the carmaker, especially because of the fact that its implications go beyond a mere technology-based engineering fraud.

The seventh principle—*Consistency and comparability*—is very difficult to be applied in everyday business interaction, because organizations are showing reluctance towards disclosing more details about their strategy. Due to this reason, the organizations adopt a mimetic behavior sometimes when decide to provide information which can be comparable over time and which could turn into an accurate industry benchmark. Most organizations seek to maintain their competitive advantage and, thus, they avoid disclosing specific information which might give a market advantage to the competitors. The DI and DK are use as tools for keeping and enhancing the competitive advantage and to reduce the level of comparability between some indicators (see Fig. 7.2). As a conclusion, organizations are using benchmark data or present quantitative indicators which are largely used by competitors.

We consider that IR contributes to the improvement of the systems and organizational processes that occur across any company. By attempting to cover the key business themes that have potential to boost the image of the company in the market, IR is interpreted as a communication tool which facilitates the organization to differentiate from its peers. IR standards create, however, an 'almost' correct and complete perception of the company because it fails to disclose a series of

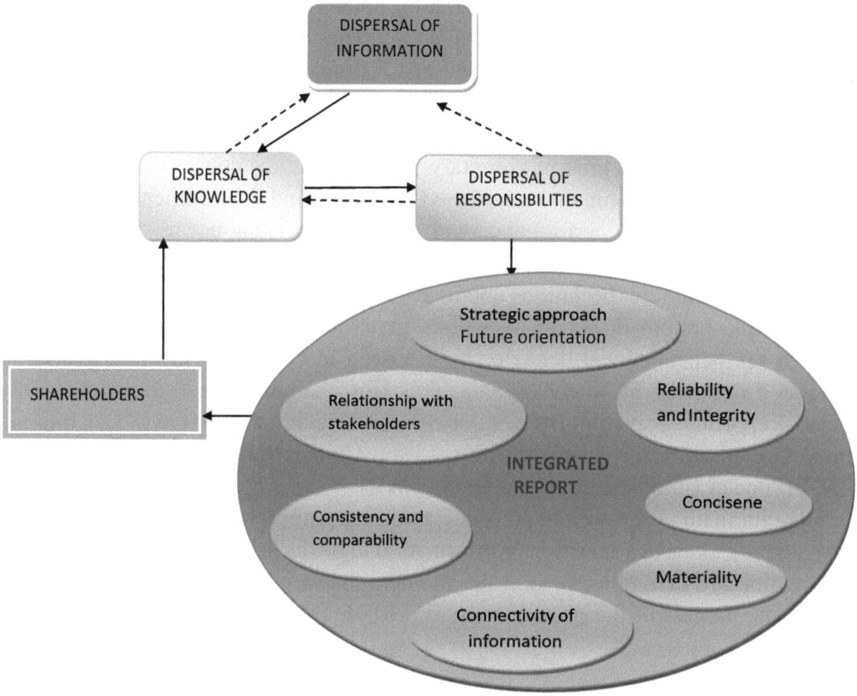

Fig. 7.2 The relationship between principles of Integrated Reporting and Triple bottom dispersal of actions. Source: Author contribution

information that might endanger its positioning in the global ever-increasing competition for market share and customer affinity.

7.7 The International Organizations as Promoters of Triple Bottom Dispersal of Actions in Integrated Reporting Approach

In 2014, Brown and Dillard reached the conclusion 'that integrated reporting, as conceived by the IIRC, provides a very limited and one-sided approach to assessing and reporting on sustainability issues' (Brown and Dillard 2014, p. 1120). This finding has lifted us the ball to the net and has given us the opportunity to build our judgment around integrated reporting.

In the last years, many research (Brown and Dillard 2014) have focused their attention on the comparison between integrated reporting and others types of reporting (i.e. sustainability reporting, financial reporting etc.) and the conclusion obtain was that integrated reporting provides plenty information about the financial capital, however the information required by other key stakeholders such as public community is often neglected (MacDonald and Hughes 2009).

IR is oriented towards protecting internal benefits (i.e. the real risks and opportunities; to disclose some financial indicators and to preserve non-financial performance in order to reveal minimal elements of its performance) and to embellish the external benefits (i.e. to present voluntary initiatives as key concern of the management strategy and to maintain undisclosed the social and environmental failures, to assess the sustainability performance with respect to existing norms or benchmarks in order to improve its reputation and to convince the external stakeholders that its true values met their expectations about sustainable development).

The triple bottom dispersal of actions gives to companies the possibilities to broaden up their unethical behavior by invoking the pluralism in the approach to integrated reporting by many organizations (for example, The International Integrated Reporting Council—IIRC, Global Reporting Initiative—GRI, The International Organisation for Standardization—ISO, United Nations Global Compact—UN, European Union—EU, World Bank—WB).

The pluralism in the approach to integrated reporting broads the opportunity for the company to make a selection of information to be disclose taking into account the aspects that are favorable for it at the moment and not the set of information that reflects the process or the phenomenon as a whole (see Fig. 7.3).

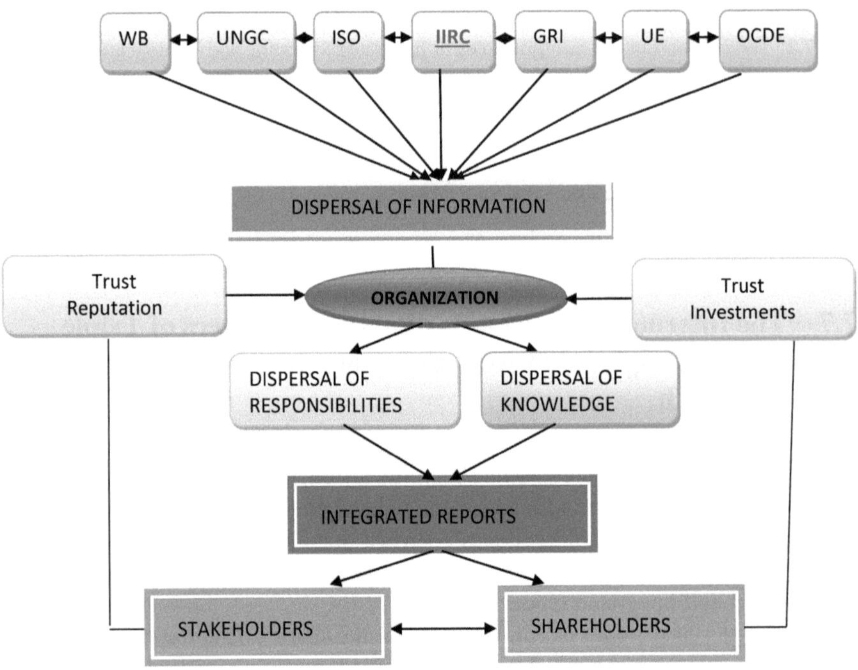

Fig. 7.3 Who is generating the triple bottom dispersal of actions in Integrated Reporting? Source: Author contribution

The GRI inspired IIRC, but GRI practices are recognized by the business environment (GRI 2015; GRI 2017a, b; IIRC 2013), while IIRC assesses the performance from an exclusive financial perspective and it does not pay attention to the relationship with stakeholders (Leach et al. 2010).

Although the triple bottom dispersal of actions is embedded in the IIRC approach because it does not support the organizational accountability, limits the stakeholders' empowerment and does not reflect a real sustainability (Boiral 2013), because 'Truthful words are not beautiful; beautiful words are not truthful. Good words are not persuasive; persuasive words are not good. The words of truth are always paradoxical.' (Sasson 2012, p. 12).

7.8 How IR Diminishes the Triple Bottom Dispersal of Actions?

Many researchers have paid increased attention to the motivations and to the internal mechanism that contributes to the formation of image of the company, and they have neglected the real performance enhanced by the IR principles (Brown and Dillard 2014; Soh et al. 2015).

The engagement processes of the companies in internal and external aspects are driven by ethical principles that must comply to the existing regulation and, for this reason companies are looking for ways to avoid the rules and to use dispersal of responsibilities as an excuse for their unethical behavior.

The vicious circle of the triple bottom dispersal of actions reflects the relationship between the quality of integrated report which influences the shareholders and stakeholders behavior toward the company, and the triple bottom dispersal of actions is used as method to improve the signals provided by shareholders and stakeholders after having reviewed the information comprised in the annual integrated report.

Standardization is not a solution as long as the main cause is not diminished, triple bottom dispersal—dispersal of responsibilities—dispersal of information—dispersal of knowledge.

In 2016, GRI (2016, p. 7) stated that 'we need to ensure the true value of sustainability data is recognized, and made accessible to all', but depending on their main activity, for any organization 'the true value of sustainability data' has a different meaning.

The organizations can claim the specific confidentiality constraints, thus avoiding to disclose part of the information. In conclusion, standardization is not viewed as a viable solution for the improvement of IR, because it does not solve the issues related to the triple bottom dispersal and does not ensure the confidentiality of the data.

The vicious cycle of triple bottom dispersal of actions is influenced by multiple and subsequent interactions between economic, social and political organizations and needs and expectancies of the civil society.

An Integrated Report, regardless of the information discloses, will be the reflection of the triple bottom dispersal of actions and the difference between the reports will be made on how the information is presented and not through their consistency. As a consequence, the information discloses within the Integrated Report must to be relevant for all stakeholders and to capture, in a synthetic way, the essence of the process or the phenomenon presented by it.

The role and the importance of IR cannot be minimized, but organizations and key stakeholders alike are required to, at least, attempt to understand the needs and expectancies of each other and to mutually help to the diminishing the triple bottom dispersal actions as follows:

Dispersal of information can be diminished through a better presentation of the strategic mission of the organization in the framework of the real needs of the key stakeholders. If the strategic mission of the organization fits the needs of the key stakeholders, the information to be disclosed will be related in relationship with internal and external challenges.

The political decisions made at national and international level have a strong impact on the organizational governance of companies, but it is not mandatory to represent a source of reducing the dispersal of responsibilities at company level (i.e. the financial and environmental scandals are one of the hidden sources of the political decisions). The relationship between political decisions and dispersal of responsibilities at the organizational level is underlined by the research conducted in the sphere of corruption (Mény and Rhodes 1997).

IR reflects an integrated thinking which means a deeper knowledge and understanding of shareholders' higher expectations and of stakeholders' requirements regarding the organizational performance outcomes. Some scholars consider that standardization is a cause of the lack of pertinent knowledge in the IR (King and Lenox 2000).

Dispersal of knowledge (DK) is one of the most important dispersal that reduces the potential of the company to create value, because 'Those who have knowledge, don't predict. Those who predict; don't have knowledge' (Sasson 2012, p. 7). Shareholders are constantly evaluating the value that organizations are able to generate, while stakeholders always differentiate between organizations.

As a conclusion, if the company wants to avoid the market risk and poor judgment, it develops a set of dispersal based on a selective disclosure of the information, on an ambiguous delimitation of responsibilities, and on a lack of relevant knowledge.

References

Alvesson, M., & Deetz, S. (2000). *Doing critical management research*. London: Sage.

Boiral, O. (2013). Sustainability reports as simulacra? A counter-account of A and A+ GRI reports. *Accounting, Auditing & Accountability Journal, 26*(7), 1036–1071.

Brown, J., & Dillard, J. (2014). Integrated reporting: On the need for broadening out and opening up. *Accounting, Auditing & Accountability Journal, 27*(7), 1120–1156. https://doi.org/10.1108/AAAJ-04-2013-1313.

Burlea Schiopoiu, A. (2010). Ethics of information in distributed business environments. In M. Pankowska (Ed.), *Infonomics for distributed business and decision-making environments: Creating information system ecology* (pp. 301–315). Hershey, PA: IGI-Global.

Burlea Schiopoiu, A. (2015, May 12–13). The impact of dispersal of responsibilities on organisational behaviour. In *Workshop on research advances in organizational behavior and human resources management* (pp. 19–22). Université Paris Dauphine, DRM – Management & Organisation.

Burlea Schiopoiu, A., & Idowu, O. S. (2016). The independence of the managers: An ethical dilemma. *International Journal of Social Entrepreneurship and Innovation, 4*(2), 152–171. https://doi.org/10.1504/IJSEI.2016.076688.

Burlea Schiopoiu, A., & Remme, J. (2017). The dangers of dispersal of responsibilities. *Amfiteatru Economic, 19*(45), 464–476.

Busco, C., Frigo, M. L., Riccaboni, A., & Quattrone, P. (Eds.). (2013). *Integrated reporting: Concepts and cases that redefine corporate accountability*. Cham: Springer.

Cooper, D. J., & Morgan, W. (2013). Meeting the evolving corporate reporting needs of government and society: Arguments for a deliberative approach to accounting rule making. *Accounting and Business Research, 43*(4), 418–441.

Del Baldo, M. (2015). Is it time for integrated reporting in small and medium-sized enterprises? Reflections on an Italian experience. In S. O. Idowu, C. S. Frederiksen, A. Y. Mermod, & M. E. J. Nielsen (Eds.), *Corporate social responsibility and governance: Theory and practice* (pp. 183–209). Heidelberg: Springer.

Dembinski, P. H., Lager, C., Cornford, A., & Bonvin, J.-M. (Eds.). (2006). *Enron and world finance a case study in ethics*. New York: Palgrave Macmillan.

Eccles, R. G., & Krzus, M. P. (2010). *One report: Integrated reporting for a sustainable strategy*. New York: Wiley.

Ferrell, O. C., & Ferrell, L. (2010). The responsibility and accountability of CEOs: The last interview with ken lay. *Journal of Business Ethics, 100*, 209–219.

Flower, J. (2015, March). The international integrated reporting council: A story of failure. *Critical Perspectives on Accounting, 27*, 1–17.

Frias-Aceituno, J. V., Rodríguez-Ariza, L., & García-Sánchez, I. M. (2014). Explanatory factors of integrated sustainability and financial reporting. *Business Strategy and the Environment, 23*, 56–72.

Global Reporting Initiative - GRI. (2015). *Second analysis paper sustainability and reporting trends in 2025*. Accessed June 15, 2016, from https://www.globalreporting.org/resourcelibrary/Sustainability-and-Reporting-Trends-in-2025-2.pdf

Global Reporting Initiative - GRI. (2016). *The next era of corporate disclosure: Digital, responsible, interactive*. Accessed October 15, 2017, from https://www.globalreporting.org/resourcelibrary/The-Next-Era-of-Corporate-Disclosure.pdf

Global Reporting Initiative - GRI. (2017a). Accessed March 2, 2017, from https://www.globalreporting.org/information/current-priorities/integrated-reporting/Pages/default.aspx

Global Reporting Initiative - GRI. (2017b). *Reporting principles and standards disclosure*. Accessed June 1, 2017, from https://www.globalreporting.org/resourcelibrary/GRIG4-Part1-Reporting-Principles-and-Standard-Disclosures.pdf

Hess, D. (2014). The future of sustainability reporting as a regulatory mechanism. In D. R. Cahoy & J. E. Colburn (Eds.), *Law and the transition to business sustainability* (pp. 125–139). Cham: Springer.

International Integrated Reporting Council - IIRC. (2013). *The international IR framework*. London: IIRC.

King, A. A., & Lenox, M. J. (2000). Industry self-regulation without sanctions: The chemical industry's RC program. *Academy of Management Journal, 43*(4), 698–716.

KPMG International. (2013). *The KPMG survey of corporate responsibility reporting 2013*. Amsterdam: KPMG International.

Leach, M., Scoones, I., & Stirling, A. (2010). Dynamic sustainabilities: Technology, environment. In *Social Justice*. London: Earthscan.

MacDonald, B. S., & Hughes, E. J. (2009). *Separating fools from their money: A history of American financial scandals* (2nd ed.). Abingdon: Routledge.

Mény, Y., & Rhodes, M. (1997). Illicit governance: Corruption, scandal and fraud. In M. Rhodes, P. Heywood, & V. Wright (Eds.), *Developments in west European politics*. London: Palgrave.

Milne, M. J., & Gray, R. (2013). W(h)ither ecology? The triple bottom line, the global reporting initiative, and corporate sustainability reporting. *Journal of Business Ethics, 118*(1), 13–29.

Rost, K., & Osterloh, M. (2010). Opening the black box of upper echelons: Drivers of poor information processing during the financial crisis. *Corporate Governance: An International Review, 18*(3), 212–233.

Ruiz-Lozano, M., & Tirado-Valencia, O. (2016). Do industrial companies respond to the guiding principles of the integrated reporting framework? A preliminary study on the first companies joined to the initiative. *Revista de Contabilidad – Spanish Accounting Review, 19*(2), 252–260.

Sasson, R. (2012). *The amazing cotes of Lao Tzu*. Accessed March 15, 2017, from https://www.successconsciousness.com/free_ebooks/lao_tzu_quotes.pdf

Soh, D., Leung, P., & Leong, S. (2015). *The development of integrated reporting and the role of the accounting and auditing profession: Social audit regulation* (pp. 33–57). Cham: Springer.

Thomson, I. (2015, March). But does sustainability need capitalism or an integrated report a commentary on the international integrated reporting council: A story of failure by flower. *Journal of Critical Perspectives on Accounting, 27*, 18–22.

Tschopp, D., & Huefner, R. J. (2015). Comparing the evolution of CSR reporting to that of financial reporting. *Journal of Business Ethics, 127*(3), 565–577.

Volkswagen. (2014). *The Volkswagen Sustainability Report 2014*. Accessed March 22, 2016, from www.sustainabilityreport2014.volkswagenag.com

Wong, R., & Millington, A. (2014). Corporate social disclosures: A user perspective on assurance. *Accounting, Auditing & Accountability Journal, 27*(5), 863–887.

Chapter 8
The Mimicry of Integrated Reporting: An Analysis of the Principles-Based Approach

Adriana Tiron-Tudor, Tudor Oprisor, and Gianluca Zanellato

8.1 Introduction

Organizational reporting is subject to constant dynamic, mainly due to the fact that reporting entities and standard setters are actively trying to meet the needs of most (if not all) information users. As a result of this dynamic, we notice the emergence of new reporting formats, developed with the purpose of enhancing the disclosure level and improve the le~~vel of~~ understanding when it comes to presenting information about an organization's activity. One such reporting model which gained a lot of attention in the recent years is integrated reporting (commonly abbreviated <IR>).

Developed as an instrument of disclosure which takes the holistic view (or "the bigger picture") into account, <IR> has been subject to scrutiny in academia, as well as intense debates involving professional organizations and standard setters. However, taken constructively, these debates have only attempted to improve the implementation process by ensuring the development of a well drafted framework, suited to the needs of information users in line with the constituting elements of such a report.

Within our paper, we focus on the "principles-based approach" undertaken by the International <IR> Framework. According to the governing body (the IIRC), this approach is intended to provide "*balance between flexibility and prescription that recognizes the wide variation in individual circumstances of different organizations while enabling a sufficient degree of comparability across organizations to meet relevant information needs*" (IIRC 2013b). Furthermore, our analysis goes even further and aims to gather an insight on the extent to which the involved standard

A. Tiron-Tudor (✉) · T. Oprisor · G. Zanellato
Babes-Bolyai University, Cluj Napoca, Romania
e-mail: adriana.tiron@econ.ubbcluj.ro; tudor.oprisor@econ.ubbcluj.ro; gianluca.zanellato@econ.ubbcluj.ro

S. O. Idowu, M. Del Baldo (eds.), *Integrated Reporting*, CSR, Sustainability, Ethics & Governance, https://doi.org/10.1007/978-3-030-01719-4_8

setters and professional organizations influence the drafting of the <IR> principles (thus, outlining a mimicry process between <IR> and traditional reporting principles).

To this point in time in the development of <IR>, it is clear that this new reporting set does not intend to *substitute* the traditional reporting formats (whether financial or non-financial, by nature), but merely to *ensure complementarity* and *enhance understandability* (through the "user-friendly" drafting of the report itself and the avoidance of over-technical provisions). Also, to make this reporting format attractive and its implementation accessible, the framework needs to provide a frame where the constituting elements of the report can be gathered from existing sources of data, to the highest extent possible (in order to avoid additional reporting costs and burdens). In this respect, our underlying cause for investigation (and the main starting point of our analysis) for this paper is that the main <IR>—specific constructs (including the guiding principles) are developed in line with those of existing traditional reporting formats (drafted by the main international standard setters).

To achieve our research goal in this paper, we mainly use a documentary research approach, analyzing the literature focused on <IR> and outlining how this reporting model evolved in the recent years and what is the current status and development of the <IR> constructs (including the principles). Also, we gather some insights from the literature regarding the principles-based approach in other reporting formats, developed and/or outlined by standard setters and professional organizations which were involved in the development of <IR> (to have an understanding on the possible sources of inspiration for the principles' drafting in the <IR> Framework). Last, but not least, we analyze the provisions from the <IR> Framework itself and outline the connections and matches with the former mentioned principles, as well as explain and emphasize the new instances which are introduced in this reporting system.

8.2 Constructing the Integrated Reporting Frame: Insights from the Literature

The evolution of "integrated thinking" and the transformation of sustainability reporting (by adopting an integrated approach) have been promoted in the field of corporate reporting by practitioners and theorists alike (Vaz et al. 2016; Velte and Stawinoga 2016; Rowbottom and Locke 2016; Lai et al. 2014; Eccles and Saltzman 2011). The topic even generated fiery debates over the success or failure of <IR> (Flower 2015 versus Adams 2015). However, early adopters of <IR> have presented a lot of interest within this field as they would have preliminary evidence regarding the institutionalization of <IR> (Stubbs and Higgins 2014; Higgins et al. 2014). The findings of these studies generally show that "while the organizations that are producing some form of integrated report are changing their processes and structures, or at least talking about it, their *adoption of integrated reporting has not necessarily stimulated new innovations in disclosure mechanisms*". These findings hint that although we are discussing a new reporting model in terms of development, the wheels that power the mechanism are not something really innovative, but the system rather re-uses existing principles and elements to produce a holistic output (thus sustaining our argument in favor of principles mimicry).

The following step has been to outline a template for the integrated report which would take into account "*the story about an organization's journey towards reaching its vision*", aligning to the core principles and content elements of <IR>. Ultimately, <IR> is seen as an opportunity for the summation of (reporting) parts to be included in a holistic construct, reporting about the "*web of interactions and implications of financial, social, environmental, and governance-related organizational activities for stakeholders*" (Abeysekera 2013). This summation of parts works as a synergy between components from the triple bottom line and is a rebranded synthesis of reporting components, using information from existing sources (or reports).

Furthermore, the use of <IR> has deeper implications on an organization's activity than merely a way of providing broader view on the created value and the business model. <IR> is perceived also as a proxy for the overall quality of the management (pinpointing the increasing focus on intangible assets and emphasizing the "*externalities*" on the environment and society). The concept of "*integrated thinking*" is supposed to achieve "balance between short-term business imperatives and ongoing value creation". Hence, <IR> should provide a reliable way to point out high-quality businesses (Churet and Eccles 2014).

The main research directions for <IR> have been towards a cost-benefit analysis (in the perspective of the concept's implementation), as well as the collaboration opportunities between private, public and nongovernmental sectors in order to establish a global movement for sustainable actions (Eccles and Saltzman 2011). In close connection, many discussions have been focused on the mandatory/voluntary status of <IR> for companies in order to create added-value to corporate reporting (Eccles et al. 2010), as well as emphasizing and analyzing connections with principles and content elements—such as materiality (Eccles et al. 2012; Mio and Fasan 2013), capitals (Karatzimas 2016; Coulson et al. 2015; Bartolini et al. 2013) or governance (Wulf et al. 2014) external factors, such as the cultural system (Frías-Aceituno et al. 2013; Garcia-Sanchez et al. 2013).

Given the fact that integrated reporting is an emerging topic, many studies seem to focus on the emphasizing the development of the field of <IR>, by synthetizing most of the relevant academic work and insights. Some comprehensive studies acknowledge the antecedents of <IR> (precursor tools and the evolution of sustainability reporting), the pioneers in the field (early adopter entities) and the regulatory developments (mainly, from the IIRC). Ultimately, this leads to an interesting series of research questions regarding the agenda for <IR> development and future perspectives for implementation (such as the study developed by de Villiers et al. 2014). In a similar manner, Cheng et al. (2014) have pinpointed several key issues that need to be resolved, respectively: the focus on financial capital providers (in detriment of other stakeholders), the meaning of "overall stock of capital" and tradeoffs between capitals, as well as the problem of assurance of integrated reports.

The IIRC has undertaken considerable efforts in the recent years to issue guidelines for the proper implementation of <IR>. In this respect, critical perspectives have emerged stating that <IR> needs to "broaden up" and "open up" dialogue, creating the assumptions for a debate on how reporting standards might assist or obstruct the construct of "sustainable business practices". Therefore, the view of the IIRC which business cases as a primary logic is fairly limited and there is need for a more pluralistic approach, as well as new accounting technologies and engagement practices (Brown and Dillard 2014).

In the same research direction, Haller and van Standen (2014), following an extensive literature review, propose "value added statements" as a practical and effective reporting instrument for <IR>. The proposed "value-added statement" not only meets the guiding principles of <IR>, but also reports on the monetary effects of different types of capital included in <IR> and in this way complements and represents the concept of <IR> very well.

The vast majority of the studies conducted in this field have the aim to ensure the proper comprehension of the <IR> concept and construct, but also to provide an adequate level of guidance for implementing regulations and methodologies. In this respect, the IIRC has launched the "Integrated Reporting Framework" project. By doing so, the Council has set the grounds for a restructuring process in the field of corporate reporting.

As we can observe in Table 8.1, we can conclude that the development of the IIRC has been intense in the recent years, as it has become one of the major governing bodies in accounting in merely 4 years. Also, the concept of <IR>, due to its popularity, has come—in terms of guidance issuance—from a Discussion Paper in 2011 to an official Framework Draft at the end of 2013 (and has further initiatives and development phases for the following years).

The <IR> Framework—whether in its draft form or the final version—has also been discussed and analyzed in the accounting literature (Loska 2011; Reuter and Messner 2015). The main focus of these studies is to assess a level of understanding regarding the political nature of standard-setting in the context of <IR> and to reveal different concerns (for instance, the scope of audience of <IR>; issues of materiality and the relationship between <IR> and other existing reporting frameworks; assurance provision, etc.). However, it is still under debate whether all types of entities (from all sectors) are not prone to implementing an <IR> system which is encompassed by the same conceptual model (including, on all accounts, the guiding principles from the current framework).

Table 8.1 Development timeline for <IR>

Date	Development	Notes
September 2011	Discussion paper towards integrated reporting—Communicating value in the twenty-first century published	Comment deadline 14 December 2011
October 2011	Initial organizations participating in the IIRC pilot programme announced	The pilot programme is a 2-year programme for a group of companies who will work as a network of peer group organizations, exchanging knowledge and sharing experiences on integrated reporting
July 2012	Draft outline of the integrated reporting framework	Not part of the formal consultation process, but the IIRC welcomed feedback
November 2012	Prototype of the international (<IR>) framework	Not part of the formal consultation process, but the IIRC welcomed feedback
April 2013	Consultation draft of the international <IR> framework published	Comment deadline 15 July 2013
December 2013	International <IR> framework published	Available online: http://integratedreporting.org/resource/international-ir-framework/
November 2014	Integrated reporting public sector Pioneer network launched	Initial discussions regarding the <IR> public sector Pioneer network started in April 2014. The official launching conference took place in Washington, in November 2014
2014–2017	The breakthrough phase	Moving from the creation of the framework and testing phase to the development and early adoption across reporting entities
April 2017	Consultation process—a new round of feedback	Comment deadline: 30 April 2017

Source: http://www.iasplus.com/en/resources/sustainability/iirc (IIRC 2011), with enhancements

8.3 Matching Pieces from Other Puzzles: An Emphasis on the Principles-Based Approach in Other Reporting Frames

The globalization phenomenon created the prerequisite for requiring (financial) reporting standards on an international level. This idea is very well summed up by Hughes (2007), in the Financial Times, stating that "*the goal of a single worldwide accounting language has long been a dream. Today is fast becoming a reality – and the pace is picking up*". This idea is empowered by the outcome of the hard times occurring after the financial crisis (as the economic system became a worldwide web of interactions and the risk of contagion is quite elevated). In this respect, the need for a reduction in reporting standard disparities between different regions and different types of entities is of paramount importance. For this desiderate, we notice the emergence of the requirement for a set of commonly accepted principles governing the reporting mechanisms.

However, this is not a problem of novelty, but a recurrent one, as the need for principles to standardize financial reporting occurred long ago, since the time of the Great Depression in 1929 (the crash of the New York Stock Exchange), when the American Institute of Accounts AIA regulated the form and contents of listed societies' reports. In time, the attempts to provide proper standardization became more prominent and, in 1973, the International Accounting Standards Committee was founded, from the will and with the assistance of the professional organizations of different countries like: Australia, France, Germany, Canada, Japan, USA, Netherlands, Great Britain. This Committee later became the International Accounting Standards Board (in 2001), the main actor of financial reporting standardization and promoter for accounting and reporting standards harmonization, as well as convergence—with other standard setters' initiatives.

As a front-runner in the field, the IASB gives the tone for accounting and reporting standardization, hence the prime objective is to enable a principles-based system, which benefits from widespread appraisal and application. A good example in this sense is the European Union Commission through the regulation number 1606/ 2002 which "*requires all EU-listed companies, including banks and insurance companies, from 2005 onwards, to prepare their consolidated accounts in accordance with the IAS/IFRS*" (www.eur-lex.europa.eu). Through this regulation, all the EU-listed companies are required to adopt the IAS, being one of the most important signs that shows there is an interest to apply (at least within the European Union) the same financial reporting standards, with the same principles all over the European Union.

The idea behind the principles based approach is that to give a clear idea of the economic value or results of the company or the transactions, as exposed by the Financial Accounting Standards Committee from the American Accounting Association, which "*believes that a principles-based approach is more likely to result in transactions that reflect their true economic substance*" (Shotridge and Myring 2004). However, this system is not entirely reliable as "*accountants are required to make a considerable number of estimates for which they are responsible* [...]" (Carmona and Trombetta 2008), and this raises many questions over the fairness of the accountants' assessments.

Furthermore, the principles-based approach enabled many discussions, being opposed to rules-based accounting (Mano et al. 2006; Cunningham 2007; Collins et al. 2012). This is an ongoing debate and a conceptual dichotomy encompassing the accounting and reporting discussions for a long period of time. The key difference between them is that "principles-based accounting provides a conceptual basis for accountants to follow instead of a list of detailed rules" (Shotridge and Myring 2004). Moreover, in the case of principles-based systems, professional judgement is a prime feature and a distinctive element of accounting process (Schipper 2003).

The widespread acceptance of IAS/IFRS around the world is clearly on an increasing trend. One of the reasons which consolidates the shift towards the principles-based system is the reduction of financial reporting's so-called "aggressiveness" (Nelson 2003). This reduction is supposed to benefit all stakeholders as the reports prepared by the companies should become more reliable and more truthful, offering a much clearer idea of the health of the company. Also, principles-based systems would reduce the phenomenon of aggressiveness because, "in addition to technical skills, the accountant's role involves some ethical and legal implications that are barely discernible under a rules-based system" (Carmona and Trombetta 2008). This means that with principles-based systems, accountants will be held accountable for their work, as through a rules-based system the accountant would just apply the rules. In this way, the true value, or results of the company, will be clearer and harder to be manipulated by managers.

Within the standardization process, many "regulators" and professional organizations are involved, whether on actual guidelines issuing processes, or on the consultative side. Beside the IASB, which we already presented, we emphasize (non-exclusively) the role in standardization and guideline issuance of the International Federation of Accountants (IFAC), International Public Sector Accounting Standards Board (IPSASB)—which is an independent committee of IFAC, the Charted Institute of Public Finance and Accounting (CIPFA), International Organization for Standardization (ISO), Global Reporting Initiative (GRI), as well as professional organizations, such as: AccountAbility (AA) and Social Accountabilty International.

The former mentioned organizations have a wider scope and aim to address a wider range of issues and types of stakeholders (from all sectors and fields), but—ultimately—they work in synergy to provide the frame for best reporting practice. In this sense, we can notice influences across the standards (and works of different standard setters). A good example is the fact that the ones developed for the public sector (IPSASs) have been inspired from the existing standards from the private sector. In a similar manner, we can observe the fact that principles encompassed in every reporting frame follow similar coordinates when addressing the presentation of accounting information.

Nevertheless, the frameworks and sets of principles issued by these standard setters are not fully distinctive of each other. In this sense, evidence from the literature shows, through comparative analyses, how the corresponding delineations overlap and to what extent do they contribute to concurring goals. In some cases, the analyses approach standards placed in a similar context (Tiron-Tudor et al. 2015; Göbbels and Jonker 2003), such as accountability, through AA1000 (AccountAbility 2008) and SA8000 (Social Accountability International 2014), whereas others embed multiple standards is more complex settings, such as, the architectural overlap between the ISO 26000 (International Organization for Standardization 2017), G4 guidelines (GRI 2013) and the <IR> Framework (Idowu et al. 2016). The common denominator between these research endeavors is the perception and tracking of similarities in approaches within the developed standards, leading to a possibility of exploring the nuances in scope, the drafting and development process (whether they have similar roots), as well as the implications of their implementation.

Resuming an earlier idea, we note that the rising interest for non-financial information has put the social and environmental pillars from the triple-bottom line within the sight of standard setters (although the provisions of delineations proves to be quite challenging). Whereas the environmental perspective includes more technical measurements and indicators, the social perspective embeds a larger and more qualitative extension. The development of ISO 26000 generates evidence in this sense (Castka and Balzarova 2007), leading to an increase in assessment accuracy concerning social impact, as well as placing the measurement in the context of quality management. The common perception is that the social dimension (with its corresponding manifestations, such as CSR) enables a myriad of interactions and an emphatic impact on stakeholders, thus in need of an increased attention from standard setters and the accounting profession within the standardization and implementation processes (Tilt 2010).

Another point of view is the context of new reporting trends. The emergence of sustainability and <IR> has proven to be a catalyst for the development of frameworks and sets of principles which enclose their architecture and practical implementation. Also, the breakdown of the triple-bottom line into multiple instances (e.g. the capitals), as well as the accountability setting acting as a liaison, have given the prerequisite of an in-depth analysis of the emergence process in case of the guiding principles (tracing the forms of mimicry with other precursor frameworks).

From Table 8.2, we can notice that each mentioned standard setter and professional organization has issued a set of principles which suits the disclosure needs for its specific type of reporting. From the International Accounting Standards (IAS) (IASB 2011) as well as from the International Public Sector Accounting Standards—IPSAS (which exhibits a pronounced form of mimicry to IAS) (IPSAS 2006), we can observe that the principles provisions are likely prone to financial accounting and reporting. Also, we can notice a different prioritization and the fact that the accrual based accounting is not present in the IPSAS, as there is still a heterogeneity between adopters and an ongoing debate on cash versus accrual accounting in the public sector.

On the other hand, the principles for effective business reporting processes issued by the Professional Accountants in Business Committee (PAIB)—a committee within IFAC—seem to be a bit more methodological and have a wider extension than the other standards (PAIB 2013), offering a stepwise method towards good reporting practices. The Global Reporting Initiative (GRI) principles take into consideration non-financial information and forward-looking orientation, as well (by addressing stakeholders and sustainability) and the quality of the reported information (principles 5–10 from Table 8.2, GRI column). The social dimension is very emphatic within the ISO 26000 and SA8000 standards (with key issues addressed), whereas AA1000 (although simpler in terms of construction) focuses more on accountability and stakeholder relationships.

To advance further with our analysis, we now pose the following research questions:

RQ1. *How is <IR> influenced by these organizations (standard setters)—as a principles-based system?*

RQ2. *How does principles mimicry manifest itself?*

Table 8.2 Synthesis of the main standard setters' guiding principles

Standard setter	IASB	IPSAS Board (independent organ of IFAC)	PAIB (committee within IFAC)	GRI	ISO	AccountAbility	Social accountability international
Main principles	1. Fair presentation and compliance with IFRSs 2. Going concern 3. Accrual basis of accounting 4. Materiality and aggregation 5. Offsetting 6. Frequency of reporting 7. Comparative information 8. Consistency of presentation	1. Fair presentation and compliance with IPSASs 2. Going concern 3. Consistency of presentation 4. Materiality and aggregation 5. Offsetting 6. Comparative information	1. Committing to effective reporting practices 2. Determining roles and responsibilities 3. Planning and controlling the reporting processes 4. Engaging stakeholders 5. Defining the reporting content 6. Selecting frameworks and standards 7. Determining reporting processes 8. Using reporting technology 9. Analyzing and interpreting reported information 10. Obtaining assurance and providing accountability	1. Stakeholder inclusiveness 2. Sustainability context 3. Materiality 4. Completeness 5. Balance 6. Comparability 7. Accuracy 8. Timeliness 9. Clarity 10. Reliability	1. Accountability 2. Transparency 3. Ethical behavior 4. Respect for stakeholder interests 5. Respect for the rule of law 6. Respect for international norms of behavior 7. Respect for human rights	1. Inclusivity 2. Materiality 3. Responsiveness	1. Child labor 2. Forced and compulsory labor 3. Health and safety 4. Freedom of association and right to collective bargaining 5. Discrimination 6. Disciplinary practices 7. Working hours 8. Remuneration
Reference	IAS 1: Presentation of financial statements	IPSAS 1—Presentation of financial statements	PAIB—Principles for effective business reporting processes	G4—Reporting principles and standard disclosures	ISO 26000—Social responsibility	AA 1000—AccountAbility principles standard	SA 8000—Social accountability

Source: Authors' projection, using information from the mentioned standards and references

For a thorough analysis, we first need to provide the context in which the <IR> Framework evolved from the first consultation draft to the final (approved) version and emphasized the impact of stakeholder consultation.

8.4 Advancements Towards a Refined Construct: Development and Usefulness of the Integrated Reporting Principles

Integrated Reporting has recently become a very popular subject, with most of the discussions being focused on regulatory perspectives and reporting performance improvement. The IIRC—as the main governing body—has been putting intense efforts to promote and issue guidelines for Integrated Reporting. The first brick was set in April 2013; the Council issued a Consultation Draft for the Integrated Reporting Framework (IIRC 2013a), thus inviting companies, organizations and various stakeholders from all over the world to consult the document and provide comments.

According to the IIRC website, the initiative has been well received as the Council received 359 comment letter submissions, most of them providing a valuable contribution to the improvement of the framework. Most suggestions have been embedded in the final version of the Framework (IIRC 2013b) and further explanations, guidance and presentations regarding the responses from the comment letters have been published in additional documents (IIRC 2013c, d).

From the synthesis compiled by the IIRC (IIRC 2013d), we can notice that the question regarding the principles of <IR> generated many points of discussion (as all the 359 respondents addressed this question). As a result, 90% of the respondents supported the principles-based requirements as a whole (including 38% with qualification). Some revisions were conducted for the final version of the <IR> Framework connected to the principles as an outcome of the consultation process, with respondents expressing concern over some elements of phrasing, challenges in application (and expectations' meeting), as well as the impact of materiality assessment (by the reporting entities) on disclosure levels and the usefulness of the reports. Each type of concern was thoroughly addressed in the document, with explanations over the acceptance (for revision) or dismissal.

In the final version of the <IR> Conceptual Framework, the main constructs on which the system revolves (the guiding principles, the content elements and the fundamental concepts) are thoroughly presented (IIRC 2013b). The guiding principles (presented in the first column from Table 8.3) represent the key outliers underpinning the preparation of an integrated report, providing guidance on how the information should be presented.

From Table 8.3, we can see that most of the principles from the <IR> Framework have a correspondent in other frameworks or standards. Regarding the matches

Table 8.3 Similarity of <IR> principles with the provisions from existing frameworks

<IR> Principle	Provision from the <IR> framework (quotes from IIRC 2013b)	Similarity with other frameworks or standards
Strategic focus and future orientation	An integrated report should provide insight into the organization's strategy, and how it relates to the organization's ability to create value in the short, medium and long term, and to its use of and effects on the capitals	– IFAC principle "1. Committing to effective reporting practices" – GRI principle "2. Sustainability context" (partially)
Connectivity of information	An integrated report should show a holistic picture of the combination, interrelatedness and dependencies between the factors that affect the organization's ability to create value over time	– GRI principle "2. Sustainability context" (partially)
Stakeholder relationships	An integrated report should provide insight into the nature and quality of the organization's relationships with its key stakeholders, including how and to what extent the organization understands, takes into account and responds to their legitimate needs and interests	– IFAC principle "4. Engaging stakeholders" – GRI principle "1. Stakeholder inclusiveness" – AA1000 principles "1. Inclusivity" and "3. Responsiveness" – ISO26000 principles "1. Accountability" (partially) and "4. Respect for stakeholder interests"
Materiality	An integrated report should disclose information about matters that substantively affect the organization's ability to create value over the short, medium and long term	– IAS principle "4. Materiality and aggregation" – IFAC principle "1. Committing to effective reporting practices" (partially) – IPSAS principle "4. Materiality and aggregation" – GRI principle "3. Materiality" – AccountAbility principle "2. Materiality"
Conciseness	An integrated report should be concise	–
Reliability and completeness	An integrated report should include all material matters, both positive and negative, in a balanced way and without material error	– IAS principle "8. Consistency of presentation" (partially) – IFAC principle "10. Obtaining assurance and providing for accountability" – IPSAS principle "3. Consistency of presentation" (partially) – GRI principles "5. Balance", "4. Completeness", "7. Accuracy" and "10. Reliability" – ISO 26000 principle "2. Transparency" (partially)
Consistency and comparability	The information in an integrated report should be presented: (a) on a basis that is consistent over time; and (b) in a way that enables comparison with other organizations to the extent it is material to the organization's own ability to create value over time	– IAS principles "8. Consistency of presentation" (partially) and "7. Comparative information" – IFAC principle "10. Obtaining assurance and providing for accountability" – IPSAS principles "3. Consistency of presentation" (partially) and "6. Comparative information" – GRI principle "6. Comparability"

Source: Authors' projection, based on IIRC (2013b) and findings from Table 8.2

with the GRI principles, our findings are consistent with the findings presented by Mio (2016), which first emphasized the form of mimicry. The study undertakes an analysis on the IIRC Framework and, in the final part, conducts a comparison between GRI and IIRC on a conceptual level (with focus on the provisions regarding the principles). However, our results show that the mere comparative analysis with the GRI framework can be extended to other settings which encompass key elements specific to <IR> (such as accountability, stakeholder engagement and materiality).

One interesting finding is that that "Connectivity of information" emulates in part the "Sustainability context" principle from the GRI (by requiring the presentation of information regarding "the performance of the organization in the wider context of sustainability", which we found to be the only link to an existing standard). Our analysis adds to Mio's (2016) findings an argument that the strategic focus and future orientation are also in match with the "Sustainability context" principle, as the forward looking orientation of the <IR> system is closely connected to ensuring high levels of sustainability, taking into account the use of and effects on all the capitals. The "Sustainability context" principle (as it is delineated in the GRI guidelines) presents many similarities in this sense with the former mentioned <IR> principle (also with reference with the dynamic impact in time over value creation, preservation or depletion).

Aside from these forms of mimicry, we also found that principles such as materiality, reliability, completeness, consistency and comparability—which are ground principles in the IAS and IPSAS, are also present in the <IR> Framework. Although methodological, the IFAC provisions regarding principles for effective business reporting processes give the outline for several principles included in the <IR> Framework. The marking similarities occur in the case of "Strategic focus and future orientation" (with the strategic input required when committing to effective reporting practices), "Stakeholder relationships" (with the web of interactions involved in the process of engaging stakeholders), as well as the common principles of materiality, reliability, completeness, consistency and comparability.

Another interesting finding of our study is that "Conciseness" (as an <IR> principle) has no prominent form of mimicry and represents an innovative instance of an integrated report. Formerly expressed as the "Efficient communication" principle (different from materiality), it requires the preparers to reasonably the limit the actual length of the report, by presenting only the most important aspects of their activity, so that the users get a transparent and complete view of their activity. Basically, the integrated report works as a corporate or an organizational summary. This is a specific trait of <IR> as the traditional forms of reporting are focused on exact presentations of financial and non-financial information, in comprehensive formats (such as: financial statements with lengthy explanatory notes; financial reports; sustainability reports).

All in all, the mimicry phenomenon in present in the case of <IR> principles and the evolution is rather fair, as most of the standard setters were involved in the drafting of the <IR> Framework and their intention was to contribute to the construction of a useful complementary system (which enhances disclosure levels), not a substituting one. In this respect, their intake was (justifiably) to use markers of a working system (such as existing principles, which can be understood by preparers, matching with <IR> requirements).

8.5 Conclusion

Integrated reporting has gathered much attention in the recent years from all sorts of stakeholders interested in the system's development. In the literature, many authors attempted (and most succeeded) to tackle specific issues regarding the constructs embedded in the <IR> model, as well as conducting prospective analyses, while others positioned themselves as front-runners and conducted exploratory studies on early adopters and pinpointed practical issues. Some of these studies (including those conducted as an outcome of the public consultation process undertaken by the IIRC) were focused on the <IR> Framework and revealed an issue worth looking into, respectively the guiding principles.

Within our paper, we addressed the grounds on which the principles-based approach has developed and gain the timely upper hand on the rules-based approach. Consequently, the <IR> principles represent the outliers encompassing the preparation requirements of an integrated report, providing guidance on how the information should be presented. However, our underlying assumption is that these principles are not necessarily an innovation, but manifest a form of mimicry in comparison to other existing guiding principles (drafted by other standard setters and prone for implementation in other forms of reporting). In this respect, from a selection of major standard setters and professional organizations (IASB, IFAC, IPSASB, and GRI), we synthesized key guiding principles and conducted an overlapping with the <IR> set to see whether there are matches or similarities.

Our findings show that there are many similarities (whether partial or complete) between existing standards and frameworks in terms of principles (mainly with IAS and IPSAS for common principles, and also with IFAC and GRI for distinctive principles), in line with other existing studies (see Mio 2016). The GRI matches are the most prominent as <IR> emerged as an upgraded form of sustainability reporting, also based on the triple bottom line. Also, we find touches of innovation in the case of strategic focus and future orientation (an <IR> principle which makes a distinctive note from traditional reporting and is only partially linked to the sustainability context from the GRI), and conciseness (which has no current links to other existing principles).

Ultimately, we find the mimicry phenomenon for <IR> principles a normal prerequisite in the Framework development process. The standard setters and professional organizations are very much interested in the evolution of the delineation (fact taken from the outcome of the public consultation process) and were even involved in the drafting of the <IR> Framework, their intention being to make this reporting system a complementary one to the ones already existing, thus using a set of principles that are proven to provide a good frame for report drafting.

The main limitation of the paper is that it refrains to conducting an analysis on a conceptual level and focuses mainly on documentary research. Thus, our research perspectives include an analysis on reporting entities, aiming to verify to what extent they are in line with all these principles and how they address the same guiding principle in two (or more) different forms of reporting.

Acknowledgement This work was supported by a grant of the Romanian National Authority for Scientific Research and Innovation, CNCS—UEFISCDI, project number PN-II-RU-TE-2014-4-0265.

References

Abeysekera, I. (2013). A template for integrated reporting. *Journal of Intellectual Capital, 14*(2), 227–245.

AccountAbility. (2008). *AA1000 AccountAbility principles standard (AA1000APS) 2008*. Accessed March 14, 2017, from https://www.accountability.org/standards/

Adams, C. A. (2015). The international integrated reporting council: A call to action. *Critical Perspectives on Accounting, 27*, 23–28.

Bartolini, M., Santini, F., & Silvi, R. (2013). Performance measurement and capitals. In C. Busco et al. (Eds.), *Integrated reporting: Concepts and cases that redefine corporate accountability* (pp. 127–145). Cham: Springer.

Brown, J., & Dillard, J. (2014). Integrated reporting: On the need for broadening out and opening up. *Accounting, Auditing & Accountability Journal, 2*(7), 1120–1156.

Carmona, S., & Trombetta, M. (2008). On the global acceptance of IAS/IFRS accounting standards: The logic and implications of the principles-based system. *Journal of Accounting and Public Policy, 26*(6), 455–461.

Castka, P., & Balzarova, M. A. (2007). A critical look on quality through CSR lenses. *International Journal of Quality & Reliability Management, 24*(7), 738–752.

Cheng, M., Green, W., Conradie, P., Konishi, N., & Romi, A. (2014). The international integrated reporting framework: Key issues and future research opportunities. *Journal of International Financial Management & Accounting, 25*(1), 90–119.

Churet, C., & Eccles, R. (2014). Integrated reporting, quality of management and financial performance. *Journal of Applied Corporate Finance, 26*(1), 56–64.

Collins, D. L., Pasewark, W. R., & Riley, M. E. (2012). Financial reporting outcomes under rules-based and principles-based accounting standards. *Accounting Horizons, 26*(4), 681–785.

Coulson, A. B., Adams, C. A., Nugent, M. N., & Haynes, K. (2015). Exploring metaphors of capitals and the framing of multiple capitals: Challenges and opportunities for <IR>. *Sustainability Accounting, Management and Policy Journal, 6*(3), 290–314.

Cunningham, L. A. (2007). A prescription to retire the rhetoric of 'principles based systems' in corporate law, securities regulation and accounting. *Vanderbilt Law Review, 60*, 1–62.

de Villiers, C., Rinaldi, L., & Unerman, J. (2014). Integrated reporting: Insights, gaps and an agenda for future research. *Accounting, Auditing & Accountability Journal, 27*(7), 1042–1067.

Eccles, R., & Saltzman, D. (2011). Achieving sustainability through integrated reporting. *Stanford Social Innovation Review*, 55–61. Accessed March 8, 2014, from http://people.hbs.edu/reccles/2011SU_Features_EcclesSaltzman.pdf

Eccles, R., Cheng, B., & Saltzman, D. (2010). *The landscape of integrated reporting reflections and next steps*. Cambridge, MA: Harvard Business School.

Eccles, R., Krzus, M., Rogers, J., & Serafeim, G. (2012). The need for sector-specific materiality and sustainability reporting standards. *Journal of Applied Corporate Finance, 24*(2), 8–14.

European Union. (2002). *International accounting standards*. Regulation number 1606/2002. Accessed August 27, 2016, from http://eur-lex.europa.eu/legal-content/EN/TXT/?uri=URISERV%3Al26040

Flower, J. (2015). The international integrated reporting council: A story of failure. *Critical Perspectives on Accounting, 27*, 1–17.

Frías-Aceituno, J. V., Rodríguez-Ariza, L., & García-Sánchez, I. M. (2013). Is integrated reporting determined by a country's legal system? An exploratory study. *Journal of Cleaner Production, 44*, 45–55.

Garcia-Sanchez, I. M., Rodriguez-Ariza, L., & Frias-Aceituno, J. V. (2013). The cultural system and integrated reporting. *International Business Review, 22*, 828–838.

Göbbels, M., & Jonker, J. (2003). AA1000 and SA8000 compared: A systemic comparison of contemporary accountability standards. *Managerial Auditing Journal, 18*(1), 54–58.

GRI. (2013). *Reporting principles and standard disclosures*. Accessed August 27, 2016, from https://www.globalreporting.org/resourcelibrary/GRIG4-Part1-Reporting-Principles-and-Standard-Disclosures.pdf

Haller, A., & van Standen, C. (2014). The value added statement – An appropriate instrument for integrated reporting. *Accounting, Auditing & Accountability Journal, 27*(7), 1190–1216.

Higgins, C., Stubbs, W., & Love, T. (2014). Walking the talk(s): Organisational narratives of integrated reporting. *Accounting, Auditing & Accountability Journal, 27*(7), 1090–1119.

Hughes, J. (2007). IFRS: US stands out on worldwide language. *Financial Times*. Accessed August 27, 2016, from https://www.ft.com/content/5d41f3d0-9141-11dc-9590-0000779fd2ac

IASB. (2011). *International accounting standard 1 – Presentation of financial statements*. Accessed August 27, 2016, from http://ec.europa.eu/internal_market/accounting/docs/consolidated/ias1_en.pdf

Idowu, S. O., Dragu, I., Tiron-Tudor, A., & Fărcaş, T. V. (2016). From CSR and sustainability to integrated reporting. *International Journal of Social Entrepreneurship and Innovation, 4*(2), 134–151.

IIRC. (2011). Accessed May 26, 2016, from http://www.iasplus.com/en/resources/sustainability/iirc

IIRC. (2013a). *The consultation draft of the international <IR> framework*. Accessed March 6, 2014, from http://www.theiirc.org/wp-content/uploads/Consultation-Draft/Consultation-Draft-of-the-InternationalIRFramework.pdf

IIRC. (2013b). *The international <IR> framework*. Accessed March 6, 2014, from http://www.theiirc.org/wp-content/uploads/2013/12/13-12-08-THE-INTERNATIONAL-IR-FRAMEWORK-2-1.pdf

IIRC. (2013c). *Basis for conclusions*. Accessed March 6, 2014, from http://www.theiirc.org/wp-content/uploads/2013/12/13-12-08-Summary-of-significant-issues-IR.pdf

IIRC. (2013d). *Summary of significant issues*. Accessed March 6, 2014, from http://www.theiirc.org/wp-content/uploads/2013/12/13-12-08-Summary-of-significant-issues-IR.pdf

International Organization for Standardization. (2017). *ISO 26000 – Social responsibility*. Accessed March 14, 2017, from https://www.iso.org/iso-26000-social-responsibility.html

IPSAS. (2006). *IPSAS 1 – Presentation of financial statements*. Accessed August 27, 2016, from https://www.ifac.org/system/files/publications/files/ipsas-1-presentation.pdf

Karatzimas, S. (2016). Integrated reporting. In A. Farazmand (Ed.), *Global encyclopedia of public administration, public policy, and governance* (pp. 1–8). Cham: Springer.

Lai, A., Melloni, G., & Stacchezzini, R. (2014). Corporate sustainable development: Is 'integrated reporting' a legitimation strategy? *Business Strategy and the Environment, 25*, 1–13.

Loska, T. (2011). *Integrated reporting: Towards a framework for a sustainable international corporate reporting* (Master's Thesis). Retrieved from https://fowaseson.firebaseapp.com/aa864/integrated-reporting-towards-a-framework-for-a-sustainable-international-corporate-reporting-by-thomas-loska-b007sf1goe.pdf

Mano, R. M., Mouritsen, M., & Pace, R. (2006). Principles-based accounting. *The CPA Journal, 62* (2), 60–63.
Mio, C. (2016). Integrated reporting: The IIRC framework. In C. Mio (Ed.), *Integrated reporting: A new accounting disclosure* (pp. 1–18). London: Palgrave Macmillan.
Mio, C., & Fasan, M. (2013). The determinants of materiality disclosure in integrated corporate reporting. In *Working papers series, no. 9*. Venezia: Universita Ca' Foscari.
Nelson, M. W. (2003). Behavioral evidence on the effects of principles- and rules-based standards. *Accounting Horizons, 17*(1), 91–104.
PAIB. (2013). *International good practice guidance: Principles for effective business reporting processes*. Accessed August 27, 2016, from https://www.ifac.org/publications-resources/principles-effective-business-reporting-processes
Reuter, M., & Messner, M. (2015). Lobbying on the integrated reporting framework: An analysis of comment letters to the 2011 discussion paper of the IIRC. *Accounting, Auditing & Accountability Journal, 28*(3), 365–402.
Rowbottom, N., & Locke, J. (2016). The emergence of <IR>. *Accounting and Business Research, 46*(1), 1–33.
Schipper, K. (2003). Principles-based accounting standards. *Accounting Horizons, 17*(1), 61–72.
Shotridge, R. T., & Myring, M. (2004). Defining principles-based accounting standards. *The CPA Journal, 74*(8), 34–37.
Social Accountability International. (2014). *SA 8000 standard – Social responsibility*. Accessed March 14, 2017, from http://www.sa-intl.org/index.cfm?fuseaction=page.viewpage&pageid=1689
Stubbs, W., & Higgins, C. (2014). Integrated reporting and internal mechanisms of change. *Accounting, Auditing & Accountability Journal, 27*(7), 1068–1089.
Tilt, C. (2010). Corporate responsibility, accounting and accountants. In S. O. Idowu & W. L. Filho (Eds.), *Professionals' perspectives of corporate social responsibility* (pp. 11–32). Cham: Springer.
Tiron-Tudor, A., Dragu, I. M., Cordoş, G. S., & Oprişor, T. (2015). Defining a methodology for social audit based on the social responsibility level of corporations. In M. M. Rahim & S. O. Idowu (Eds.), *Social audit regulation* (pp. 257–279). Cham: Springer.
Vaz, N., Fernandez-Feizoo, B., & Ruiz, S. (2016). Integrated reporting: An international overview. *Business Ethics: A European Review, 25*(4), 577–591.
Velte, P., & Stawinoga, M. (2016). Integrated reporting: The current state of empirical research, limitations and future research implications. *Journal of Management Control, 26*, 1–46.
Wulf, I., Niemoeller, J., & Rentzsch, N. (2014). Development toward integrated reporting, and its impact on corporate governance: A two-dimensional approach to accounting with reference to the German two-tier system. *Journal of Management Control, 25*(2), 135–164.

Chapter 9
Integrated Report: Is it a Strong or Weak Accountability Tool?

Antonella Silvestri and Stefania Veltri

9.1 Introductory Remarks

The concept of Integrated Reporting (IR) has been proposed and discussed recently in the literature by different scholars (Eccles and Krzus 2010; Abeysekera 2013; Adams and Simnett 2011; Adams 2015; de Villiers et al. 2014; Vaz et al. 2016) and practice (KPMG 2012; PWC 2013), with the aim to improve the usefulness and efficiency of corporate reporting.

Institutionally, the discussion has been heavily influenced so far by the International Integrated Reporting Council (IIRC) and its initiatives to develop an Integrated Reporting Framework by the end of 2013. In this Framework the IIRC focuses on the "fundamental concepts", "guiding principles" and "content elements" of an integrated report and raises some issues concerning the "preparation and presentation" of such a report (IIRC 2013).

Several initiatives have emerged in different regions of the world and in different sectors to practice the "integrated thinking" promoted by the <IR> Framework (Busco et al. 2013; Frías-Aceituno et al. 2014; IIRC 2013). South Africa was the first country to direct listed companies to produce an IR (Veltri and Silvestri 2015; PwC 2013).

Integrated reporting has emerged fast therefore as a new accounting practice to help firms understand how they create value and enable them to communicate this effectively to external stakeholders. Its main feature is to combine financial and non-financial disclosures of a company's performance in one report.

Although the paper is the result of an equal joint effort by the authors, their primary individual contributions are reflected in the following of sections of the paper. Antonella Silvestri: Sects. 9.3 and 9.4; Stefania Veltri: Sects. 9.1 and 9.2.

A. Silvestri · S. Veltri (✉)
Department of Business Administration and Law, University of Calabria, Rende, Italy
e-mail: antonella.silvestri@unical.it; stefania.veltri@unical.it

S. O. Idowu, M. Del Baldo (eds.), *Integrated Reporting*, CSR, Sustainability, Ethics & Governance, https://doi.org/10.1007/978-3-030-01719-4_9

Two competing positions emerge in the literature: the first, which considers IR as the evolution of the annual report (AR); the second, seeing IR as the evolution of the sustainability report (SR), the most diffused voluntary report disclosing non-financial information.[1]

The first position (IR as evolution of AR) finds its theoretical justification in the rationale behind the IIRC framework. The IIRC framework clearly states that: "an integrated report should be prepared primarily for providers of financial capital in order to support their financial capital allocation assessments" (IIRC 2013, p. 8). Even if the subsequent paragraph states that "an integrated report and other communications resulting from the IR will be of benefit to all stakeholders interested in an organization's ability to create value over time", the indication of the IIRC framework is clear in indicating investors and providers of financial capital as the intended users of IR. Owing to this, IR is viewed as an evolution of AR, rather than of SR (Fasan 2013; Tweedie 2014).

The second position (IR as evolution of SR) starts from the consideration that organizations have come under pressure to adapt their business practices to consider the increasing public awareness of environmental, social and governance (ESG) issues (Kolk and van Tulder 2010; Seuring and Mueller 2008; GRI 2013). Stakeholders demand that an organization's ESG performance be transparent, and corporate social responsibility (CSR) reports have proved vital in meeting those demands.

The number of organizations worldwide utilizing voluntary CSR reports has increased substantially since the financial crisis in 2008/2009 (KPMG 2015), and CSR reporting has been subject to increasing standardization (e.g. Global Reporting Initiative), regulation (e.g. EU directive 2014/95) and research activity (see Huang and Watson 2015). Among the exponents of this second position, we can quote Velte and Stawinoga (2017); Giovannoni and Fabietti (2013); Stacchezzini et al. (2016) etc.

Consistently with Fasan (2013), we believe IR is a tool that has the potential to overcome the limitations of both AR (complexity, short-termism, shortage of non-financial information) and SR (low reliability and trust from investors, and disconnection with financial performance). Nevertheless, the main challenge linked to IR is the difficulty of integrating financial information derived from traditional financial reporting with non-financial information derived from voluntary reports in a single integrated report.

A recent research (Dumay et al. 2016) highlighted that there is a disconnection between academic research in IR and IR practice, as the majority of IR articles do not research practice, specific organizations or engage practitioners as fellow researchers and authors. Consistently with Dumay et al. (2016), we also believe that a mainly empirical perspective has the potential to advance knowledge in the IR field. To face

[1]Among the voluntary reports we can comprehend the Corporate Governance (CG) report, the Intellectual capital(IC) reports, the corporate social responsibility report (CSR), the environmental report, the sustainability report (SR), which includes environmental and CSR issues. In the paper, the term CSR and SR will be used interchangeably.

the problem of the gap between theory and practice, this paper aims to develop a specific research template able to support the researchers to analyze the degree of accountability of the organizational IRs. The building of this research template represents the main originality of the paper.

9.2 Literature Review

The literature on IR has had an exponential growth in recent years, therefore there is a huge number of studies focused on IR (Eccles et al. 2014).

From an academic perspective, there is a growing contemporary interest in IR, as evidenced by the circumstance that several papers have been presented in leading accounting conferences (Massaro et al. 2016). IR is a recent research field which is passing through the typical phases of an emerging research area in parallel with IC research (Petty and Guthrie 2000).

Studies belonging to the first phase focused on raising awareness of IR as a suitable tool to address stakeholders' needs (Eccles et al. 2014). Actually, the majority of studies lie in the second phase, devoted to a critical analysis of guiding principles and content elements of the IR framework (Flower 2015; Adams 2015); some studies started to investigate the effect of IR on financial performance and value creation (Baboukardos and Rimmel 2016). In the third stage of IR research, researchers should become involved with the praxis of IR inside organizations and critically analyze the existing gap between the espoused benefits of IR in the literature and the reality of organizational practices, the process of IR implementation, and the quality of IR (Dumay and Garanina 2013).

This research could be included in the third stage, as it aims to investigate the quality of IRs in terms of degree of accountability of the document towards organizational stakeholders.

In the paper, we accept Christensen and Ebrahim's (2006) conceptualization of accountability as "being answerable to stakeholders for the action of the organization" (Gray et al. 2014). Within the accountability paradigm, there are two different positions (Del Baldo and Baldarelli 2015). According to the first position, the organizations have to be financially sustainable and socially responsible to all of their multiple stakeholders, in order to create social value and achieve their organizational goals (Andreaus 2007; Costa et al. 2011). This identifies a "strong" form of accountability. According to the second position, the organizations voluntary self-report on their trustworthiness as part of a reputation-building process (Owen et al. 2001), to give a positive signal regarding the reputation of the firm (Simaens and Koster 2013). This second position identifies a "weak" form of accountability (Owen et al. 2001).

In the strong accountability perspective organizations are ethical, meaning that they accept social responsibility as an ethical obligation, integrating in their social accountability a responsibility towards an efficient and effective use of their resources (Surroca et al. 2010). Ethical, political, and integrative theories of CSR suggest that firms/managers have an incentive to be honest, trustworthy, and ethical in their business processes, and thus tend to adhere to a high standard of behavior (Garriga and Melé 2004). On the other hand, under the weak accountability perspective an organization communicates non-financial information with the only aim to a positive signal regarding its reputation (e.g., Simaens and Koster 2013).

Depending on the approach followed, the meaning of IR changes for the organizations: according to the first approach (strong accountability), IR is used as an accountability tool; under the second approach (weak accountability), IR is used as an image-building and reputational tool (Veltri and Bronzetti 2014).

Empirical research on the IR quality appears fragmented, dispersed and conducted by a large variety of theoretical models (Velte and Stawinoga 2017). Among these studies, only the study of Pozzoli and Gesuele (2016) intends quality in the sense of accountability of an IR, analyzing the degree of IR accountability of 11 Italian public utilities belonging to the IIRC pilot program. Several studies measure the IR quality through an IR quality score/index. Among these, we can quote the studies of Stent and Dowler (2015), Lipunga (2015) and Clayton et al. (2015) investigated the quality of IR of eight companies listed on the Johannesburg Stock Exchange by constructing an IR quality index based on elements such as length of report and structure. Stent and Dowler (2015) create an IR quality score based on four "best practice reporting" firms in New Zealand. Lipunga (2015) builds up an IR quality index based on the disclosure items and IR headings of 12 companies' IRs listed on the Malawi Stock Exchange. Clayton et al. (2015) construct an IR quality index based on the IRs disclosed of eight companies listed on the Johannesburg Stock Exchange taking into consideration elements such as length of IR and its structure. Other empirical studies focused on IR quality are centered on the adherence to the principles and/or content elements to a framework considered a benchmark. Among these we can quote the study of Marx and Mohammadali-Haji (2014) and the study of Veltri and Silvestri (2015). Marx and Mohammadali-Haji (2014) recognize that the quality of IR is very heterogeneous in South Africa and that some firms merely renamed their annual reports as integrated reports to create the appearance of compliance with the developments of IR. Veltri and Silvestri (2015) analyse the IR disclosed by the University of Free State (UFS) in terms of content and guiding principles according to the IIRC framework.

Based on the above literature review, the research aims to contribute to the debate on the IR quality by proposing a research template able to guide the researchers in evaluating the quality of IR disclosed by organizations in terms of accountability.

9.3 The Research Template

The research template we intend to use to read and understand the information provided by IR comprehends three dimensions: a) the attitude of organization towards its stakeholders; b) the way in which the organization discloses its value creation process through the business model representation; c) the type of integration originating IR.

The three dimensions have been selected analyzing the literature linking each of them to the degree of accountability of the voluntary reports disclosed by organizations.

The references related to the three dimensions are specified in the following subsections devoted to each of them.

The research template is a tool allowing researchers to express a circumstantial judgment on the approach (strong or weak) used by the organization in disclosing their IR and to compare different organizations in terms of IR quality. In other words the research template allows unveiling whether the organization is guided by an accountable perspective (ceremonial) or by a legitimation (substantive) perspective in disclosing its IR (Haji and Anifowose 2016). The use of IR at a strong rather than at a weak accountability level, implies more responsiveness on the part of organizations towards stakeholders and requires technical and cultural organizational changes (Doni and Gasperini 2014).

9.3.1 Stakeholders Involvement

The primary IR stakeholders come from the corporate, investment, accounting, securities, regulatory, academic, civil society and standard-setting sectors (IIRC 2013). In a preliminary version, the IIRC Framework stated that the target audience for IR was providers of financial capital (Stubbs and Higgins 2014). This position was criticized by Flower (2015) for its emphasis on "value for investors" and not "value for society", and argues that there is no obligation on firms to report harm inflicted on entities outside the firm (such as the environment) where there is no subsequent impact on the firm. In fact, according to the *stakeholder theory*, an organization should create wealth for all participants (or stakeholders), in contrast to the traditional financial model based on creating value for the principal agent or shareholder (García-Sánchez et al. 2013). In a further version of IIRC Framework was rectified in favor of accountability towards all stakeholders.

The stakeholder responsiveness is a relevant guiding principle of IR on the basis of IIRC Framework (Table 9.1), which underlines the relevance of the relationship between organisation and stakeholders, as value is not created by the organisation alone, but through relationships with third parties (IIRC 2013).

As the basic proposition of the stakeholder theory is that the firm's survival depends on its successful management of relationships with stakeholders, the latter need to be informed of the economic, social and environmental impact of corporate performance if they are to continue this waiving of resources, or to penalize inadequate performance by revoking the support provided. Since the information provided in financial statements is usually insufficient, the concept of corporate transparency must be expanded to other areas, such as the social and environmental

Table 9.1 The IIRC guiding principles

Guiding principle	Meaning
Strategic focus and future orientation	An IR should provide insight into the organization's strategy, and how it relates to the organization's ability to create value in the short, medium and long term, and to its use of and effects on the capitals
Connectivity of information	An IR should show a holistic picture of the combination, interrelatedness and dependencies among the factors that affect the organization's ability to create value over time
Stakeholder relationships	An IR should provide insight into the nature and quality of the organization's relationships with its key stakeholders, including how and to what extent the organization understands, takes into account and responds to their legitimate needs and interests
Materiality	An IR should disclose information about matters that substantively affect the organization's ability to create value over the short, medium and long term
Conciseness	An IR should be concise
Reliability and completeness	An IR should include all material matters, both positive and negative, in a balanced way and without material error
Consistency and comparability	The information in an integrated report should be presented: (a) on a basis that is consistent over time; and (b) in a way that enables comparison with other organizations to the extent that it is material to the organization's own ability to create value over time

Source: Adapted from IIRC (2013)

aspects of corporate behaviour and must be presented in an integrated form (Azcarate et al. 2011; Frías-Aceituno et al. 2013).

According to the stakeholder theory literature (Waddock 2002), there are three levels of stakeholder involvement: *stakeholder mapping* (first level), in which the corporation maps its stakeholders, if possible distinguishing between primary and secondary; *stakeholder management* (second level), in which the corporations tries to manage stakeholders' expectations, balancing different positions; and *stakeholder engagement* (third level), in which corporations involve their stakeholders in decision-making processes, sharing information, having dialogue with them and creating a model of mutual responsibility (Manetti 2011; Rinaldi 2013).

On the basis of the above consideration, we believe that the organizations following a *stakeholder engagement* approach achieve a strong accountability towards their stakeholders, while the propensity for *stakeholder management* highlights an opportunistic and strategic approach to the stakeholder theory (Freeman 1984). We intend to evaluate the attitude of an organization towards its stakeholders by reading throughout its IR, and analyzing a series of elements indicative of the quality of its stakeholder involvement, that is, whether stakeholders are included in collaboratively designing IR performance metrics with organizations, and whether they are regularly approached to receive or provide feedback on subsequent IR.

9.3.2 Business Model

The business model is a key concept in the managerial literature (Beattie and Smith 2013; Aprile and Magnaghi 2014). During the last 20 years, many academics and practitioners have focused their attention on the concept of the business model, without finding a shared definition (Zott et al. 2011).

IIRC provided its own definition of business model as "the organization's chosen system of inputs, business activities, outputs and outcomes that aims to create value over the short, medium and long term" (IIRC 2013, p. 6) (see Fig. 9.1).

The business model is one of the most important content elements of IR on the basis of the IIRC Framework (Table 9.2), illustrating how the firms' business activities create/destroy value by processing their inputs, that is, the six forms of capital, with specific interconnections among them.

As regards the representation of the business model, which is the core of IR, it is difficult to make the firm's value creation process explicit and understandable, both owing to the capability of managers to prepare it and to the stakeholders' ability to read it. Moreover, a further relevant problem related to the business model lies in the opportunity to unveil the firm's value creation dynamics, because it is commercially sensitive information (Tweedie 2014).

The IIRC believes that IR is a more effective reporting approach because it focuses on value creation through the lens of the six forms of capital (*financial, manufactured, intellectual, human, social and relationship,* and *natural*) rather than sustainability reporting's focus on environmental and social impacts through the lens of stakeholder materiality (Nugent 2015). Table 9.3 illustrates the six forms of capital and their meaning according to IIRC Framework.

Not all the above-mentioned forms of capital are owned by the organization, but each form of capital is considered as an "input" since it constitutes or affects the business process. The business model draws from the six forms of capital (included in the IR framework) as its main inputs. These inputs are then transformed into outputs through business activities. Both business activities and outputs will generate a range of outcomes that, in turn, will eventually affect the various capitals.

In order to show the organizational value creation process, the business model should be interpreted and represented as the key element linking an organization's strategy, governance and performance. Moreover, in order to be an effective tool, which is useful for the decision-making process, business model should develop the following steps: a) start from organizational strategy; b) deeply understand the relationships between its elements and with the external environment; c) take into consideration the information feedback in terms of recommendation for future changes (Bernabè and Giorgino 2013). Conversely, a business model which is decoupled from the strategy, not tailored to the firm specificity, and not able to show the relationships with its environment, does not address the IIRC aim, which is to show stakeholders the business value creation process.

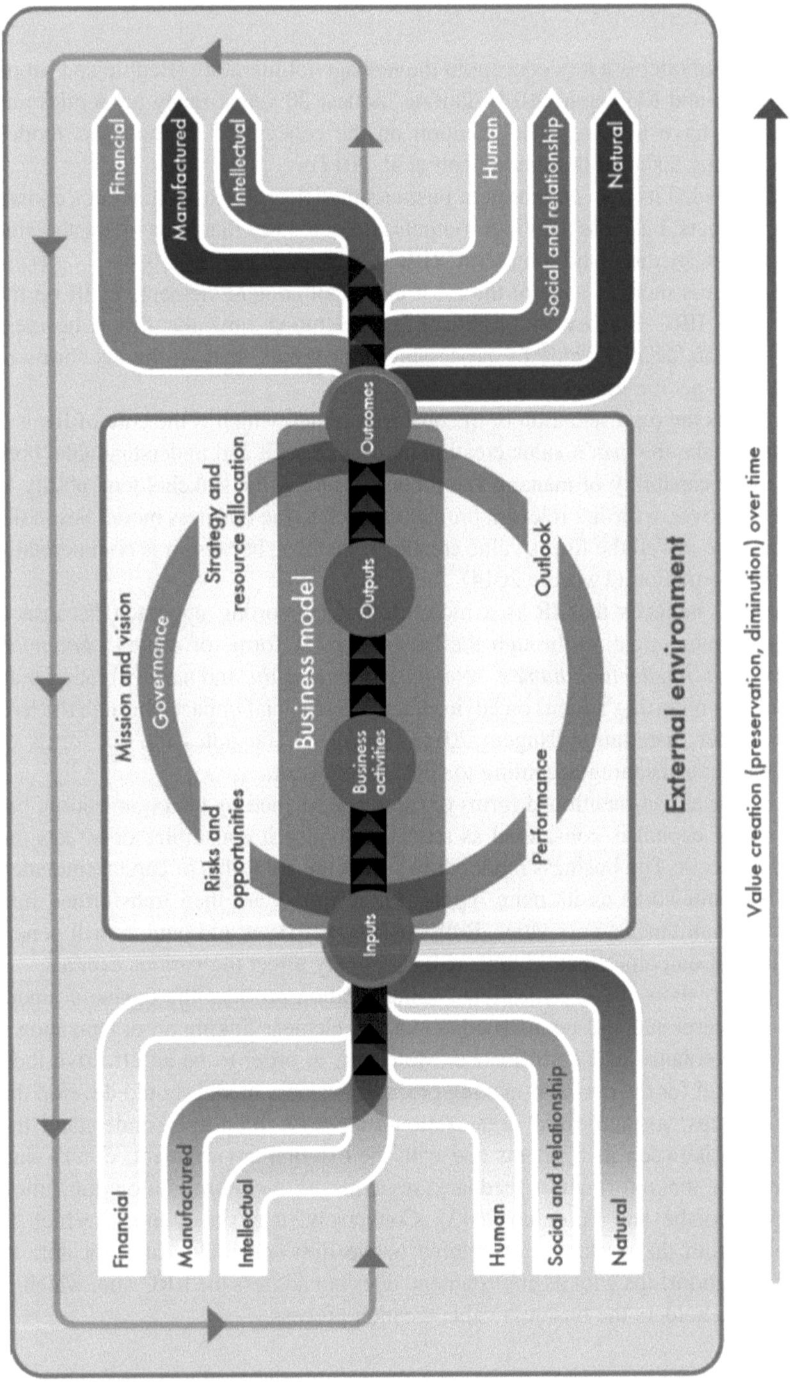

Fig. 9.1 The IIRC business model. Source: IIRC (2013)

Table 9.2 The IIRC content elements

Content element	Question to answer	Including
a. Organizational overview and external environment	What does the organization do and what are the circumstances under which the organization operates?	The organization's mission and vision; key quantitative information; significant factors affecting the external environment and the organization's response
b. Governance	How does the organization's governance structure support its ability to create value in the short, medium and long term?	Organizations' leadership structure; specific process and particular actions; remuneration and incentives
c. Business model	What is the organization's business model?	Inputs, business activities; outputs, outcomes
d. Risks and opportunities	What are the specific risks and opportunities that affect the organization's ability to create value over the short, medium and long term, and how is the organization dealing with them?	The specific source of risks and opportunities; the organization's assessment of risks; the specific steps taken to manage risks
e. Strategy and resource allocation	Where does the organization want to go and how does it intend to get there?	The organization's strategic objective; the resource allocation plan; the linkage between them
f. Performance	To what extent has the organization achieved its strategic objectives for the period and what are its outcomes in terms of effects on the capitals?	Quantitative indicators on targets and risks; the organization's effects on capitals; the state of key stakeholders relationships; links with past and future performance
g. Outlook	What challenges and uncertainties is the organization likely to encounter in pursuing its strategy, and what are the potential implications for its business model and future performance?	The organization's expectations and how the organization is equipped to face them; the discussions of potential implications for future financial performance
h. Basis of preparation and presentation	How does the organization determine what matters to include in the integrated report and how are such matters quantified or evaluated?	The organization materiality process; the description of reporting boundary; frameworks and methods used to quantify or evaluate material matters
i. General reporting guidance		Disclosure of material matters; disclosures about the capitals; time for short-, medium- and long-term aggregation and disaggregation

Source: Adapted from IIRC (2013)

Table 9.3 The six capitals of business model

Capital type	Meaning in the IIRC framework
Financial	The pool of funds available to an organization for use in the production of goods or the provision of services obtained through financing or generated by firm
Manufactured	Manufactured physical objects (as distinct from natural physical objects) that are available to an organization for use in the production of goods or the provision of services
Intellectual	Organizational, knowledge-based intangibles
Human	People's competencies, capabilities and experience, and their motivations to innovate
Social and relationship	The institutions and the relationships within and between communities, groups of stakeholders and other networks, and the ability to share information to enhance individual and collective well-being
Natural	All renewable and nonrenewable environmental resources and processes that provide goods or services that support the past, current or future prosperity of an organization

Source: Our elaboration based on the IIRC (2013)

9.3.3 Integration

The relevance of integrated thinking in producing an IR is underlined also by the second guiding principle, the *connectivity* of information (Table 9.1). Under a theoretical profile, according to the ways in which the partial reports are combined, three different kinds of integration can be achieved: *weak* aggregation, *strong* aggregation and integration in a *narrow sense* (Paternostro 2013).

Within the weak aggregation approach, the IR is constructed starting from a main partial report, which the organisation "enhances" simply by adding other information perceived as secondary. The IR can derive both from the Financial Statement including social and environmental information, or instead from the Social and Environmental Report aggregating financial information. The IR resulting from this aggregation, is not a real IR for two reasons: 1) the focus remains the same as the original main report; 2) there is no connectivity between information.

Within the strong aggregation approach, instead, the IR results from the aggregation of a number of partial reports, all maintaining their identity. Unlike the first approach, there is no "main" report to which secondary information is added, but instead there is equilibrium among the different reports making up the IR. Even in this second kind of integration, the information is not interconnected and the IR does not provide a comprehensive synthesis.

Finally, within the third approach, the partial reports are not identifiable in the IR, just providing information that merges within the IR. Only in this approach could the IR be identified as defined in the IIRC framework, as an "enhanced main report" derives from the weak aggregation and an "aggregated report" from the strong aggregation. In this third approach, the IR respects both the materiality principle (as it includes not all information, but only that relevant to drive the organizational value creation process) and the conciseness principle (the information is assimilated in IR without overloading and without identifying the original source), so realizing the IR role for IIRC framework, which is to provide a global evaluation of organizational value creation ability (Paternostro 2013).

Table 9.4 summarizes the three dimensions included within the research template together with the degree of accountability that dimension could assume.

9.4 Considering Conclusions, Limits and Future Research Directions

The production and presentation of an IR extends the information contained in traditional financial statements. Nevertheless, providing integrated information understandable for the stakeholders is difficult for all kinds of organizations (Du Toit 2017). In the paper, we research IR practice from a reporting perspective, with the aim to investigate the IR quality, that is whether IR is used as a weak or a strong accountability tool. In the first meaning, organizations use IRs to be answerable towards their stakeholders, while in the second meaning organizations use IRs as reputational tools.

The purpose of the paper is thus addressed to create a research template on the bases of literature analyses, then applying it to best practice IR case studies to verify their accountability degree in disclosing IR, in other words whether the organizations follow a legitimatization or an accountable approach (Haji and Anifowose 2016).

The implementation of the research template illustrated in Table 9.4 requires to analyze the organizational IR in terms of the three selected dimensions. First of all, it is necessary to position the investigated IR within the research template for each dimension, from low to high. The positioning of IR on each dimension is a function of the degree assumed by each dimension along a continuum going from a minimum to a maximum. The final evaluation on the degree of accountability of the

Table 9.4 The research template

	Degree of accountability		
Dimensions	Low	Medium	High
Attitude towards stakeholders	Stakeholders mapping	Stakeholders management	Stakeholders engagement
The vision of business model	Business model decoupled from the strategy	Business model weakly related to the strategy	Business model derived from the strategy
Integration of the financial and non-financial reports	Weak integration	Strong integration	Integration in the narrow sense

Source: Own elaboration

comprehensive IR is thus obtained by mediating in qualitative terms the judgements expressed for each dimension.

The use of IR at a strong accountability level, rather than at a weak level, implies more responsiveness on the part of organizations towards stakeholders (Stubbs and Higgins 2014) and requires technical and cultural organizational changes (García-Sánchez et al. 2013).

The study adds knowledge to the IR empirical literature stream addressed to analyze IR from an accountability perspective (Wild and Van Staden 2013; Ruiz-Lozano and Tirado-Valencia 2016). This framework contributes to the developing field of IR, fostering further research and debate on this reporting phenomenon (de Villiers et al. 2017).

The created research template presents both strength and a weakness points. As for the strength point, it gives an original contribution to the IR literature simplifying the elements to consider to express an evaluation about the degree of accountability followed by the organization in disclosing IR, an issue on which there is a concern in IR literature (Haji and Anifowose 2016; Stent and Dowler 2015). As for the weakness point, the created research template introduce subjectivity elements both in the selection of the key dimensions included in the research framework, and because observations and conclusions are based on the authors' analysis under an interpretative approach.

A future research direction could be directed to apply the research framework to a real case study to verify the usefulness of the research framework in evaluating the degree of accountability of a company's IR (Silvestri et al. 2017). Moreover, the implementation of the same research framework to other organizational IR documents could allow different organizations to be compared in terms of the quality of their IR disclosed (external comparison) and the same organization to be compared in different periods of time (internal comparison).

A further research direction, coherently with Doni and Gasperini (2014) and Fasan (2013), could be addressed to verify whether and how the internal implementation of IR has been able to modify the organizational management control systems and whether IR information is used by managers for decision-making purposes, so completing the reporting perspective with an internal approach.

References

Abeysekera, I. (2013). A template for integrated reporting. *Journal of Intellectual Capital, 14*(2), 227–245.

Adams, C. A. (2015). *Understanding integrated reporting: The concise guide to integrated thinking and the future of corporate reporting*. Oxford: Do Sustainability.

Adams, S., & Simnett, R. (2011). Integrated reporting: An opportunity for Australia's not-for-profit sector. *Australian Accounting Review, 21*(3), 292–301.

Andreaus, M. (2007). Quale modello di rendicontazione per l'impresa sociale? *Impresa Sociale, 76*, 59–68.

Aprile, R., & Magnaghi, E. (2014). Business model and integrated reporting: A first analysis. In *Knowledge and management models for sustainable growth*, e-book of proceedings of IFKAD – International forum on knowledge asset dynamics (pp. 3818–3836). Matera: IKAM.

Azcarate, F., Carrasco, F., & Fernandez, M. (2011). The role of integrated indicators in exhibiting business contribution to sustainable development: A survey of sustainability reporting initiatives. *Spanish Accounting Review, 14*, 213–240.

Baboukardos, D., & Rimmel, G. (2016). Value relevance of accounting information under an integrated reporting approach: A research note. *Journal of Accounting and Public Policy, 35* (4), 437–452.

Beattie, V., & Smith, S. J. (2013). Value creation and business models: Refocusing the intellectual capital debate. *The British Accounting Review, 45*, 243–254.

Bernabè, F., & Giorgino, M. C. (2013). "Integrating" business model and strategy. In C. Busco, M. L. Frigo, A. Riccaboni, & P. Quattrone (Eds.), *Integrated reporting. Concepts and cases that redefine corporate accountability* (pp. 111–126). Cham: Springer.

Busco, C., Frigo, M. L., Riccaboni, A., & Quattrone, P. (Eds.). (2013). *Integrated reporting. Concepts and cases that redefine corporate accountability*. Cham: Springer.

Christensen, R. A., & Ebrahim, A. (2006). How does accountability affect mission? The case of a nonprofit serving immigrants and refugees. *Nonprofit Management & Leadership, 17*(2), 195–209.

Clayton, A. F., Rogerson, J. M., & Rampedi, I. (2015). Integrated reporting vs. sustainability reporting for corporate responsibility in South Africa. *Bulletin of Geography, 29*, 7–17.

Costa, E., Ramus, T., & Andreaus, M. (2011). Accountability as managerial tool in non-profit organizations: Evidence from Italian CSVs. *VOLUNTAS: International Journal of Voluntary and Nonprofit Organizations, 22*, 470–492.

de Villiers, C., Rinaldi, L., & Unerman, J. (2014). Integrated reporting: Insights, gaps and an agenda for future research. *Accounting, Auditing and Accountability Journal, 27*(7), 1042–1067.

de Villiers, C., Hsiao, P.-C., & Maroun, W. (2017). Developing a conceptual model of influences around integrated reporting, new insights and directions for future research. *Meditari Accountancy Research, 25*(4), 450–460.

Del Baldo, M., & Baldarelli, M. G. (2015). From weak to strong CSR: The experience of the EoC (Economy of Communion) industrial parks in Germany and Italy. *UmweltWirtschaftsForum, 3*, 1–16.

Doni, F., & Gasperini, A. (2014). The internal integrated reporting and the value creation: A case study approach. In *Knowledge and management models for sustainable growth*, e-book, of proceedings of IFKAD – International forum on knowledge asset dynamics (pp. 2666–2704). Matera: IKAM.

Du Toit, E. (2017). The readability of integrated reports. *Meditari Accountancy Research, 25*(4), 629–653.

Dumay, J., & Garanina, T. (2013). Intellectual capital research: A critical examination of the third stage. *Journal of Intellectual Capital, 14*(1), 10–25.

Dumay, J., Bernardi, C., Guthrie, J., & Demartini, P. (2016). Integrated reporting: A structured literature review. *Accounting Forum, 40*(3), 166–185.

Eccles, R. G., & Krzus, M. P. (2010). *One report: Integrated reporting for a sustainable strategy*. New York: Wiley.

Eccles, R. G., Krzus, M. P., & Ribot, S. (2014). *The integrated reporting movement: Meaning, momentum, motives, and materiality*. New York: Wiley.

Fasan, M. (2013). Annual reports, sustainability reports and integrated reports: Trends in corporate disclosure. In C. Busco (Ed.), *Integrated reporting* (pp. 41–58). Cham: Springer.

Flower, J. (2015). The international integrated reporting council: A story of failure. *Critical Perspectives on Accounting, 27*, 1–17.

Freeman, R. E. (1984). *Strategic management: A stakeholder approach*. Boston, MA: Pitman.

Frías-Aceituno, J. V., Rodríguez-Ariza, L., & García-Sánchez, I. M. (2013). The role of the board in the dissemination of integrated corporate social reporting. *Corporate Social Responsibility and Environmental Management, 20*(4), 219–233.

Frías-Aceituno, J. V., Rodríguez-Ariza, L., & Garcia-Sánchez, I. M. (2014). Explanatory factors of integrated sustainability and financial reporting. *Business Strategy and the Environment, 23*(1), 56–72.

García-Sánchez, I. M., Rodríguez-Ariza, L. B., & Frías-Aceituno, J. V. (2013). The cultural system and integrated reporting. *International Business Review, 22*, 828–838.

Garriga, E., & Melé, D. (2004). Corporate social responsibility theories: Mapping the territory. *Journal of Business Ethics, 53*(1–2), 51–71.

Giovannoni, E., & Fabietti, G. (2013). What is sustainability? A review of the concept and its applications. In C. Busco (Ed.), *Integrated reporting* (pp. 21–40). Cham: Springer.

Gray, R., Adams, C., & Owen, D. (2014). *Accountability, social responsibility and sustainability: Accounting for society and the environment*. Harlow: Pearson.

GRI. (2013). *The sustainability content of integrated reports. A survey of pioneers*. GRI, Amsterdam. Accessed March 8, 2016, from https://www.globalreport-ing.org/resourcelibrary/GRI-IR.pdf

Haji, A. A., & Anifowose, M. (2016). The trend of integrated reporting practice in South Africa: Ceremonial or substantive? *Sustainability Accounting, Management and Policy Journal, 7*(2), 190–224.

Huang, X. B., & Watson, L. (2015). Corporate social responsibility research in accounting. *Journal of Accounting Literature, 34*, 1–16.

IIRC – International Integrated Reporting Council. (2013). *Integrated reporting: The IIRC*. Accessed April 16, 2015, from www.theiirc.org/

Kolk, A., & van Tulder, R. (2010). International business, corporate social responsibility and sustainable development. *International Business Review, 19*, 119–125.

KPMG. (2012). *Integrated reporting. Performance insight through better business reporting*. Issue 2. Accessed April 10, 2015, from www.kpmg.com/AU/en/IssuesAndInsights/ArticlesPublications/Better-Business-Reporting/Documents/integrated-reporting-issue-2.pdf

KPMG. (2015). *The KPMG survey of corporate responsibility reporting 2015*. Accessed March 24, 2016, from http://www.kpmg.com/CN/en/IssuesAndInsights/ArticlesPublications/Documents/kpmg-survey-of-corporate-responsibility-reporting-2015-O-201511.pdf

Lipunga, A. M. (2015). Integrated reporting in developing countries. Evidence from Malawi. *Journal of Management Research, 7*, 130–156.

Manetti, G. (2011). The quality of stakeholder engagement in sustainability reporting: Empirical evidence and critical points. *Corporate Social Responsibility and Environmental Management, 18*(2), 110–122.

Marx, B., & Mohammadali-Haji, A. (2014). Emerging trends in reporting: An analysis of integrated reporting practices by South African top 40 listed companies. *Journal of Economic and Financial Sciences, 7*, 231–250.

Massaro, M., Dumay, J., & Guthrie, J. (2016). On the shoulders of giants: Undertaking a structured literature review in accounting. *Accounting, Auditing & Accountability Journal, 29*(5), 767–801.

Nugent, M. (2015, April 20). *Integrated reporting: The current state of play*. Melbourne: Presentation to the Association of Chartered Certified Accountants.

Owen, D., Swift, T., & Hunt, K. (2001). Questioning the role of stakeholder engagement in social and ethical accounting, auditing and reporting. *Accounting Forum, 25*, 264–282.

Paternostro, S. (2013). The connectivity of information for the integrated reporting. In C. Busco (Ed.), *Integrated reporting* (pp. 59–77). Cham: Springer.

Petty, R., & Guthrie, J. (2000). Intellectual capital literature review: Measurement, reporting and management. *Journal of Intellectual Capital, 1*(2), 155–176.

Pozzoli, M., & Gesuele, B. (2016). From theory to practice: First adoption of integrated reporting by the Italian public utilities. In D. Mancini, R. P. Dameri, & E. Bonollo (Eds.), *Strengthening information and control systems* (pp. 121–132). Cham: Springer.
PWC. (2013). *The value creation journey*. A survey of JSE top-40 companies' integrated reports. Accessed March 6, 2014, from www.pwc.com/corporatereporting
Rinaldi, L. (2013). Stakeholder engagement. In C. Busco, M. L. Frigo, A. Riccaboni, & P. Quattrone (Eds.), *Integrated reporting. Concepts and cases that redefine corporate accountability* (pp. 95–110). London: Springer.
Ruiz-Lozano, M., & Tirado-Valencia, P. (2016). Do industrial companies respond to the guiding principles of the integrated reporting framework? A preliminary study on the first companies joined to the initiative. *Revista de Contabilidad, 19*(2), 252–260.
Seuring, S., & Mueller, M. (2008). Core issues in sustainable supply chain management—A Delphi study. *Business Strategy and the Environment, 17*, 455–466.
Silvestri, A., Veltri, S., Venturelli, A., & Petruzzelli, S. (2017). A research template to evaluate the degree of accountability of integrated reporting: A case study. *Meditari Accountancy Research, 25*(4), 675–704.
Simaens, A., & Koster, M. (2013). Reporting on sustainable operations by third sector organizations. *Public Management Review, 15*(7), 1040–1062.
Stacchezzini, R., Melloni, G., & Lai, A. (2016). Sustainability management and reporting: The role of integrated reporting for communicating corporate sustainability management. *Journal of Cleaner Production, 136*(A), 102–110.
Stent, W., & Dowler, T. (2015). Early assessments of the gap between integrated reporting and current corporate reporting. *Meditari Accountancy Research, 23*, 92–117.
Stubbs, W., & Higgins, C. (2014). Integrating reporting and internal mechanisms of change. *Accounting, Auditing & Accountability Journal, 27*(7), 1068–1089.
Surroca, J., Tribò, J. A., & Waddock, S. (2010). Corporate responsibility and financial performance: The role of intangible resources. *Strategic Management Journal, 31*, 463–490.
Tweedie, D. (2014, July 7–9). Integrated reporting: Symptom or cure of new capitalism's ills?. Proceedings of the Critical Perspectives on Accounting Conference, Toronto.
Vaz, N., Fernandez-Feijoo, B., & Ruiz, S. (2016). Integrated reporting: An international overview. *Business Ethics: A European Review, 25*(4), 577–591.
Velte, P., & Stawinoga, M. (2017). Integrated reporting: The current state of empirical research, limitations and future research implications. *Journal of Management Control, 28*(3), 275–320.
Veltri, S., & Bronzetti, G. (2014). Intellectual capital reporting in the Italian nonprofit sector. An image-building or an accountability tool? In E. Costa, L. D. Parker, & M. Andreaus (Eds.), *Accountability and social accounting for social and non-profit organizations*. Bingley: Emerald.
Veltri, S., & Silvestri, A. (2015). The Free State university integrated reporting: A critical consideration. *Journal of Intellectual Capital, 16*(2), 443–462.
Waddock, S. A. (2002). *Leading corporate citizens: Vision, values, value added*. Boston, MA: McGraw-Hill.
Wild, S., & van Staden, C. (2013). *Integrated reporting: Initial analysis of early reporters–an institutional theory approach* [Available on line]. Accessed May 4, 2015, from http://www.apira2013.org/
Zott, C., Amit, R., & Massa, l. (2011). The business model: Recent developments and future research. *Journal of Management, 37*(4), 1019–1042.

Chapter 10
Integrated Reporting, Integrated Thinking and Gaming Companies: Myths and Paradoxes

Maria-Gabriella Baldarelli

10.1 Introduction

The chapter takes its ideas from the concept that there are some situations in which it is very difficult to speak about integrated thinking (Mella 2012) and integrated reporting (Eccles and Krzus 2010; Paternostro 2012; Busco et al. 2013; Mio 2016; Del Baldo and Nesheva-Kiosseva 2017). Especially when there are economic mechanisms that try to follow the patterns of development and money drain.

Moreover, in the transition from weak to strong sustainability (Baldarelli 2009; Baldarelli and Del Baldo 2017) an important role is played by the tools of accountability (Matacena 2017), especially by those tools adopted to measure and to account for corporate sustainability, that have a fundamental function of company legitimisation (Gray et al. 1993) within its own territory. In this chapter, we wish to analyse from the accounting perspective, the myths and paradoxes that can be found in the tools of accountability (Gray et al. 2014) in gambling companies (Cefaloni 2014).

The approach is based on the analysis of literature contributions regarding the relationship between sustainability development and accountability, and especially the transition from weak to strong sustainability (Bebbington and Contrafatto 2006). Furthermore, it is important that we include the role of accounting and social reporting, and integrated reporting too, in this transition (Gray et al. 1993), with particular reference to the role of "social emancipation", which accounting and reporting may carry out in the aspects of eco-justice (Gray et al. 1993: 307) and therefore, it is especially important to underline legitimacy theory as the base framework for the empirical analyses. Following our view in this "emancipation process" Integrated Reporting (IR) is involved, because may change how to measure

M.-G. Baldarelli (✉)
University of Bologna, Bologna, Italy
e-mail: maria.baldarelli@unibo.it

S. O. Idowu, M. Del Baldo (eds.), *Integrated Reporting*, CSR, Sustainability, Ethics & Governance, https://doi.org/10.1007/978-3-030-01719-4_10

and to account the value that the company produces. Then IR may pursue to a more comprehensive concept of company as we can read in the IR Framework: *"Strategic focus and future orientation: An integrated report should provide insight into the organization's strategy, and how it relates to the organization's ability to create value in the short, medium and long term, and to its use of and effects on the capitals"* (IR Framework: 6). These aspects sum up the relationship between company culture and anthropological culture which foresees a helicoidally relationship (Catturi 2004; Lai 2004).

The chapter is based on the analysis of one significant research case (Naumes and Naumes 2006) related to the SISAL company.

The case analysis helps us understand the stakeholder's expectations and such reflections can be drawn in the transition from weak to strong eco-justice, so that our idea is at the base of the building of integrated thinking, where gambling companies are involved.

The chapter involves the myths and paradoxes that can be found in the accountability instruments in certain sectors, such as gambling (Cefaloni 2014). These reflections can be considered from the point of view of their relationship between integrated reporting and integrated thinking.

About Integrated thinking derives from the following words: "*...But the parts of the world are all so related and linked to one another that I believe it impossible to know one without the other and without the whole... I hold it equally impossible to know the parts without knowing the whole and to know the whole without knowing the parts in details." (Mella* 2012*: V. Ed. or: Pascal* 1901 *(translation): 72).*

This work is divided in the following sections: Sect. 10.2 describes literature review about legitimacy theory, weak and strong eco-justice and integrated thinking; Sect. 10.3 presents methodology of case analysis; Sect. 10.4 contains SISAL case and finally Sect. 10.5 is about discussion and conclusion.

In the next section, we are going to present literature regarding the passage from weak and strong eco-justice that consider based on legitimacy theory.

10.2 From Weak to Strong Eco-Justice, Integrated Thinking and the Legitimacy Theory: Some Notes

The company can survive where It is legitimised by society, since, in this way, It can: educate stakeholders, modify their perception of the business, direct attention towards (or distract attention away from) the negative aspects and finally and, eventually, change external expectations (Gray et al. 1996, Ch. 2.6; Lindblom 1993).

Another strand derives from Political Economy theory by Cooper and Sherer (1984) and also by Guthrie and Parker (1990). They affirm that classic Economic Policy, which originates from the thinking of Marx is based upon conflicts between labour and capital while the State plays a role in contributing in resolving such conflicts.

To complete the thought highlighted above, for other Authors legitimacy theory does not cover all aspects that may be investigated, therefore we need to pause more upon the aspects, which concern the role played by the State. Especially, what is the role of the State in creating conditions that are suitable for the possible development of a greater incentive towards the opening up to information relating to those aspects of social accounting (Archel et al. 2009) as we may read: "...*although the state is not necessary conceived as a neutral arbiter of the common good, the forces that shape state interventions are presumed to operate in a pluralistic setting. Furthermore, it is presumed that the state itself does not have ideological inclinations and material needs that predispose it towards some of those groups and not others*" (Archel et al. 2009: 1287).

According to the same Authors, legitimacy theory is based upon the social contract and It is legitimised to use the information which results from social accounting. As regards communication of the social and environmental variables, this is based upon social expectations of the companies and upon the aspects which concern legitimisation. Often however, there are some gaps between expectations and communication which create conflicts. Such conflicts depend on: behaviour of the media, size of the company and ownership layout. In the thought of the authors we are talking about, it is affirmed that whenever there is a conflict between company and society, the company may (strategically) put one of the following action plans into play: the first one regards the modification of the behaviour trying to fill this gap. The second action plan consists in acting upon the expectations of society without modifying its behaviour. The third action plan, instead, consists in manipulating the expectations distracting attention away from the gap. The fourth action, instead, consists in manipulating society, modifying the expectations in order to direct them towards the goals of the company. According to our point of view, the case analysed in this chapter comes precisely within these two final aspects. Such aspects are "mediated" by an important ongoing trend, about which we will talk later, which enters within the dynamics that permit the transition from weak sustainability to strong sustainability, amongst which there is the "pact" (Baldarelli and Del Baldo 2017), that goes beyond strictly contractual aspects.

Regarding Legitimacy theory applied at the opening of company information towards the environment, this neutral vision of the State is not to be taken for granted so much, because many differences between various stakeholder groups persist in society (pluralistic Legitimacy theory).

Within the field of social accounting, various scholars, indeed, criticise the neutrality of the State (Patten 1992; Darrell and Schwartz 1997), they also affirm that the State does not remain neutral with respect to the "common good", but it intervenes via various factors (pluralistic setting) (Archel et al. 2009: 1287).

Agreeing with this theory, the State intervenes by way of the dominant classes to legitimise the strategies of certain successful companies. They affirm that the financial statement is not neutral or a passive document, yet the numbers may be manipulated in order to orientate and to legitimise actions that are directed by power positions. The same process, in our opinion may happen about integrated reporting.

In this sense, even Political Economy Theory considers the State responsible for maintaining situations of conflict and therefore, situations of inequality (since there is no pact?) (Dillard 1991: 9). These dynamics are inserted, from our point of view, in the transition from weak to strong sustainability (Baldarelli and Del Baldo 2017).

Sustainable development may be defined as follows: "*The key question inherent to sustainable development therefore, seems to interest the ways through which we organise the economic system, in such a way that development is carried through without damaging the environmental heritage, upon which every present thing and any future development is based*" (Bebbington and Contrafatto 2006: 217).

The second viewpoint is that of the company, as evidenced in other works (Baldarelli and Del Baldo 2017), to which we refer, which adapts the behaviour of its particular business to the context previously described and it is translated into a trend aimed at sustainability, which may be implemented by way of actions of differing intensity.

Sticking to the theme of sustainability, the other viewpoint instead considers the integral development of the company, where sustainability takes on connotations linked with the anthropological culture (Sorci 2006; Zadek 2006; Alford et al. 2006).

Within the same strand, we underline Catturi (2009), Who distinguishes between the concepts of: evolution, growth and development. Evolution is defined by later states of programmed perfection (Catturi 2009: 4). So, a growth, even if it is only a quantitative one which is harmonious as well, corresponds to development, Catturi (2009: 20).

As may be noted, the characteristics that sustainable development takes on are ever more numerous but we can read an evolution going on from those, which are merely quantitative and concern distribution of wealth to those of a qualitative-relational and inclusive character, which are currently emerging as those which are the hardest to implement.

Therefore, in relation to the intensity with which the same sustainability we want to focus on the passage from weak to strong sustainability following the dimension of eco-justice.

Weak sustainability directed towards eco-justice consists, especially, of considering the intra- and inter-generational distribution of resources and wealth produced.

As regards "strong" eco-justice, the company is called to arrange certain rather sophisticated reporting tools and to ask to undergo a social and environmental auditing of a certain standard (Bebbington 2007).

This logic is based on some processes that are involved in the legitimacy theory that let us to understand the dynamics to pass from weak to strong eco-justice using different accountability tools.

The question that we pose is, therefore, to single out the dynamics which allows for the transition from weak eco-justice to strong eco-justice and in the following section we propose reflecting upon methodology to analyse SISAL case.

10.3 Methodology

The case of SISAL (Naumes and Naumes 2006; Matacena 2017) is interesting because:" *In a short number of years, the gambling market has grown exponentially. Total revenues have broken the ceiling of 100 billion Euro per year . . .. Italy holds the record in Europe and third place in the world ... 47% of the poor play and 56% of those belonging to the lower middle class ... The urban backdrops change. Food shops disappear under the pressure of shopping centres in order to cede their place to "buy gold" shops or rooms for gambling, bars become small gambling dens ... a part of the population lives as if deported and without any links to protection. In front of so many doors which are closed and which are never knocked on out of shame, remains the extreme attempt to try one's luck. Challenging destiny is all part of the human condition, but the pathological adrift is just a stone's throw away ... according to estimates, from 2 to 4% of the population"* (Cefaloni 2014: 11–1).

Results will be presented and are relative to accountability documents: financial statements from 2007 to 2016, social reporting and sustainability reporting (2011 and 2016). We chose 2016, because SISAL in this year celebrated its 70th year of activities. At the end of 2017 SISAL incorporated Schumann Ltd and created a new group that is named: SISAL group spa. Furthermore, the analysis of websites of the company and relative social networks has been carried out for the same periods.

We decided to begin the analysis of SISAL from the social report of 2011, since it is particularly important for the time of economic and financial crisis which was slightly abating in western countries and, besides, the publication of the Social Report in 2011 is in its third year and, therefore, assumes an informative worth which is more consolidated than its initial years.

However, the work's main limitation regards the empirical approach based on only one case, which hinders the generalisation of results and requires further research steps that aimed at considering more cases to be investigated in depth.

10.4 Accountability in the Case of SISAL and Integrated Reporting: Some Missing Links?

SISAL had been founded in 1946 to have sports gaming activities and it is the first Italian Company to operate in the gaming sector as a Government Licensee. Progressively from the field of sport, it changed its activities to include various typologies of gambling, but we want to start with the words that appear on the web site:

> Over the years, SISAL has created a succession of popular, high-quality games, including the Totocalcio, the Totip and, more recently, SuperEnalotto and Win for Life. SISAL today is a major Italian group operating in the Gaming and Payment Services sector and in 2015 reported a turnover of about €15.1 billion, over 1700 employees and a network of more than 45,000 points of sale; manages a broad offering of on- and offline games: lotteries, betting, casino games and gaming machines; offers the public in excess of 500 payment services, through its network of retailers and over 80 partners (www.SISAL.com/eng).

The mission of SISAL is: "*SISAL's widespread presence across Italy represents a social network that allows the Company to offer innovative and responsible gaming experiences. For 65 years, SISAL has played a key role in Italy's economy, mirroring its growth and keeping pace with socio-cultural developments and the changing habits of Italians from the post-war era to date.*" (Social Report 2011: 7).

And the objective of 2011 had been: "*SISAL's aim is to become a leader in sustainability by 2016. Its ethical commitment has also resulted in the introduction of a Responsible Gaming Program, in accordance with the guidelines of the Autonomous Administration of State Monopolies and with best practice and international standards. Furthermore, as part of its stakeholder engagement, the SISAL Group has intensified its dialogue with stakeholders and has developed a reporting system—in other words, a documentation and measurement system—that allows it to 'report' on the activities carried out, while also encouraging stakeholders within the Company to 'be aware' of their actions and consciously quantify them.*" (Social Report 2011: 8).

Governance (Table 10.1) is developing as described: "*SISAL Holding Istituto di Pagamento (SHIP) S.p.A. The SISAL Group holding company responsible for the management and coordination of companies in the group, offering more than 300 payment services with 70 partners, both private and public, authorised by the Bank of Italy to operate as a payment institution. Its registered office is situated at 13 via di Tocqueville, Milan. It is wholly owned by Gaming Invest Sàrl. The Group's total profits in 2011 were €13.3 billion, up by 18% compared with 2010. SISAL S.p.A. Controlled by SHIP S.p.A., this manages a network of 41,659 online merchants offering numerous games including SuperEnalotto, SiVinceTutto SuperEnalotto, Vinci per la vita—Win for Life and Eurojackpot. It also sells a wide range of third-party products and services, such as national and international telephone cards and PINs for mobile phone top-ups for all the major Italian mobile operators, as well as pay-per-view cards for digital terrestrial television*" (Social Report 2011: 15).

Table 10.1 SISAL structure and organisation charts

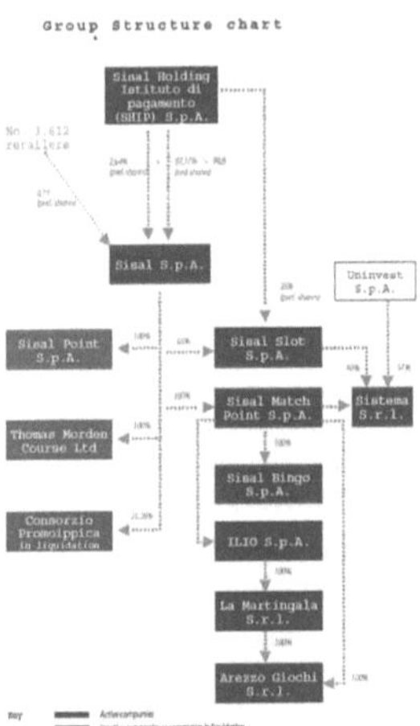

Source: Social Report 2011: 15

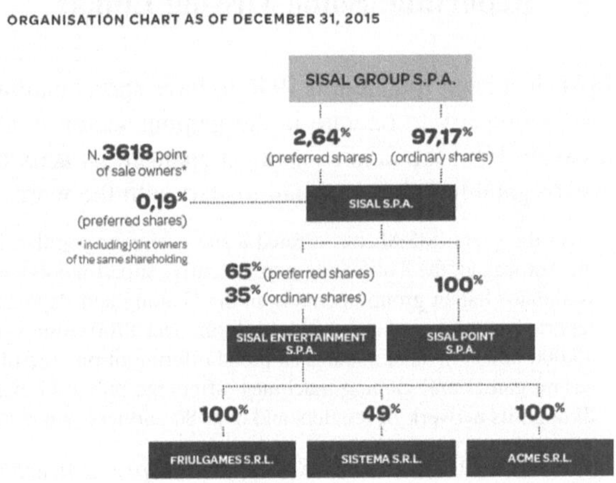

Sustainability Report 2016: 46

We especially underline that ACME srl which is responsible for the slot machine distribution has increased its revenues greatly.

Source: Social Report 2011: 19

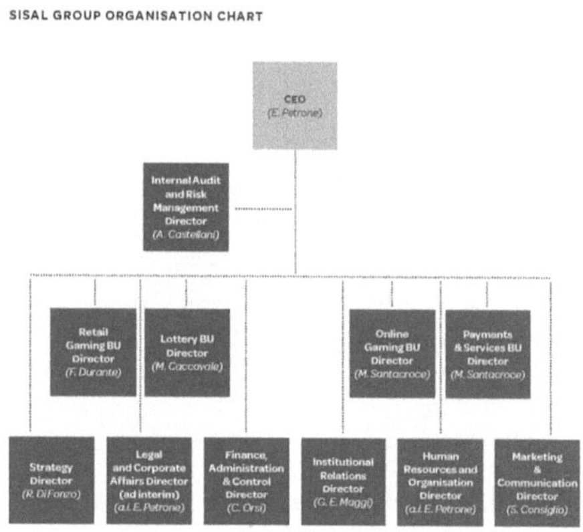

Sustainability Report 2016: 48

Group Brands

Source: Social Report 2011

Concerning financial information we can show some figures, that are very interesting in underlining the development of gambling (Tables 10.2, 10.3 and 10.4).

Table 10.2 Financial statement- trend of Profit in millions of €

	2007	2008	2009	2010	2011	CAGR 2007/2011
Lotteries	16,042	17,635	18,876	18,122	19,421	4.9%
Betting and pools*	5,558	6,357	6,142	6,228	5,294	-1.2%
AWP (slot + VLT)	18,072	21,685	25,525	31,534	44,615	25.3%
Bingo	1,753	1,636	1,512	1,954	1,850	1.4%
Skill & Card Games	0	242	2,348	3,146	8,418	226.2%**
Total profit	41,425	47,555	54,402	60,984	79,597	17.7%

Source: Social Report 2011: 22

Table 10.3 Total revenue detail

Dettaglio raccolta totale	2013	2014	2015	2016	CAGR 2013/2016
Lotterie	17.321	17.258	17.195	18.669	2,5%
Scommesse e CPS	4.651	6.285	7.836	10.063	29,3%
ADI	47.391	46.744	48.161	49.440	1,4%
Bingo	1.664	1.624	1.598	1.702	0,8%
Skill, Card & Casinò Games	13.281	12.318	13.229	15.986	6,4%
Totale Mercato dei Giochi	84.308	84.229	88.019	95.859	4,4%

Source: SISAL GROUP-Consolidated Financial statement 31/12/2016(Italian language): 13 Sustainability Report, 2016

The message that shines through from the very beginning concerns the highlighting of a responsible gaming, when, as we have seen a little bit beforehand, the increase in gambling has gone through the roof over the years.

That which, in 2011, was declared, and that is: "*The SISAL Social Report, now in its third edition, is the Group's main tool for reporting and communicating responsible corporate governance and social responsibility. It is, therefore, a final statement of corporate social responsibility activities addressed to all SISAL stakeholders that shows, with absolute transparency, all of the activities already carried out and the guidelines for future development. It thus represents, for SISAL, its affiliated companies and stakeholders, a public monitoring and planning tool for improving the Company's performance*" (Social Report 2011: 9).

Table 10.4 Composition of revenues 2016 (millions of €)

Revenues (in thousands of Euro)	2016
GNTN (Number Games with National Totalizator)	5,383
Gaming Machines	34,764
Horse Racing Betting	777
Big Bets Revenues	2
Virtual Races Revenues	2,444
Betting Contests	39
Online Gaming	3.159
Total	46.568

Source: Sustainability Report, 2016

SISAL policy is to drive in the direction of responsible gaming. But one can wander: "What is the meaning of responsible gaming if can create a sort of "addiction" and become a damage for the development of civil society?"

The results of accountability that are presented before are showing a progressive increase of gambling that create "addiction" and the terms that the company use to name this activity is: "collection", that means a term similar to that used by banks to collect money.

From data emerges that each operation is aimed to produce money and profit and focus on "the business model" that is a little bit strange in the respect of its original objectives and mission.

So, what is the role of accountability? May we find integrated thinking in this document about accountability? The answer is negative, because everything, under the umbrella of responsible gaming, is orientated only to gain.

About integrated reporting and especially in the different ways of representation of value created by companies, that is represented by different capitals: social, relational, etc., perhaps It is a good occasion to let possible to communicate the loss of relations and the loss of social capital that companies like SISAL procure.

After the presentation of the case, we are going to show, in the next section, discussion and conclusion.

10.5 Discussion and Conclusion

In the previous paragraphs We highlighted certain data to underline that from a company which started its operations to introduce positive "games" and however to develop, if we may say so, our country, it has transformed into a company which is completely different and which tries to involve all stakeholders in this process. Above and beyond what is written in the analysed documents, the presence of a "movement", among the Italian people, which tries, from the bottom, to face this overflowing of "chances" of addiction as does the association we mentioned previously: "EconomiaFelicità" makes us reflect.

This association includes, among its various activities, the Slot-mob campaign too which was established in 2013 to face up to the flood of compulsive gambling: "(http://www.economiafelicita/slotmob.it)" "*The Slotmob campaign was set up in July 2013, promoted by various exponents of Civil Society, to combat the problem of legalised gambling which is inundating all Italy. The gambling business is enormous in size: in 2012 in Italy €80 billion were bet, for a net intake of €8 billion for the Italian State. However, the social costs associated with this business are no less: more than 800,000 people at risk of addiction (GAP—Gioco d'Azzardo Patologico [Compulsive Gambling]), families destroyed, numerous cases of suicides due to too many debts, without counting Mafia involvement which launders money via Slot machine halls as well as the cases of usury which are ever increasing. The State, practically, fills in the holes in the financial statement by promoting gambling through a short-term, short-sighted vision, without considering the social impact this brings. And, as usual, should the profits go into the hands of the businesses operating in this sector, the costs fall upon the general public (costs for Local Health Authorities for the care of gamblers, for the fight against crime). The idea is that a bar without a slot machine holds more space for people. The basic idea is always the same: as consumers always reward the bars which chose to give Slot machines up by consuming hundreds of breakfasts and cocktails, this in order to train ourselves to vote with our pockets and to make people aware of the matter. In the end, if from today we chose to buy only in those bars that have no slot machines, the issue would already be solved: no bartender would be willing to offer a product that nobody wants onto the market. It is market logics, but used for ethical purposes. To this we add gaming, any gaming, as long as it is healthy and leads to human relationship, and as long as it is not the ill gaming of the slot machines which leads to isolation and addiction. This is the reason why, in every Slotmob a table-football tournament is organised. Within the space of a few months, the associations which joined have reached nearly 90 in number and this number grows with the passing of each day and the Slotmob campaigns continue, week after week, to win enormous uptake throughout all Italy.*"

It is not our intention to analyse specifically such an association rather, take a very initial conclusion upon the role played by accountability and upon the role of the transparency of information, on the one hand, recalling legitimacy theory and, on the other, considering the transition from weak to strong eco-justice in order to pursue sustainability matters. These considerations reflect upon the construction of a system thinking, which has the construction of positive and not "pathological", social relationships as its central aim.

Actually, the attempt at these activities puts into doubt that certain aspects highlighted in SISAL accountability documents consider every stakeholder and that a reading map of that which is "good" for stakeholders has developed, which does not correspond to the culture that emerges from the Civil Economy, expressed by the countless initiatives of Slot-mob. Indeed, these players strongly affirm that they prefer not to have slot machines in the bars where they have their coffee and that they want a "healthy" economy, which places the person and not "addiction" at its core. This apparently "banal" aspect renders itself carrier of an anthropological/economic culture, which questions companies like SISAL as regards the future.

In that, the relationship between anthropological culture and company culture is very important (Catturi 2009) and shows that the engaging message coming from documents published by SISAL is not very effective and that it cannot totally manipulate a collective thinking towards a profit at the expense of "addiction" as shown in legitimacy theory (Patten 1992; Darrell and Schwartz 1997; Archel et al. 2009) (Table 10.5).

Thus, those we have named "myths" are highlighted: accountability advances a relational improvement within the economic system, between companies, and especially the presence of an "integrated" reporting system would lead to an integral development and to a system thinking towards a non-generic, but specific, common good where all "poverties" are eliminated.

We are spectators, instead, to the development of "new and silent poverties", like that of gambling addiction, which, in the case analysed, is masked up in the points

Table 10.5 Social Report 2011—accountability and stakeholder engagement

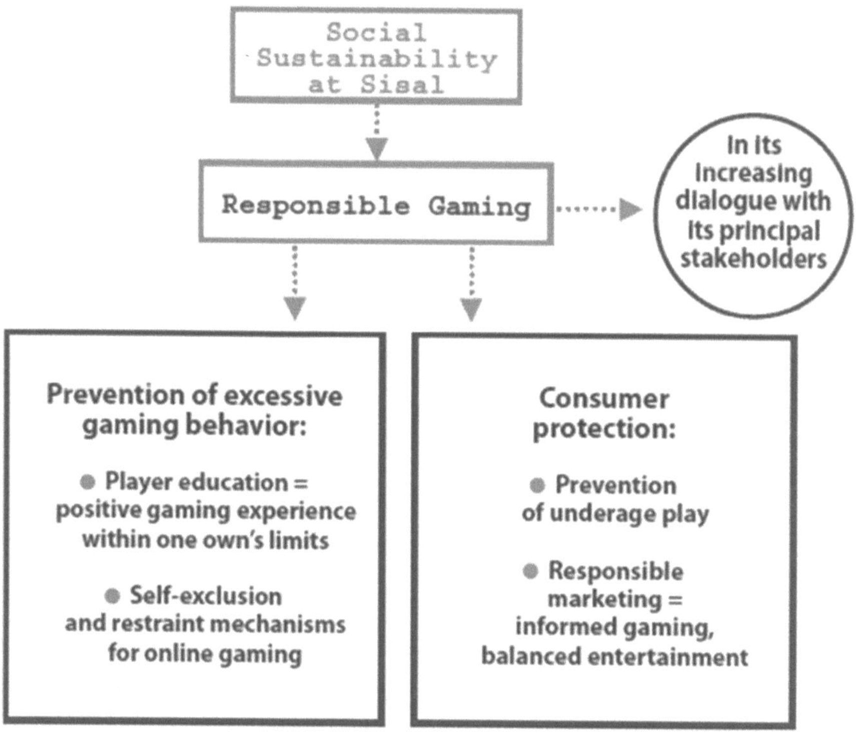

where with engaging and charming words, the mirage of «gaming» which «produces wealth» is introduced.

The paradoxes we may bring to the fore are that, through this type of accountability, maybe even any other typology: sustainability reporting and integrated reporting (Eccles and Krzus 2010; Paternostro 2012; Busco et al. 2013; Mio 2016; Del Baldo and Nesheva-Kiosseva 2017), an improvement of the economic conditions and the elimination of poverties is not obtained. Otherwise, social/civil phenomena, such as the Slot-mob associations, could not be explained.

To sum up, the transition from weak to strong ec-justice seems impossible to us in certain sectors of business and, therefore, also the "mirage" of a pathway towards a system thinking, supported by mere tools of accountability, which must be integrated and investigated carefully, in that process of eco-justice we spoke above.

The case of SISAL represents an example that falls within those actions of manipulation while distracting attention from the "void"/gap. In this case, the void becomes an "abyss" of words and "misleading" information (Archel et al. 2009) which, at their base, have no positive relational circuits that obstacle the building of system thinking.

In this case, we are assisting to the "game" to create information, not surely "integrated information" to make confusion and to create misunderstanding showing that is positive what in reality is negative that is good what in reality is bad.

Scientific researches and publications about integrated reporting and integrated thinking, are good occasions to try to see accountability using other eyes and to try to begin a process to record and to account the real responsibility of enterprise using new tools and new words, that are more and more important in the sustainability world, that each one of us want to contribute to build.

References

Alford, H., Clark, M. A. C., Cortright, S. A., & Naughton, M. (Eds.). (2006). *Rediscovering abundance interdisciplinary essays on wealth, income, and their distribution in the Catholic social tradition*. Indiana: Notre Dame.

Archel, P., Husillos, J., Larrinaga, C., & Spence, C. (2009). Social disclosure, legitimacy theory and the role of the state. *Accounting, Auditing and Accountability Journal, 22*(8), 1284–1307.

Baldarelli, M. G. (2009). La responsabilità sociale nel terzo millennio: Prospettive economico-aziendali. In A. Matacena & M. Del Baldo (Eds.), *Responsabilità sociale d'impresa e territorio. L'esperienza delle piccole e medie imprese marchigiane* (pp. 62–98). Milano: Angeli.

Baldarelli, M. G., & Del Baldo, M. (2017). New categories for responsible corporate governance starting from the "Unity in Multiplicity". In M. Aluchna & S. O. Idowu (Eds.), *Responsible corporate governance, CSR, Sustainability, Ethics & Governance* (pp. 131–149). Cham: Springer.

Bebbington, J. (2007). *Accounting for sustainable development performance*. Burlington: Elsevier.

Bebbington, J., & Contrafatto, M. (2006). Sviluppo sostenibile: una rivisitazione della letteratura. In G. Rusconi & M. Dorigatti (Eds.), *Impresa e responsabilità sociale*, Collana: Persona, Imprese e Società 6 (pp. 206–235). Milano Angeli.

Busco, C., Frigo, M. L., Riccaboni, A., & Quattrone, P. (Eds.). (2013). *Integrated reporting: Concepts and cases that redefine corporate accountability*. Cham: Springer.

Catturi, G. (2004). *La determinazione del "valore creato"* (Vol. II). CEDAM: Padova.
Catturi, G. (Ed.). (2009). *L'armonia della complessità dalla crescita allo sviluppo aziendale*. Padova: CEDAM.
Cefaloni, C. (2014). *Vite in gioco. Oltre la slot economia*. Roma: Città Nuova.
Cooper, D. J., & Sherer, M. J. (1984). The value of corporate accounting reports: Arguments for a political economy of accounting. *Accounting, Organizations and Society, 9*(3–4), 207–232.
Darrell, W., & Schwartz, B. N. (1997). Environmental disclosures and public policy pressure. *Journal of Accounting and Public Policy, 16*(2), 125–154.
Del Baldo, M., & Nesheva-Kiosseva, N. (2017). Toward the future perspectives of business integrated measurement and communication. In M. G. Baldarelli, M. Del Baldo, & N. Nesheva-Kiosseva (Eds.), *Environmental accounting and reporting: Theory and practice* (pp. 91–161). Cham: Springer.
Dillard, J. (1991). Accounting as a critical social science. *Accounting, Auditing, Accountability Journal, 4*(1), 122–136.
Eccles, R. G., & Krzus, M. P. (2010). *One report: Integrated reporting for a sustainable strategy*. Hoboken, NJ: Wiley.
Gray, R., Bebbington, J., & Walters, D. (1993). *Accounting for the environment*. London: Chapman Publishing.
Gray, R., Owen, D., & Adams, C. (1996). *Accounting and accountability*. London: Prentice-Hall.
Gray, R. U., Adams, C., & Owen, D. (2014). *Accountability, social responsibility and sustainability: Accounting for society*. Harlow: Pearson.
Guthrie, J., & Parker, L. D. (1990). Corporate social disclosure practice: A comparative international analysis. *Advances in Public Interest Accounting, 3*, 159–175.
IR Framework. https://integratedreporting.org/wp-content/uploads/2013/12/13-12-08-THE-INTERNATIONAL-IR-FRAMEWORK-2-1.pdf
Lai, A. (2004). *Paradigmi interpretativi dell'impresa contemporanea. Teorie istituzionali e logiche contrattuali*. Milano: Angeli.
Lindblom, C. K. (1993). *The implication of organizational legitimacy for corporate social performance and disclosure*. Chapter presented at the Critical Perspective on Accounting Conference, New York.
Matacena, A. (2017). The subject we are dealing with: MGA model of the enterprise (§ 1.1). In M. G. Baldarelli, M. Del Baldo, & N. Nesheva-Kiosseva (Eds.), *Environmental accounting and reporting, theory and practice*. Cham: Springer.
Mella, P. (2012). *System thinking: Intelligence in action*. Springer –Verlag Italia. (English translation from the original Italian edition "Guida al system thinking", Il sole 24ore 2007).
Mio, C. (Ed.). (2016). *Integrated reporting, a new accounting disclosure*. London: Palgrave.
Naumes, W., & Naumes, M. J. (2006). *The art and craft of case writing* (2 nd ed.). London: M.E. Sharp.
Pascal, B. (1901). *The thoughts of Blaise Pascal* (translated from the text of M. Auguste Molinier by C. Kegan Paul. London: George Bell and Sons).
Paternostro, S. (2012). *L'accountability nell'aziendal integrale. Prime riflessioni sulla valenza informativa del bilancio integrato Collana Quaderni di ricerca economico-aziendali: teoria e casi/5*. Roma: Aracne.
Patten, D. M. (1992). Intra-industry environmental disclosures in response to the Alaskan oil spill: A note on legitimacy theory. *Accounting, Organizations and Society, 17*(5), 471–475.
SISAL GROUP-Consolidated Financial statement 31/12/2016 (Italian language).
SISAL Social Report. (2011).
SISAL Sustainability Report. (2016).
Sorci, C. (2006). *Responsabilità e sviluppo integrale delle aziende, SIDREA, Appunti per un dibattito sulla cultura aziendale* (pp. 85–94).
Zadek, S. (2006). Responsible competitiveness reshaping global markets through responsible business practices. *Corporate Governance., 6*(4), 332–348.

Chapter 11
Assurance on Integrated Reporting: A Critical Perspective

Miriam Corrado, Paola Demartini, and John Dumay

11.1 Introduction

In December 2013, the International Integrated Reporting Council (IIRC) released the Integrated Reporting <IR> Framework, which provides, through a principle-based approach, the guidelines and main concepts for issuing an integrated report. The <IR> Framework aims to combine a holistic view of financial and non-financial information to improve the quality of corporate disclosure, by promoting integrated thinking and a better understanding of how different 'capitals' interact in creating value (IIRC 2013).

According to Simnett and Huggins (2015, p. 30), this innovative form of communication differs from current financial and sustainability frameworks that are both unable to provide an integrated view of organisations. While annual reports have become progressively more complex and longer, the release of stand-alone sustainability reports requires stakeholders to examine multiple corporate documents. Consequently, redundant and overlapping information can decrease users' trust (Simnett and Huggins 2015, p. 31; Monciardini et al. 2016, p. 2).

The <IR> Framework emerged in a context in which stakeholders require a transparent and reliable disclosure of the whole business. According to Adams et al. (2016, p. 284), the <IR> Framework "provides a mechanism to address the

M. Corrado
Università Politecnica delle Marche, Ancona, Italy
e-mail: m.corrado@pm.univpm.it

P. Demartini (✉)
University of Roma TRE, Rome, Italy
e-mail: paola.demartini@uniroma3.it

J. Dumay
Macquarie University, Sydney, NSW, Australia
e-mail: john.dumay@mq.edu.au

S. O. Idowu, M. Del Baldo (eds.), *Integrated Reporting*, CSR, Sustainability, Ethics & Governance, https://doi.org/10.1007/978-3-030-01719-4_11

non-financial information needs of providers of financial capital", by enabling organisations to tell their "value creation story". The IIRC intends for <IR> to become the corporate reporting "norm" and an effective instrument in providing both financial and non-financial information.[1]

European legislators have also perceived combining financial and non-financial information a relevant initiative. Directive 2014/95/EU now requires all European listed companies, banks and insurers to disclose, in their management report, an environmental and social statement including employee, human rights, anti-corruption and bribery matters. According to Monciardini et al. (2016, p. 11), EU law emphasizes "*the failure of existing accounting rules to respond to broader societal, environmental and economic needs that go beyond shareholder interests.*" However, critics still perceive the <IR> framework to maintain and increase companies' focus on shareholders' needs in a financial and traditional reporting construct, instead of being addressed to broader social or environmental needs (*ib.* pp. 14–15).

Finally, while the <IR> framework aims to offer a concise and holistic view of corporate disclosure, critics argue that its underpinning model mainly addresses the needs of financial capital providers (Milne and Gray 2013; Flower 2015; Dumay et al. 2016). This is an important difference from the integrated reporting adopted in South Africa (King Report on Governance for South Africa-2009, IDSA, 2009), which addresses a more comprehensive range of stakeholders and considers assurance as a feature of accountable governance (de Villiers et al. 2014, p. 1049; Tweedie and Martinov-Bennie 2015, p. 56).

Another important question in prior research is whether the information <IR> provides is sufficient and reliable enough to enhance users' trust. Unlike the financial reports, sustainability and integrated reports are adopted on a voluntary basis. Managers have wide discretionary power in these reports, particularly in <IR>, which directs managers to adopt a strategic and forward-looking approach (Simnett et al. 2016). Thus, many researchers believe that without an independent third-party or other credibility-enhancing mechanisms able to ensure the information provided is reliable, <IR> becomes another "marketing" document or form of rhetorical storytelling (Park and Brorson 2005; Simnett and Huggins 2015; Mori Junior et al. 2014).

Within <IR> literature, more skeptical perspectives argue that assurance practices cannot overcome problems embedded in the <IR> framework. Conversely, others believe that the <IR> Framework has the potential to play a relevant role in internal and external decision-making processes and move capital markets towards a longer and more sustainable view (Tweedie and Martinov-Bennie 2015, p. 56). Mainstream research considers assurance as a fundamental mechanism for enhancing user trust (Cohen and Simnett 2015, p. 68).

[1] As the IIRC expresses in its Mission: *"The IIRC's mission is to establish integrated reporting and thinking within mainstream business practice as the norm in the public and private sectors."* http://integratedreporting.org/the-iirc-2/

In the light of these main critical points, in this chapter, we try to understand whether the <IR> assurance process can represent a useful tool to enhance users' trust. Our findings, based on a pilot study involving a group of experts engaged in <IR> assurance debates, are in line with the more critical social, environmental and sustainability accounting and reporting (SEA) literature (O'Dwyer and Owen 2005; Milne and Gray 2013). Our explorative research reveals how assurance is perceived mostly as a tool for improving managers' confidence in their own non-financial reporting rather than a credibility-enhancing instrument for stakeholders.

The chapter is structured into six sections: The literature review presented in Sect. 11.2 outlines the main assurance issues identified in both past CSR research and the most recent <IR> literature. The latter was used to develop the key themes for interviews with experts engaged in the <IR> assurance discussions, as detailed in the methodology Sect. 11.3. Findings follow in Sect. 11.4. Section 11.5 discusses the findings and, finally, Sect. 11.6 highlights research limitations and an agenda for further research.

11.2 Literature Review

Numerous prior studies examine the value of, and demand for, assurance in both financial and sustainability reports. These studies adopt different theoretical perspectives, ranging from the institutional (Jensen and Berg 2012) to the more traditional stakeholder (Bepari and Mollik 2016; Wong and Millington 2014) and legitimacy theories (O'Dwyer et al. 2011; Perego and Kolk 2012). In particular, the procedure for assuring non-financial information has been well developed throughout the last decade with reference to assurance practices in Sustainability Reporting (SR) (Velte and Stawinoga 2017). Although SR and <IR> differ, both incorporate non-financial information. Therefore, the issues past research raises about SR assurance offer a starting point for interrogating the relatively new process of <IR> assurance.

Opinions on the role of non-financial information assurance are not converging. Eccles et al. (2012a, b, p. 12) state that assurance on SR has an important function in making reports more comparable and reliable, and consequently increasing the trust of users who know an independent third party reviews the information and the underlying reporting processes. On the other hand, O'Dwyer and Owen (2005) question the independence of the assurance exercise, due to a large degree of management control over the assurance process.

According to Dumay et al. (2018), legitimacy theory is widely used to explain social and environmental accounting practices, and a widely accepted version of legitimacy theory is that outlined by Suchman (1995). The advantage of Suchman's (1995) account of legitimacy theory is that he distinguishes between the strategic and institutional perspectives and outlines how organisations maintain, gain or repair legitimacy. Thus, legitimacy theory is an appropriate lens to understand how organisations either adopt or justify SR practices, such as assurance, because assurance helps to develop trust between stakeholders and managers and confirm that information about material issues is reliable (Dumay et al. 2015).

Suchman (1995, p. 574) proposes an inclusive and broad-based definition of legitimacy as "a generalized perception or assumption that the actions of an entity are desirable, proper, or appropriate within some socially constructed systems of norms, values, beliefs, and definitions." Beyond the broad definition of legitimacy, Suchman (1995) identified that studies of legitimacy have assumed two distinctive dimensions of the pursuit of organizational legitimacy, being the strategic and institutional approaches. In broad terms, the strategic perspective "adopts a managerial perspective and emphasizes the ways in which organizations instrumentally manipulate and deploy evocative symbols" (Suchman 1995, p. 572). Thus legitimacy is seen as an operational resource assuming a "high level of managerial control over legitimating processes" (p. 576). Alternatively, the institutional view assumes the perspective that "cultural definitions determine how the organization is built, how it is run, and, simultaneously, how it is understood and evaluated" (Suchman 1995, p. 576). Within the context of reporting and assuring non-financial information, drivers of strategic legitimacy originate from the organization's management, influencing the boundaries of the whole process of communication with the organization's stakeholders in support of their goals, whereas drivers of institutional legitimacy originate from the organization's external environment, where management decides to adopt a voluntary assurance to comply with culturally defined norms of behavior.

Studies based on the legitimacy theory demonstrate how external pressures influence management decisions in adopting assurance practices in SR (Perego and Kolk 2012), often with the aim to enhance reputation and legitimacy (Michelon et al. 2015; Gürtürk and Hahn 2015). From a legitimacy theory perspective, organisations seek to survive in their social context by implementing procedures that communities expect and desire, and take remedial actions to improve external perception (Suchman 1995; Guthrie et al. 2007). According to Dowling and Pfeffer (1975, p. 127), organisational communication represents a tool through which a company can "become legitimate", being identified by its values. Therefore, companies use corporate disclosure to influence external expectations and perceptions about the organisation (Tregidga et al. 2006, p. 4; Guthrie et al. 2007, p. 6; Beck et al. 2017).

Prior SR assurance studies using the legitimacy theory concur that assurance practices in SR play a strategic role in influencing users' perceptions and creating a positive corporate image. In particular, it is possible to distinguish the assurance practices as a mechanism to enhance corporate legitimacy (Cohen and Simnett 2015, p. 65) from the assurers' strategies adopted to legitimise the assurance practices themselves and create "a need and demand for assurance" (O'Dwyer et al. 2011). In the second case, practitioners adopt a consulting and a "co-producing" approach, undermining the independence of the assurance process and favoring the management that controls the report content (Bepari and Mollik 2016; O'Dwyer et al. 2011).

Since <IR> is a means for communicating the organisation's value-creation process to capital providers (IIRC 2013), and "potentially" also to other stakeholders (de Villiers et al. 2014, p. 1050), <IR> assurance might represent a positive process that increases credibility of the organisations' "value creation story" and clarifies how it is defined (Adams 2013, 2015, p. 52). Consequently, this research will answer the following research question:

RQ 1) Can <IR> assurance increase users' trust?

11.3 Literature Review: <IR> Assurance

In this section, we identify the key issues emerging from the IIRC consultation paper (2015) and literature on <IR> assurance. In recognising the growing importance of the role of assurance in <IR>, the IIRC (2015) collected the significant concerns from the previous consultation paper, which was addressed not only to academics and assurance providers but also to "all those with an interest in building the credibility of and trust in corporate reporting" (IIRC 2014a, b, p. 2). The report highlights three main considerations: the awareness that assurance practice in <IR> can increase the reliability of information disclosed and stakeholders' trust; the need for innovation in the development of assurance, since traditional criteria may not be suitable in <IR>; the framework is still evolving, and it's too early to require assurance.

Although the IIRC is not interested in giving particular directions or providing standard-setters for the role of assurance in <IR> ("*The IIRC does not aspire to be a leader in assurance*" (IIRC 2015, p. 6), they encourage research and innovative approaches to resolve the identified concerns.

Despite the recent emergence of several studies on <IR>, IR assurance is under-researched (Dumay et al. 2016, p. 9). Academics usually refer to assurance concerns within their broader study on the <IR> Framework, and the opinions on the role and the usefulness of <IR> assurance are conflicting. Some authors claim assurance is necessary to ensure "complete, correct and comparable information" and contrast the discretion entrusted to managers to hide negative information that can damage the company's reputation (Flower 2015, p. 10). Other researchers call for a better understanding of the most challenging assurance characteristics necessary to ensure information reliability and users trust (Simnett and Huggins 2015; de Villiers et al. 2014). Conversely, some authors question whether there is enough interest in integrated reporting from users to require assurance (Cheng et al. 2014, p. 101; Rensburh and Botha 2014, p. 151). However, other studies underline the inability of assurance providers to apply reliable methodologies and procedures to narrative and looking-forward information, arguing that the lack of non-financial and generally accepted standards makes it impossible to develop a complete materiality process and comparative analysis across organisations (Adams 2015; Eccles et al. 2012a).

Table 11.1 Unsolved issues and main concerns in <IR> assurance

IR assurance issues	Main concerns
The role of assurance (Simnett and Huggins 2015; Maroun and Atkins 2015; de Villiers et al. 2014; Zhou et al. 2016)	• Assurance as a managerial instrument to influence investors' perception (i.e. a 'greenwashing' mechanism) • Uncertain added value of assurance process in integrated reporting • Impacts of assurance on users' behaviour and potential benefits on financial market • Lack of broader social accountability and stakeholder-inclusivity in assurance process
Materiality and completeness in non-financial information (Maroun and Atkins 2015; Simnett and Huggins 2015; Mio 2013)	• Completeness vs. conciseness <IR> principles • High degree of managerial discretion in defining material aspects • Subjectivity of materiality process, which relies more on strategic views than factual aspects • Establish material thresholds for narrative and future-oriented information
Lack of global regulation and specific standards (Adams 2015; Eccles et al. 2012b)	• How to ensure more narrative and future-oriented information • Lack of generally accepted criteria for non-financial information • Need for innovative assurance standards that combine financial and non-financial assurance • Presence of different types of assurance standards and guidelines (ISA 720, ISAE 3410, ISAE 3000, ISAE 3400, AA 1000, GRI)

From the perspective of a consulting firm, integrated reporting assurance represents a competitive market for several assurance providers, which have important economic interests at stake (Cohen and Simnett 2015, p. 65; O'Dwyer et al. 2011). Since listed companies in South Africa adopt integrated reports on an "apply or explain" basis, the most significant auditing firms have already faced some of the most critical issues. From their perspective, a combination of internal and external assurance could represent an efficient strategy for finding innovative solutions that ensure the reliability of <IR> metrics (Chartered Institute of Internal Auditors 2015, p. 9; KPMG 2012, p. 12; Ahmed Haji and Anifowose 2016; Abdifatah and Mutalib 2016). However, the IAASB (2014) considers <IR> assurance "*at an early stage of development and [it needs] further development based on user needs*". In addition, the existing standards and guidelines are unsuitable for covering the broader information that the <IR> Framework requires, and there is a need for innovative forms of assurance methodologies able to enhance the credibility of reports (Simnett et al. 2016, p. 10; IAASB 2016).

The few studies focusing on integrated reporting highlight some relevant challenges in <IR> assurance regarding the following three main aspects: the role of assurance; materiality and completeness of non-financial information; and, a lack of global regulation and specific standards (Maroun and Atkins 2015; Simnett and Huggins 2015; Mio 2013; Demartini and Trucco 2017; Maroun 2018). Table 11.1 shows the main concerns about assurance practices in <IR>, which we used to structure the interviews with <IR> experts.

11.4 Methodology

Our investigation is an exploratory study, which aims to uncover key themes and issues that can develop a research agenda for future research (Van Teijlingen and Hundley 2001). The research adopts a qualitative approach, using semi-structured interviews (see annex I) conducted in Australia with a group of experts engaged in <IR>. Interviews incorporate three main perspectives on <IR> assurance from practitioners, academics and report users.

The research design had two phases. First, we analysed prior SR and more recent IR literature to identify key issues to interrogate in interviews. Second, we collected data through in-depth interviews and other sources of evidence, which included <IR> meetings and documents research participants provided, to determine the extent to which these issues recur in IR assurance.

Interviews represent one of the most important data sources and are designed as a kind of "guided conversation" (Yin 2014, p. 110). According to Qu and Dumay (2011, p. 246), "*the semi-structured interview involves prepared questioning guided consistently and systematically by identified themes interposed with probes designed to elicit more elaborate responses*". Semi-structured interviews are beneficial because they are a flexible and insightful tool, which allows us to obtain a deep understanding of different individual perceptions (Qu and Dumay 2011, p. 246; Ritchie and Lewis 2003, p. 141), as well as a deeper interrogation of themes the IIRC's consultation and prior research has only briefly discussed.

Consequently, our interviews are based on three principal themes previously identified in our literature review (see Table 11.1):

- The role of assurance in the <IR> context;
- Materiality and completeness of non-financial information
- Lack of global regulation and specific standards

The themes, as illustrated in Table 11.1, are linked to some concerns that potentially undermine the trustworthiness/reliability of information reported by organisations. Consequently, discussing the above-mentioned topics with experts allows us to think rationally about whether the assurance mechanism is able to increase users' trust and confidence of <IR>.

The research provides different perspectives to help us understand how assurance can represent a reliable instrument in enhancing trust and confidence in <IR>. The eight participants involved in this research can be classified as follows: practitioners, investors and academics/institutional members (Table 11.2). Each selected participant is directly involved in the discussion of assurance practices in <IR> through participating in institutional meetings (e.g., AASB meetings), providing feedback in the IIRC's consultation processes or taking part in the IIRC's Networks.

Practitioners can offer a more practical view of the methodologies adopted and the technical assurance challenges. They can propose possible solutions for the typical features of the <IR> Framework, providing a comparison of current methodologies in non-financial disclosure, and underline new concerns that require

Table 11.2 Categorisation of interviews

Perspective	Data sources	Participant's role	Data categories
Practitioners	P1	Senior sustainability manager and consultant	• Role of assurance in <IR> • Materiality and completeness • Assurance regulation and standards
	P2	Consultant and institutional member	
	P3	Financial controller and accountant	
Academics	A1	Professor and institutional member	
	A2	Researcher	
	A3	Professor	
Investors	I1	ESG investors' advisor	
	I2	ESG analyst	

additional guidance and considerations. Academics or institutional members are informed experts who can analyse IR in light of prior assurance initiatives in SR. Concerning investors, they represent the primary users of <IR> and can offer a better understanding of how the assurance practice is perceived and how it represents an effective mechanism for ensuring reliable information for the investment decision-making process.

11.5 Findings

This section provides an overview of our findings in light of the <IR> assurance debate. The findings have been grouped into three different themes previously identified in the literature review (Table 11.3).

11.5.1 The Role of Assurance

Practitioners agree that assurance has an important role in quality and confidence, but have reservations about the new technical challenges the <IR> format introduces (Bepari and Mollik 2016; Simnett and Huggins 2015; Cohen and Simnett 2015). None of the practitioners address the instrumental use of assurance in detail. Contrary to Michelon et al.'s (2015) claims, practitioners perceived the choice of implementing assurance in SR reports and <IR> as *"best practice"* (P2), and as adding value for companies rather than as a "symbolic" instrument to influence stakeholders' perception. However, some practical gaps to fully assure what the <IR> Framework requires emerge.

Practitioners perceive significant technical challenges to assuring <IR> information, underpinning the business model and business strategy. P1 maintains that

Table 11.3 Summary of main <IR> assurance concerns from interviews

Assurance issues	Main concerns from interviewees
Role of assurance	• Assurance as managerial and internal instrument to enhance reputation and legitimacy • Lack of impacts of assurance on users' behavior and potential benefits on financial market • Effective added value brought by assurance in integrated reports • Lack of broader social accountability
Materiality and completeness of non-financial information	• High degree of managerial discretion and subjectivity in defining material aspects • Business centric view of materiality process • Lack of stakeholder engagement • Completeness vs. conciseness <IR> principle • Need to establish material thresholds for narrative information
Lack of global regulation and specific standards	• Lack of compulsion and comparability • Need for innovative assurance standards or guidelines • Effectiveness of current assurance standard • Difficulties in auditing strategic information

<IR> requires the assurance of *"not codified statements"* in terms of business strategies, and it is difficult to see how these can be assured. In the same way, the assurers *"would not know how to provide assurance in a considerable manner moving into more forward-looking and value-based reports*". Moreover, as stated by P2, the <IR> Framework "*is always around the purpose of the organisation with the focus on the value that had been created rather than just a report [on what is] achieved"*.

The practitioners interviewed for this study called for more practical guidance to assure the kind of information <IR> seeks to provide in particular, the non-financial capitals—and to provide objective and evidenced-based conclusions. Hence, according to P1, the only risk is the adoption of poor quality assurance, which does not provide the confidence that stakeholders require or would expect. Conversely, *academics* underlined the 'marketing' latent purpose to adopt assurance since companies intend to *"enhance their reputational capital and be seen [as] good corporations"* to attract investors (A2).

The lack of compulsion and the narrative approach of <IR> leaves extensive discretion to management, not ensuring an adequate level of confidence for broader societal accountability purposes. In particular, adopting the storytelling approach within integrated reports, companies are more likely to tend to *"tell you the best story rather than tell you something negative"* (A2). Consequently, assurers need to be able *"to identify what has not been reported and then should be flagging if there are inconsistencies between the strategies of the company and what has been reported"* (A3). This is in line with the approach already adopted by auditors in assuring that the information released in the management report is consistent with the audited annual report (ICAEW 2008, p. 9).

Investors were generally skeptical of the usefulness of both <IR> and <IR> assurance. This is consistent with Slack and Campbell's (2016) finding of investors' unfamiliarity with <IR> that "express concerns over the measurability and connectivity of the capitals model" (p. 9). Similar to prior academic findings, I1 states that *"it is part of account board credibility communication to have the assurance. It is probably less about making sure the information is accurate and reliable for investment purposes"*. Assurance becomes a practice within organisations to feel confident about the disclosure released in the markets, rather than increase users' trust. Furthermore, as the academics A1 and A2 also suggest, *"investors do not typically spend a lot of time looking at the assurance report, given that they can read one"* (I1).

If on the one hand, <IR> represents an "*SR evolution*" (I1) for increased transparency and its supposed investor focus, on the other hand, as the interviewee I2 states—*"I don't really trust any narrative that a company provides. I always use the numbers that they provide, [. . .], but I don't really pay any attention to the story that goes along with them"*—the qualitative and narrative nature of information limits the report's trustworthiness and reliability.

11.5.2 Materiality and Completeness

Our investigation of materiality aimed to provide a deeper understanding of how, if at all, assurance can enhance confidence in the <IR> Framework. Ideally, materiality processes should ensure complete and reliable disclosure for the decision-making process and that material information is defined and selected for the assurance process. Prior CSR research finds materiality processes in non-financial disclosure not well-defined (Jones et al. 2016b).

Practitioners view <IR> materiality as especially challenging for assurers that do not have detailed knowledge of the business and are therefore not able to assess quantitative thresholds of non-financial information (Mio 2013, p. 87–91). According to P3, if "<IR> is considered as an extension of the financial report for a better understanding of the value that the organisation is creating and investment decision purposes, then materiality has to be expressed in financial terms". Thus, despite the potential of <IR> to move markets towards a longer-term perspective (Tweedie and Martinov-Bennie 2015, p. 57; Mio 2013, p. 85), a large number of investors still have short-term investment horizon and "therefore short-term returns can only be [expressed as] financial results" (P3).

Some interviewees confirm <IR>'s ability to provide enough information to investors without becoming "meaningless and lengthy" (P2, P3). However, on the other side, the lack of a robust and independent materiality process "moves away from the report being focused on what is most important, complete and transparent [for a wide range of stakeholders] to being a report that is tailored to what is most important just to the business" (P2). This seems coherent with the <IR> Framework focus on what is most important to the business rather than what is most relevant for a wider range of stakeholders in terms of environmental and social sustainability (GRI 2013a, b; Tweedie and Martinov-Bennie 2015; Jones et al. 2016a).

The academics interviewed particularly addressed the materiality challenges to the wide discretion <IR> entrusts to management in deciding which information they want to report and which they do not. Researchers already question the voluntary basis approach of non-financial reports and the six capitals in <IR> Framework (Flower 2015, p. 4; Adams 2015, p. 26; Dumay et al. 2017, pp. 461–480). In particular, it has been argued that there are unclear boundaries and a lack of prescription of <IR> capitals. As stated, in determining the materiality process, "the problem is that the company is prioritising some of the capitals over the others in <IR>; they are actually more interested in financial impacts because their strategy is to make money" (A3). Moreover, according to A2, "in <IR> you do not have to disclose everything because it is not regulated yet. [...] The CEO, chairperson, is not going to overly disclose details or sensitive information, they are vague enough to be aspirational. [...] <IR> is trying to obtain a more scientific methodology and statistics measure, but that doesn't mean to say that everything is being disclosed. They can still cherry pick information to tell the best story". These statements lead to the widespread belief that management adopts voluntary reporting with the aim to enhance reputation influenced by the external pressure in being perceived as legitimate (Perego and Kolk 2012) (Michelon et al. 2015; Gürtürk and Hahn 2015).

In particular, the participant investors were not interested in any form of investors engagement or inclusiveness within companies in determining materiality for <IR>. They appear more interested in the potentialities of <IR> to provide additional and more concise information for the investment decision-making process, rather than in the role of assurance as a mechanism to enhance information reliability. "<IR> is not going to be the entirety but it is a useful road map, a contextualization of the total story. [...] A good <IR> is like an external expression of good board reporting [...] as an early indicator of a successful strategic initiative and that is the kind of information that it is potentially new and really helpful for investors" (I1).

Thus, there seems to be the awareness that the companies do not provide complete information in their qualitative disclosure and that the concise information provided in <IR> represents the top-level view of what is going in the business.

11.5.3 Lack of Global Regulation and Specific Standards

In the debate on non-financial information assurance, many researchers highlight the lack of standardized methodologies and specific assurance standards (Gürtürk and Hahn 2015; Perego and Kolk 2012; Manetti and Becatti 2009). Furthermore, the IIRC questions whether there is a need for innovation in developing assurance methodologies in integrated reports (IIRC 2015, p. 14) due to the presence of forward-looking and strategic information (Simnett et al. 2016). Among the practitioners and academics interviewed, it is not possible to delineate a single opinion.

Some deem that current methodologies and standards, such as ISAE 3000 and AA1000, are sufficient. Others believe that it is necessary to develop further guidance and additional assurance standards for the integrated reports released according to the <IR> Framework. In particular, P3 suggests that an additional and special guidance is required for the <IR> Framework on a principles-based approach "because you can never anticipate what new non-financial information companies want to report".

In addition, most academics consider that until <IR> processes become more advanced, it is impossible to think that assurance will guarantee comparable and reliable information to investors (Stubbs and Higgins 2014). As stated by A1, "<IR> needs to be developed a lot further and become more mature before we can even start thinking about standardizing assurance. Until it is not mandatory or unless it is much more mature, I do not think assurance can lead reporting; reporting has to lead assurance". This is consistent with Dumay et al. (2017), who advocate "the need for further development and evidence to help inform improvements both from a policy and a practice perspective".

11.6 Discussion

In the existing literature on the <IR> Framework, few studies focus on the topic of assurance and analyse <IR> assurance practices. This research aims to contribute to the emerging field of <IR> assurance research using the legitimacy theory lens (Suchman 1995). In analyzing the opinions of expert practitioners, academics and investors, the paper sought to answer the following research question:

RQ 1) Can <IR> assurance increase users' trust?

Under the *legitimacy perspective lens*, previous studies on SR assurance have already demonstrated how external pressure, also coming from legislation in social and environmental contexts, acts as a driver for companies to adopt assurance practices (Perego and Kolk 2012, p. 184). On the other hand, other authors argue that assurance can become a symbolic practice driven by a management agenda to enhance reputation and maintain legitimacy (Michelon et al. 2015, p. 75; Gürtürk and Hahn 2015).

Concerns arising from the interviews have been reconsidered in the light of the two dimensions envisaged by Suchman (1995). As shown in Table 11.4, we highlight how the voluntary and non-standardized nature of <IR> reporting and assurance lead organizations to adopt a strategic legitimacy approach (pp. 275).

In fact, it emerges that managers have full discretion to decide what information to report and what to assure, as already stated by Flower (2015). This aspect undermines the independence of assurance practitioners, who tend to work very closely with managers and directors in the assurance process of non-financial data and can affect the objectivity and robustness of the whole assurance process.

Table 11.4 Dimensions of legitimacy (Suchman 1995) and <IR> assurance concerns from interviews

Dimensions of legitimacy	<IR> assurance concerns from interviewees (Table 11.3)	Interview questions (see annex I)
Strategic perspective	• Assurance as managerial and internal instrument to enhance reputation and legitimacy • High degree of managerial discretion and subjectivity in defining material aspects • Business centric view of materiality process • Effective added value brought by assurance in integrated reports • Lack of broader social accountability	Some studies suggest that companies only use assurance to enhance their reputation and legitimacy. Can you provide your view/opinion?
Institutional perspective	• Lack of global regulation and specific standards do not support an institutional legitimacy perspective	

These results are consistent with recent studies (O'Dwyer et al. 2011; Bepari and Mollik 2016) that show how assurers tend to provide their services to improve the image of organizations, rather than to make organizations more accountable. Hence, a decline in the confidence of assurance practices with non-financial information is confirmed by the sceptical perspective of academics and investors involved in our research, who believe that assurance is a strategic tool for companies to enhance their reputation. As supported by Dowling and Pfeffer (1975) and Suchman (1995), the aim to gain and maintain legitimacy leads organizations to focus their attention on the communication process with their stakeholders and, in particular, on the reporting practice identified and perceived as a way to be responsive to their larger interests (Beck et al. 2015).

Furthermore, as stated by some professionals, our research also demonstrates how assurance is perceived as a tool for improving managerial confidence in their non-financial reports. Assurance becomes necessary to enhance confidence, especially for directors, who are principally responsible for corporate disclosure and feel more confident if a third party reviews the whole process leading up to the integrated report.

Finally, some of the interviewees reveal that overall investors do not put much faith in narrative and "story-telling". Investors from our interviews believe that companies do not provide a complete and fair image of the business and are used to gather non-financial information from different sources to make informed decisions. They rely more on financial reports, which already present a high level of assurance since it is a more mature and standardized practice. In fact, as Adams et al. (2016, p. 294) also suggest, so far <IR> does not ensure the complete dimensions of businesses are adequately reported. Or at least, <IR> is not any more effective than existing traditional forms of reporting are representing businesses' overall activities.

The IIRC's proposal that <IR> should become the corporate norm does not seem to be supported by our findings, given that mandatory requirements in traditional financial reports provide more accurate and comprehensive information for investors than a concise, integrated report (de Villiers et al. 2014, p. 1046). In fact, investors in the interviews outline the need for additional details and links to other documents, comparing the integrated report to a "conceptualization map".

11.7 Conclusions

This study contributes to the emerging <IR> literature by assessing the assurance <IR> raises, and how these challenges affect users' trust. However, since the study is based on a small sample, and is inherently exploratory, there is a need for further research involving a wider sample of participants in a global context Prior research has found that investors, who are supposedly the main users (stated by IIRC), seem to have different expectations of <IR> assurance (Slack and Campbell 2016).

Our findings are consistent with this observation, and also question whether the <IR> Framework meets investors' needs. First, it is necessary to understand how much interest and demand there is, in practice, from providers of financial capitals in the <IR> Framework (IAASB 2015; Cheng et al. 2014). For this reason, we believe future research should conduct more detailed studies incorporating a wider sample of investors and promoters of <IR> on the 'other side' or <IR> reporting practice (e.g., IIRC institutional members or <IR> supporters). It is necessary to see how <IR> is going to evolve and attempt to obtain wider legitimacy, how widely it will be accepted into the market and how we can develop new assurance methodologies.

Second, practitioners called for research into additional and special guidance for assuring qualitative information. In particular, practitioners have to face further issues related to their technical skills in assuring forward-looking and strategic information at the basis of integrated reports (Maroun 2018). They require principle-based guidance for all forms of non-financial information, for both mandatory (e.g., information included in the management report) and voluntary reporting. In August 2015, IAASB started moving in this direction by releasing the paper "*Supporting Credibility and Trust in emerging forms of External Reporting: ten key challenges for Assurance Engagements*" in response to the demand for supporting credibility and trust and the need for more flexibility in assurance services (IAASB 2015, p. 6). Further investigation of the respondents' opinions is an important part of the <IR> research agenda moving forward.

Acknowledgement We would like to thank Dale Tweedie from Macquarie University for his help, guidance and insightful suggestions.

Appendix 1: Questionnaires

Practitioner and Academic Perspective

Introduction questions:

1. Interviewee's background on years of experience and his role in current position
2. Interviewee's experience in Integrated Reporting and sustainability field.

1. **Understand the role of assurance in IR context and non-financial disclosure**
   (a) What role does assurance play in non-financial reporting?
   (b) Who are considered principal users of assurance on non-financial reporting? How can assurance impact on their decision?
   (c) Some studies suggest that companies only use assurance to enhance their reputation and legitimacy. Can you provide your view/opinion? What are the impacts on companies that adopt assurance on non-financial and voluntary reporting?
   (d) Companies in <IR> report supposedly disclose their value-creation story adopting a more strategic-focus and forward-looking approach. Which consequences could this approach have for assurance practice? How auditors and consultants themselves determine most appropriate audit methodologies in relation to narrative and forward-looking information?
   (e) How these approaches ensure reliable and useful information for investors' decision making processes?
2. **Materiality and reliability of information provided in IR and SR**
   (a) What material information IR should provide compared to Financial and Sustainability Reporting?
   (b) Who should be the principal addressee of assurance process for non-financial reporting and <IR>?
   (c) Which are the current practises in non-financial reporting about stakeholders' (included investors) engagement?
   (d) The IR differs from other traditional reports for its conciseness. Which consequences does conciseness have in relation to providing complete information to capital providers? Are they considered sufficient for investment decisions?
3. **Assurance standards for non-financial information**
   (a) Which audit standards are used for non-financial information in Sustainability Reporting? Do you see these as sufficient to ensure the reliability of information and enhance the trust of report users?
   (b) Is there a need for new audit standards for IR? If so, what should these standard include?
   (c) Who should the standards be applicable to?

Conclusion

1. What do you see as the key challenges in the IR assurance moving forward? How can they be resolved?
2. Any issues you think we should have covered but haven't?

Investor Perspective

1. **Role of non-financial reporting in investors' decision-making process and role of assurance**
   (a) What are investors' primary sources of corporate disclosures for non-financial information?
   (b) Some studies suggest that companies only use assurance in SR and IR to enhance their reputation and legitimacy. Can you provide your view/opinion?
   (c) How much faith and confidence do they put in non-financial reporting?
   (d) Companies in <IR> report supposedly adopt a "storytelling" and a forward-looking approach. Which consequences could this approach have for investors' decision making processes?
   (e) How could assurance of non-financial disclosure affect investor's decisions?
2. **Materiality and reliability of information provided in IR and SR**
   (a) From investors' perspective, what information is material to IR compared to Sustainability and Financial Reporting?
   (b) How important is it to include investors in IR materiality process?
   (c) As part of IIRC Investors Network, which are the expectation of your company in being included in this network? What are the purposes of the network? The IR differs from other traditional reports in requiring conciseness. What is investors' view of whether concise integrated reports can provide useful and complete information?
3. **Assurance standards for non-financial information**
   (a) What are investors' perceptions of the current assurance methodologies and standards in non-financial disclosure? Are these sufficient to guarantee reliable information?

References

Abdifatah, A. H., & Mutalib, A. (2016). Audit committee and integrated reporting practice: Does internal assurance matter. *Managerial Auditing Journal, 31*(8/9), 915–948. https://doi.org/10.1108/MAJ-12-2015-1293.

Adams, C. (2013). Sustainability reporting and performance management in universities: Challenges and benefits. *Sustainability Accounting, Management and Policy Journal, 4*(3), 384–392.

Adams, C. A. (2015). The international integrated reporting council: A call to action. *Critical Perspectives on Accounting, 28*, 23–28. https://doi.org/10.1016/j.cpa.2014.07.001.
Adams, C. A., Potter, B., Singh, P. J., & York, J. (2016). Exploring the implication of integrated reporting for social investment (disclosure). *The British Accounting Review, 48*, 283–296. https://doi.org/10.1016/j.bar.2016.05.002.
Ahmed Haji, A., & Anifowose, M. (2016). Audit committee and integrated reporting practice: Does internal assurance matter? *Managerial Auditing Journal, 31*(8/9), 915–948.
Beck, C., Dumay, J., & Frost, G. (2015). In pursuit of a "Single source of truth": From threatened legitimacy to integrated reporting. *Journal of Business Ethics, 141*(1), 191–205.
Beck, C., Dumay, J., & Frost, G. (2017). In pursuit of a 'single source of truth': From threatened legitimacy to integrated reporting. *Journal of Business Ethics, 141*(1), 191–205.
Bepari, M. K., & Mollik, A. T. (2016). Stakeholders' interest in sustainability assurance process: An examination of assurance statements reported by Australian companies. *Managerial Auditing Journal, 31*(6/7), 655–687. https://doi.org/10.1108/MAJ-06-2015-1208.
Chartered Institute of Internal Auditors. (2015). *The role of internal audit in non-financial and integrated reporting*. London: Chartered Institute of Internal Auditors.
Cheng, M., Green, W., Conradie, P., Konishi, N., & Romi, A. (2014). The international integrated reporting framework: Key issues and future research opportunities. *Journal of International Financial Management & Accounting, 25*(1), 90–119.
Cohen, J. R., & Simnett, R. (2015). CSR and assurance services: A research agenda. *Auditing: A Journal of Practice & Theory, 34*(1), 59–74. https://doi.org/10.2308/ajpt-50876.
de Villiers, C., Rinaldi, L., & Unerman, J. (2014). Integrated reporting: Insights, gaps and an agenda for future research. *Accounting, Auditing & Accountability Journal, 27*(7), 1042–1067. https://doi.org/10.1108/AAAJ-06-2014-1736.
Demartini, C., & Trucco, S. (2017). *Integrated reporting and audit quality*. Cham: Springer.
Dowling, J., & Pfeffer, J. (1975). Organizational legitimacy: Social values and organizational behavior. *The Pacific Sociological Review, 18*, 122–136.
Dumay, J., Frost, G., & Beck, C. (2015). Material legitimacy blending organisational and stakeholder concerns through non-financial information disclosures. *Journal of Accounting & Organizational Change, 11*(1), 2–23. https://doi.org/10.1108/JAOC-06-2013-0057.
Dumay, J., Bernardi, C., Guthrie, J., & Demartini, P. (2016). Integrated reporting: A structured literature review. *Accounting Forum, 40*(3), 166–185.
Dumay, J., Bernardi, C., Guthrie, J., & La Torre, M. (2017). Barriers to implementing the international integrated reporting framework: A contemporary academic perspective. *Meditari Accountancy Research, 25*(4), 461–480.
Dumay, J., De Villiers, C., Guthrie, J., & Hsiao, P.-C. (2018). Thirty years of AAAJ: A critical study of the journal's articles impact. *Accounting, Auditing & Accountability Journal, 31*(5), 1510–1541.
Eccles, R. G., Krzus, M. P., & Watson, L. A. (2012a). *Integrated reporting requires integrated assurance*. Retrieved from http://www.financepractitioner.com/auditing-best-practice/integrated-reporting-requires-integrated-assurance?page=1
Eccles, R. G., Krzus, M. P., Rogers, J., & Serafeim, G. (2012b). The need for sector-specific materiality and sustainability reporting standards. *Journal of Applied Corporate Finance, 24*(2), 65–71.
Flower, J. (2015). The international integrated reporting council: A story of failure. *Critical Perspectives on Accounting, 27*, 1–17. https://doi.org/10.1016/j.cpa.2014.07.002.
GRI. (2013a). *G4 sustainability reporting guidelines*. Global Reporting Initiative.
GRI. (2013b). *The external assurance of sustainability reporting*. Global Reporting Initiative.
Gürtürk, A., & Hahn, R. (2015). An empirical assessment of assurance statements in sustainabilityreports: Smoke screens or enlightening information? *Journal of Cleaner Production*, 1–12. https://doi.org/10.1016/j.jclepro.2015.09.089.
Guthrie, J., Cuganesan, S., & Ward, L. (2007). Legitimacy theory: A story of reporting social and environmental matters within the Australian food and beverage industry. *5th Asian Pacific Interdisciplinary Research in Accounting (APIRA) Conference*.

IAASB. (2014). *Response to the IIRC's discussion papers "Assurance on Integrated Reporting".*

IAASB. (2015). *Working group publication: Exploring assurance on integrated reporting and other emerging developments in external reporting.* International Auding and Assurance Standards Board.

IAASB. (2016). *Supporting credibility and trust in emerging forms of external reporting.* International Auditing and Assurance Standards Board (IAASB).

ICAEW. (2008). *Assurance on non-financial information existing practices and issues.* Institute of Chartered Accountants in England and Wales, Audit and Assurance Faculty.

IIRC. (2013). *The international <IR> framework.* International Integrated Reporting Council.

IIRC. (2014a). *Assurance on <IR>. An exploration of issues.*

IIRC. (2014b). *Assurance on <IR>: An introduction to the discussion.*

IIRC. (2015). *Assurance on <IR>. Overview of feedback and call to action.*

Jensen, J. C., & Berg, N. (2012). Determinants of traditional sustainability reporting versus integrated reporting. An institutionalist approach. *Business Strategy and the Environment, 21*, 299–316. https://doi.org/10.1002/bse.740.

Jones, P., Comfort, D., & Hillier, D. (2016a). Managing materiality: A preliminary examination of the adoption of the new GRI G4 guidelines on materiality within the business community. *Journal of Public Affairs, 16*(3), 222–230. https://doi.org/10.1002/pa.1586.

Jones, P., Hillier, D., & Comfort, D. (2016b). Materiality and external assurance in corporate sustainability reporting: An exploratory study of Europe's leading commercial property companies. *Journal of European Real Estate Research, 9*(2). https://doi.org/10.1108/JERER-07-2015-0027.

KPMG. (2012). *Integrated reporting. Performance insight through better business reporting.* Retrieved from kpmg.com/integratedreporting

Manetti, G., & Becatti, L. (2009). Assurance services for sustainability reports: Standards and empirical evidence. *Journal of Business Ethics, 87*, 289–298. https://doi.org/10.1007/s10551-008-9809-x.

Maroun, W. (2018). Modifying assurance practices to meet the needs of integrated reporting: The case for 'interpretive assurance'. *Accounting Auditing & Accountability Journal.* https://doi.org/10.1108/AAAJ-10-2016-2732.

Maroun, W., & Atkins, M. (2015). *The challenges of assuring integrated reports: Views from the South African auditing community.* ACCA (Association of Chartered Certified Accountants).

Michelon, G., Pilonato, S., & Ricceri, F. (2015). CSR reporting practices and the quality of disclosure: An empirical analysis. *Critical Perspectives on Accounting, 33*, 59–78. https://doi.org/10.1016/j.cpa.2014.10.003.

Milne, M. J., & Gray, R. (2013). W(h)iter ecology? The trible bottom line, the global reporting initiative, and corporate sustainability reporting. *Journal of Business Ethics, 118*, 13–29. https://doi.org/10.1007/s10551-021-1543-8.

Mio, C. (2013). Materiality and assurance: Building the link. In C. Busco, M. L. Frigo, A. Riccaboni, & P. Quattrone (Eds.), *Integrated reporting: Concepts and cases that redefine corporate accountability* (pp. 79–94). Cham: Springer.

Monciardini, D., Dumay, J., & Biondi, L. (2016). *Integrated reporting and EU law. Competing, converging or complementary regulatory frameworks*? SMART conference 'Life-cycle based management and reporting for sustainable business'. 2017–23. University of Oslo Faculty of Law Legal Studies Research Paper Series.

Mori Junior, R., Best, P. J., & Cotter, J. (2014). Sustainability reporting and assurance: A historical analysis on a world wide phenomenon. *Journal of Business Ethics, 120*, 1–11. https://doi.org/10.1007/s10551-013-1637-y.

O'Dwyer, B., & Owen, D. L. (2005). Assurance statement practise in environmental, social and sustainability reporting: A critical evaluation. *The British Accounting Review, 37*, 205–229. https://doi.org/10.1016/j.bar.2005.01.005.

O'Dwyer, B., Owen, D., & Unerman, J. (2011). Seeking legitimacy for new assurance forms: The case of assurance on sustainability reporting. *Accounting, Organizations and Society, 36*, 31–52. https://doi.org/10.1016/j.aos.2011.01.002.

Park, J., & Brorson, T. (2005). Experiences of and views on third-party assurance of corporate environmental and sustainability reports. *Journal of Cleaner Production, 13*, 1095–1106. https://doi.org/10.1016/j.jclepro.2004.12.006.

Perego, P., & Kolk, A. (2012). Multinational accountability on sustainability: The evolution of third-party assurance of sustainability reports. *Journal of Business Ethics, 110*, 173–190. https://doi.org/10.1007/s10551-012-1420-5.

Qu, S. Q., & Dumay, J. (2011). The qualitative research interview. *Qualitative Research in Accounting & Management, 8*(3), 238–264. https://doi.org/10.1108/11766091111162070.

Rensburh, R., & Botha, E. (2014). Is integrated reporting the silver bullet of financial communication? A stakeholder perspective from South Africa. *Public Relation Review, 40*, 144–152. https://doi.org/10.1016/j.pubrev.2013.11.016.

Ritchie, J., & Lewis, J. (2003). *Qualitative Research Practice*. London: Sage.

Simnett, R., & Huggins, A. L. (2015). Integrated reporting and assurance: Where can research add value? *Sustainability Accounting Management and Policy Journal, 6*(1), 29–53. https://doi.org/10.1108/SAMPJ-09-2014-0053.

Simnett, R., Zhou, S., & Hoang, H. (2016). Assurance and other credibility enhancing mechanisms for integrated reporting. In C. Mio (Ed.), *Integrated reporting: A new accounting disclosure* (pp. 269–284). London: Palgrave Macmillan.

Slack, R., & Campbell, D. (2016). *Meeting users' information needs: The use and usefulness of integrated reporting*. London: Association of Chartered Certified Accountants (ACCA).

Stubbs, W., & Higgins, C. (2014). Integrated reporting and internal mechanism of change. *Accounting, Auditing & Accountability Journal, 27*(7), 1068–1089. Retrieved from http://hdl.handle.net/10536/DRO/DU:30066414.

Suchman, M. C. (1995). Managing legitimacy: Strategic and institutional approaches. *Academy of Management Review, 20*(3), 571–610.

Tregidga, H., Milne, M., & Kearins, K. (2006). *Organisational legitimacy and social and environmental reporting research: The potential of disclosure analysis*. 5th Australian Conference on Social and Environmental Accounting Research.

Tweedie, D., & Martinov-Bennie, N. (2015). Entitlements and time: Integrated reporting's double-edged agenda. *Social and Environmental Accountability Journal, 35*(1), 49–61. https://doi.org/10.1080/0969160X.2015.1007466.

Van Teijlingen, E. R., & Hundley, V. (2001). *Social research update: The importance of pilot studies* [online]. Retrieved from March 2018 http://sru.soc.surrey.ac.uk/SRU35.html

Velte, P., & Stawinoga, M. (2017). Empirical research on corporate social responsability assurance (CSRA): A literature review. *Journal of Business Economics, 87*(8), 1017–1066.

Wong, R., & Millington, A. (2014). Corporate social disclosures: A user perspective on assurance. *Accounting, Auditing & Accountability Journal, 27*(5), 863–887. https://doi.org/10.1108/AAAJ-06-2013-1389.

Yin, R. (2014). *Case study research*. London: Sage.

Zhou, S., Simnett, R., & Hoang, H. (2016). *Combined assurance as a new assurance approach: Is it beneficial to analysts?* In 26th Audit and Assurance conference - Thursday 5 May 2016. Said Business School, University of Oxford, Oxford, UK.

Chapter 12
Value-Oriented and Rental Approach in <IR> of Private Water Utilities Companies

Ninel Nesheva-Kiosseva

12.1 Introduction

The integrated report is a model of accountability in a unified information space where on the basis of unity of information and with the help of the possibilities that contemporary information technologies give; many aspects of the company's activities may be represented in synergy. The report serves not only management purposes but also keeps external users such as investors, state authorities, clients, and regulators informed.

The integrated report allows the use of combinations of different methods and tools known to us from economic theory, statistics and accounting, and their development. It allows the introduction of combinations of diffuse information data and techniques to create an adequate and changing information model that is flexible and capable of responding to the new and changing needs of creating value for both the organization and other interested parties.

In this essay the aim is to introduce theses that can be discussed in relation to the integrated accountability of water utility companies as a specific business. Such a discussion is also relevant in connection with the fact that there are already water utility companies that prepare Integrated Reports. The article looks at the issue of representing the value of water in the field of water extraction, processing and supply companies. Water is a natural resource on which their business activity is based. In most cases, they occupy the position of "natural monopolies" in the respective supply region, which raises a number of questions as to the characteristics of their profits, the added value and their business modelling. The paper does not address directly the question of the relationship between value and price, as this is a separate research issue.

The theory of rent was applied to the hydroelectric sector (Rothman 2000; Amundsen et al. 1992), but was not applied to the WUC. WUC's business differs

N. Nesheva-Kiosseva (✉)
Department of Business Administration, New Bulgarian University, Sofia, Bulgaria

S. O. Idowu, M. Del Baldo (eds.), *Integrated Reporting*, CSR, Sustainability, Ethics & Governance, https://doi.org/10.1007/978-3-030-01719-4_12

significantly from that of energy, based on waterpower. As with hydroelectric plants, however, as with WUC, water is the basis of their profits. Moreover, both types of business have a different economic and social significance. While electricity can be obtained from many sources other than water, there is no substitute for water when it comes to thirst, food production and industry, as a whole. That is why the water treatment and supply business has a special vital and social significance.

12.2 The Value

Integrated reporting has several important as well as specific objectives, including presentation and emphasis on value added created by the "six capitals", the application and link between the value added generated by the different types of capital and the business model (or how the organization generates its profits) and its strategic development. The integrated report should also show for which stakeholders are creating value in the organization and what value added is created for them.

In view of this, the methodological guidelines for integrating reporting define the six capitals of the organization that add value for it and its stakeholders (International Framework of <IR> 2013).

<IR> has the ambition to provide the stakeholders (above all the investors) with a more complete assessment of the value of companies.

Yet there is no tool to evaluate the full cost of the companies, to register all value drivers and the levels at which value is created in the company. In the future it will be necessary for an integrated reporting approach to be developed in this direction.

The disclosure of companies' full cost and value added requires the complex task of coping with value in economic science.

Theoretical and applied methods for assessing the economic value of natural resources, including water, have been developed by a number of scientific studies, guides and standards.

In the extremely important work "*The Measurement of Environmental and Resource Values. Theory and Methods* "by Myrick Freeman III, Herriges and Kling a classification has been made of the existing methods of measuring the values of nature and resources recovered for human use, including water. The authors analyze the entire palette of methods and tools that have been developed over the years and classify these methods in the following groups: Welfare Measures (40–126); Valuing Changes in Risk (127–170); Aggregation of Values Across Time (171–189); Valuing Longevity and Health (190–236); Environmental Quality Measures for Valuing Changes in Productivity of Natural Resource Systems (257–268); Recreation Demand Models of Valuation (269–309); Property Value Models (310–359); Hedonic Wage Models (360–381); Stated Preference Methods for Valuation, which is the preferred tool in the present article (383–418) (Myrick Freeman III et al. 2014). The models listed vary greatly. Some of the models are market-based; others are models of non-market valuation. Some of these methods, in fact, try not to produce value, but a price, which in the approach of the subjective economic schools is equal to value. Due to the fact that these are economic methods, they have the tendency and/or make the effort to monetize value. In addition to systematizing and detailing all developed methods and tools for assessing nature and resource values, the study also shows their shortcomings as well as the directions in which scientists should work for their development.

There are also a large number of studies whose main subject is water valuation. Their detailed listing is impossible within the limited scope of this paper and is beyond the scope of its specific purpose. A common example is the methodology of Systems of National Accounts and System of Environmental and Economic Accounts (SNA 2008; SEEA 2012) that recommend a methodology for assessing the economic value of water supply as a monetary assessment of the water stocks owned by net water vendors based on the Net Present Value (NVP) and being calculated with bank interest rate (SEEA 2012, p. 23). Other authors, not abandoning this method, develop the concept of the "social discount rate" or "social rate of time preference", water projects included. This is based on the fact that in most cases water supply is a public or state-owned project (Young 2002, pp. 1, 4–6).

Methods and techniques for decoupling economic and environmental value and their assessment have been developed and appraised in various empirical cases such as *"Assessing the Environmental and Economic Value of Water"*.

Other researchers, such as Eric Plottu and Beatrice Plottu, in *"The concept of Total Economic Value of Environment"* systematize and practically apply "value in use" and "non-use values" in the concept of Total Economic Value (Plottu and Plottu 2007, pp. 52–61).

In Economics, there are many and different means of "value" and its estimate. Economics does not refer to "value" in general, but to "value" that has an adjective in front of it.

According to Green's opinion (Green 2003, p. 21) it is possible to summarize the manifold groups of "the value" in economy in two groups:

1. Value in and of itself;
2. Instrumental value.

Or, which is the same:

1. Intrinsic value ("valuable in and for itself") and
2. Instrumental value or "economic form of instrumental value"- an assessment that is made by comparison with something else, aim or purpose (Myrick Freeman III et al. 2014, p. 6).

The first group concerns the understanding of value in classical economics. Adam Smith, David Ricardo and, to a great extent, J.B. Sey, as well as Marx, accept that intrinsic value in itself can be defined by the cost of production. According to Smith, intrinsic value deviates only rarely from the "exchange value" (market value).

From this point of view, the value of water should be equal to the full cost of water (Green 2003, p. 10). Full Value of Water is the sum of the Economic and Intrinsic Values and Equal of Value in Use (Green 2003, p. 22). Intrinsic values in practical application are not estimated (Rogers et al. 1998, pp. 25, 27).

The problem of assessing phenomena that have an intrinsic value is that they cannot be replaced by anything else either as a function or as an utility (Green 2003, p.23). Such is the case with water.

The representatives of the environmental sector in economy differentiate between "economic" and "environmental" values (Dietz et al. 2005, pp. 336–365), which are defined and by "environmental ethics". The ecological values are defined as: "*Worth that a community or society places on environmental goods or services such as aesthetic and recreational facilities and resources. See also environmental value added.*" (Nash 1989). Thus, environmental value added has been defined: "*Net impact of an organization's activities on the environment over a specific period*" (Business Dictionary 2018). In fact, there are many points of view of "ecological value" and "worth" depending on the views of authors dealing with the problem. However, the value of water for life on Earth is indefinable, because it is absolute, therefore it is an absolute worth and has an absolute value.

The problem with multiple points of view and definitions does not prevent the question of the value of water being placed when it comes to water utility companies (WUC). Integrated reporting allows all reporting methods and all types of reporting to be used in "one report".

When it comes to assessing the value of water, we face a number of problems and contradictions that are rooted in water's unique nature uncomplaint when evaluating the value in its entirety, and problems in terms of accounting treatment—problems associated with water's characteristics as "the basis of life" and at the same time as a factor of production that should have a certain return on its economic use.

Water is seen as a "natural resource" and the assessment of its economic value is made in practice by costing:

- The cost of securing its natural functions (the functioning of the water cycle, water flow measures, anti-drought measures and the like);
- Collection and treatment costs;
- The cost of bringing it to the consumers—households, industry, agriculture.
- The costs for research, management and information on the status of the water basins.

That is to say that in measuring the economic value of water the cost approach is used, monetary indicators are used, and in the environmental report and water balance—physical indicators. With these are measured and quantified indicators that relate to "human water treatment" and costs for it.

It is difficult to find a solution to the other aspects of the problems related to water assessment and presented in major world water documents, such as the Dublin Declaration and the Rio de Janeiro Declaration.

The Dublin Declaration points out, first of all, the importance of water to sustain life, and then for economic development and for nature. The Dublin Declaration states: "*Principle No. 1: Fresh water is a finite and vulnerable resource,* ***essential to sustain life****, development and the environment. Since water sustains life, effective management of water resources demands a holistic approach, linking social and economic development with protection of natural ecosystems. Effective management* *links land and water uses across the whole of a catchment area or ground water aquifer.*" and "*Principle No. 4:* ***Water has an economic value in all its competing uses and should be recognized as an economic good.*** *Within this principle, it is vital to recognize first the basic right of all human beings to have access to clean water and sanitation at an affordable price. Past failure to recognize the economic value of water has led to wasteful and environmentally damaging uses of the resource. Managing water as an economic good is an important way of achieving efficient and equitable use, and of encouraging conservation and protection of water resources.*" (The Dublin Statement on Water and Sustainable Development 1992a).

In the same year as the Dublin Conference, the United Nations Conference on Environment and Development (UNCED) in Rio de Janeiro was held, continuing the development of the 1972 Stockholm Conference (Report Of The United Nations Conference on the United Nations Environment Conference 1972).

The United Nations Conference on Human Environment, meeting in Stockholm from 5 to 16 June 1972, proclaims that: "*Both aspects of man's environment, both natural and man-made, are essential to his well-being and to the enjoyment of basic human rights-even the right to life itself.*" (Report Of The United Nations Conference on the United Nations Environment Conference 1972, p. 3).

The Conference of Rio de Janeiro adopts the following in its preamble, Principle 1 and Section Two "*Conservation and management of resources: Protecting and managing fresh water*": "*Human beings are at the centre of concern for sustainable development. They are entitled to a healthy and productive life in harmony with nature*". And: "*The preamble and the following eight chapters address the challenges that adaptation of human behaviour to sustainable development poses to prevailing social and economic structures and institutions* (*Preamble, The Principal 1*). With this, the Rio Declaration recognizes that water is a social and natural good, and only after that—economic (The Rio Declaration on Environment and Development 1992b, p. 1).

Although it states that water is "*a resource, essential to sustain life*", the water value assessment in the Dublin Declaration is limited to an economic assessment in the presence of competitive uses, to a certain extent ethical—by measuring affordability, and to some extent, to its environmental assessment, but not to the assessment of sustaining life. The economic assessment of water, according to the document, is the "assessment of assessments "in an ecological, ethical (equitable use) and vital sense, based on the fact that the assessment of water as an economic asset will also provide its ecological and vital features for all living creatures on the planet. Thus, for the assessment of water as ~~a~~ life, a special and complex approach is not applied, but "economic fundamentalism".

The Rio Declaration changes this approach to "economic value fundamentalism" towards natural goods, placing first in their definition the characteristic of a "social" and "natural" good.

We should also note the forgotten definition of "aspects of man's environment" at the Stockholm conference as "natural and the man-made".

Economic theory on water valuation is incomplete. The most common is that "the value of water is the willingness to pay" once people's vital needs for water have been met. More complex models for evaluating the full value of water have been created using categories such as “value in use” and “non-used values”.

Water is a complicated “object” for economic research. It is not a commodity like any other commodity. In all research on the problems of water, both declarations state “water is life.” But in economy, the category of “value of biological human life” has not been introduced and operationalized, although without the existence of man’s biological life, “economic life” is impossible. In economic sciences, notions such as “the useful life of assets”, indicators of “the quality of life of the people”, “the standard of living” have been introduced as concepts, but not the life of human beings as biological life in itself.

For example, the standard of living is a category that R. Fogel defines as “*covered more abundant food supplies and better housing*” (Fogel 1994, p. 4).

Professor Fogel proves the link between human health and economic development. Human health is a necessary prerequisite for its ability to work and create added value, i.e. to be economically active. He has stressed that we should first be able to define the concept of “good life” (Fogel 1999).

The economic category “good life”, derived and explored by Fogel’s Climometrics, should include access by default and the right to water of all living creatures on the planet, including people, without which their biological existence is impossible.

Economic science and accountability of economic activity have inevitably reached the creation of a relation between purely economic and ethical categories, taking into account both quantitative and qualitative indicators of the performance of economic subjects, which is reflected in the IR and other forms of non-financial reporting. While combining ethical categories and ethical issues with economic ones has already been largely successful, this is not the case with the biological categories and the category of “life”.

Defining the full value of water is really complicated, because defining it needs to define and value “biological life” also, as water is the foundation of life.

The French researcher Philippe Saint-Marc rightly concludes that “*Traditional economic science, based on production and labour, ignores the worth (value) of what constitutes* “a *gift of nature*”. *But today these* “*gifts*” *have become a rarity, as a result of their waste, as well as the absurd ignorance and indifference to all biological phenomena on the part of political economy and urbanistic doctrines.*” (Saint-Marc 1977, p. 331).

It would seem right to ask the question if these “free gifts” are used by “the natural monopoly” to make a profit? It would be reasonable to ask the question: What is the “business model” of such a company (how does it create its profit)? How does it create value added for itself, its investors and other stakeholders? Or is the term “natural monopoly” an ironic term for the monopoly that uses nature-created resources, gift of nature and naturally does not show its value in reports?

The <IR> methods do not provide guidelines for water value reading by any of the developed methods, this is not done by the WUC in their accountability models either.

The integrated report, with its huge potential of incorporating all types of methodologies and tools, and its purpose—to evaluate the company's full value, the company's added value and to expose their business model cannot escape these issues despite the problems in economic theory with the issue of value.

The value of water is related to the definition of water as a vital necessity, a gift of nature, as well as capital.

12.3 Value, Capital and Business Model

Capital—these are the resources pertaining to the economic activity of the enterprise. In this sense, natural resources are also delimited in <IR>.

In <IR> manufactured capital is defined as follows: "*Manufactured physical objects which are available to an organization for use in the production of goods or provision of services, including:*

- *Buildings*
- *Equipment*
- *Infrastructure such as roads, ports, bridges etc.*
- *Waste and water treatment plants.*

Manufactured capital is often created by other organizations but includes assets produced by the reporting organization for sale or when they are retained for its own use." (International Integrated Reporting Council (IIRC) 2015, p. 12). And: "*Natural Capital—All renewable and non-renewable environmental resources and processes that provide goods or services that support the past, current or future prosperity of an organization.*" *Including:*

- *Air, water, land, minerals and forests;*
- *Biodiversity and eco-system health* (International Integrated Reporting Council (IIRC) 2015, p. 12).

Business model background paper for <IR> provides guidelines to take into account water technology projects in "*Manufactured capital*", Infrastructure (such as roads, ports, bridges and wastewater treatment plants (Business model background paper for <IR> 2013, p. 11).

In conformity with these definitions, water as a gift of nature is used by water supply companies and is:

1. Separate capital—a gift of nature,
2. Non-manufactured capital;
3. It directly relates to the past, present and beneficial development of the organization and to the creation of value for society as a whole.

These characteristics logically make water stand out as different capital from the manufactured physical capital objects.

Raw water that WUC supply to households, agriculture and industry is not a man-made resource and hence "capital" that <IR> refers to the "natural capital" group. It is "natural capital" that enables these WUC to perform their business functions.

Water is a natural, material, non-financial capital asset. It has a characteristic of a capital asset because it meets the generally accepted characteristics: a permanent asset that is used to carry out business and earn income, it is used on a daily basis, and is a worth for its ruler.

Water is a gift of nature and a "common pool resource" (at least in the part recognized as "common pool resources" such as seas, oceans, lakes, groundwater), which yield water for purification and supply (See: Ostrom et al. 1994). These circumstances make it necessary to divide the assets, product of human labor that water companies possess and water as non-man-made and non-financial capital (a capital asset). As a gift of nature and non-man-made natural capital, raw water is a capital asset of humanities. Existing stocks can increase or decrease over time. The <IR> provides the opportunity for such WUC capital treatment in order to gain a more accurate assessment of sources and means for value creation for these companies.

On the other hand, companies operating in the extraction, purification and water supply business are high tech companies that make significant investments in their business. They invest in serious science projects with slow return incurring high costs, such as all R & D spending, due to which they possess a large intellectual capital, largely shaping their market value as companies.

From here what logically follows is that the capital structure of water-supply enterprises must be divided in <IR> as "man-made capital" and "non-man-made capital" (Fig. 12.1):

1. Non-manufactured capital, Non-financial capital or Capital assets—Raw water—a gift of nature, and,
2. Manufactured capital or Capital assets—dams, pipelines, water treatment plants, equipment, Intellectual capital etc. (After purchasing or extracting raw water from the company, the water is its capital asset).

With the growth of urbanization and the world's population, the cost of extracting, purifying and supplying water to consumers is steadily increasing. The

Fig. 12.1 Division of capital assets of water utilities companies

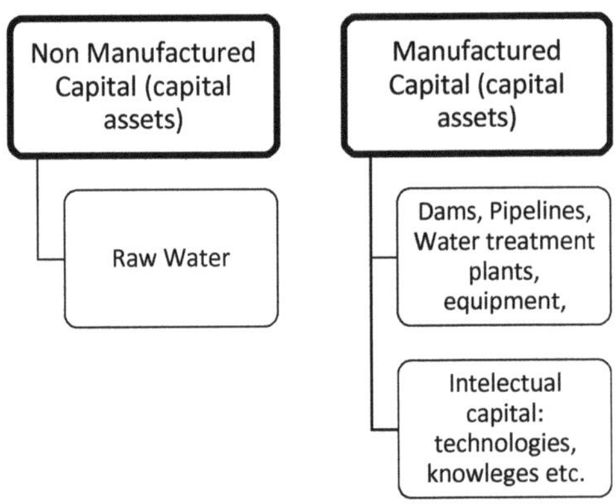

cost of water usually rises due to the investments made by the WUC. However, there is no convincing evidence that the increase in the price of water improves its quality. The separation, assessment and disclosure of companies' costs for the so-called "Environmental services" in this case is justified. For clarity on this issue, it is possible to use "modals for estimating the values of environmental services" including "Human health, Mortality; Chronic morbidity; Acute morbidity; Direct impact on humans; Economic productivity of ecological systems; Ecological, services (such as recreation); Effects on non-living systems such as materials, non-use values such as ecological stability and biodiversity." (Myrick Freeman III et al. 2014 p. 437).

Too little space is given to the Environment report in <IR> at the end of the Integrated report model. When natural resources are the base of companies' business this aspect can be developed and presented in more detail. Such is the case with water supply companies. The source of their business activities is water. The <IR> does not report by economic value (by monetary means) losses from the degradation of water resources as a result of water supply.

Different types of industries can adapt their integrated reporting models to the specific nature of their business and use different methods to analyze the status and forecasts for their future performance. This differentiated approach to the different industries is successfully applied in sustainable accountability standards such as SASB (SASB 2016).

In the integrated report of water companies, there is also a need and opportunity to create information and links between the:

1. Social aspects of water delivered—affordability; fair pricing, social return of investments;
2. Natural and vital aspects of water—the preservation of water resources and their proper distribution for the needs of households, agriculture and industry;
3. Ecological aspects—preservation, conservation of water resources and protection of the water balance;
4. Economic and financial aspects of businesses for extraction, treatment and supply of water, – Profits (Created by new investments and created by existing investments), Costs, investments and return on investment in man-made assets, rents distribution;
5. Risk to the business and investors;
6. Risk to water users for water supply, quality risk.

These aspects are related to and applicable in the content elements of <IR> in:

1. The business model of the water company;
2. The strategy of the WUC;

Complex valuation may be used in:

1. Assessment of the water use facility (lake, dam) on the basis of comparison of the values created by the property rights of the WUC, average for the region.
2. Market assessments to identify the reasons for the observed price difference between suppliers;

3. A conditional assessment approach using a poll to assess WTP for improved water and supply quality and their willingness to accept WTA to reduce water quality.

12.4 Rental Approach

With the emergence of cities and their growth, the business with water has become one of the most profitable businesses nowadays.

In 2003, a special stock index for water utilities companies, the Palisades Water Index (ZWI) was created (The Law Dictionary). This is such a lucrative business that at the end of 2012 it is superior to the banking sector in the stock exchange (Fig. 12.2).

The components of the Palisade Water Index include the following companies given in Table 12.1 These companies, included in ZWI make large investments in

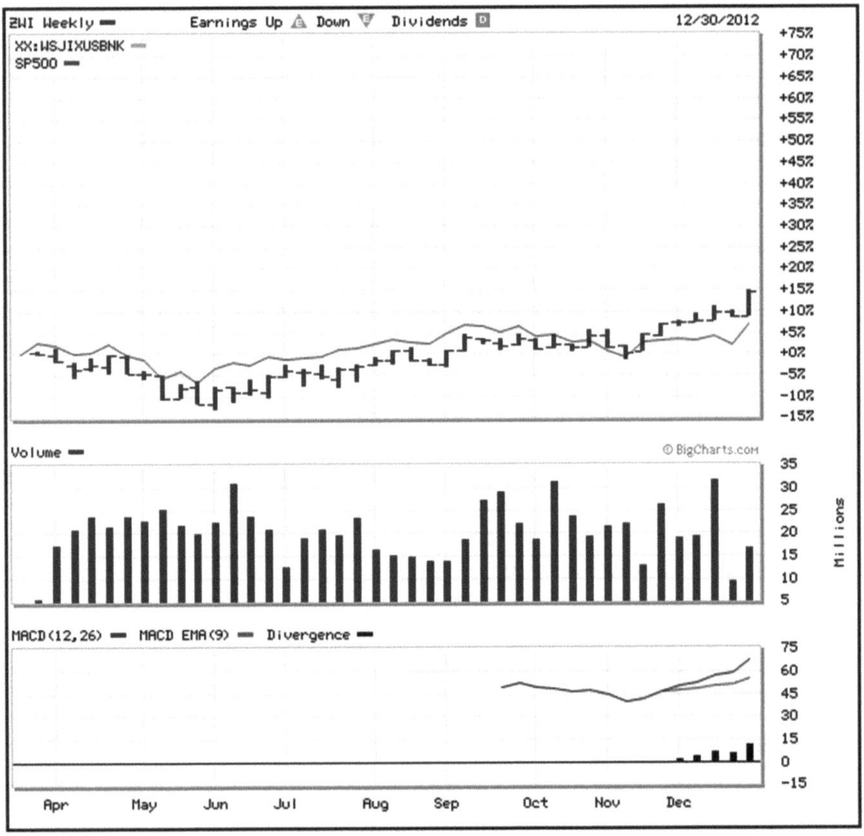

Fig. 12.2 Comparison between S&P 500 bank sector and Palisades Index. 04 March 2017; 5 years, weekly (Market Watch)

Table 12.1 List of companies-components of Palisade Water Index, ZWI (Market Watch)

Name of company and % from ZWI	<IR>/other nonfinancial reports	Description	Headquarter/ Web site
1. Tetra Tech (NASDAQ: TTEK) 4.58%	NO/ Sustainability report	Public, worldwide, Water, Environment & Infrastructure, Resource Management & Energy	US, Pasadena, http://www.tetratech.com/
2. Itron Inc. (NASDAQ: ITRI) 4.12%	NO	Public, worldwide, Water Communication modules (metrics software and others)	US, Washington http://www.itron.com/
3. URS Corporation (NYSE: URS) 4.07%	NO Sustainability and CSR report	Subsidiary of AECOM, worldwide Engendering, constructions (URS was acquired by AECOM on October 17, 2014)	US, California San Francisco, http://www.urs.com/
4. AECOM (NYSE: ACM) 4.00%	NO, sustainability report	Public, worldwide Professional services, R&D	US, California, Los Angeles, http://www.aecom.com/
5. Danaher Corp (NYSE: DHR) 3.98%	NO Sustainability and CSR reports	Public, worldwide, Conglomerate (multi-industry company) Test & measurement, industrial technologies, environmental, and Life Science & Diagnostics	US, Washington http://www.danaher.com/
6. Badger Meter (NYSE: BMI) 3.95% (flow meters)	YES	Public, Worldwide, Water meters, meter reading and analytics technologies for municipal water utilities Flow measurement and control products for water Residential water metering, Commercial water metering Water and wastewater treatment facilities	US, Milwaukee, https://www.badgermeter.com
7. Veolia Environment, (NYSE: VE) 3.92%	NO Sustainability report	Public limited company (Societe Annonime) French transnational company, Water treatment, waste management, HVAC, street lighting, facility management services	Paris, France http://www.veolia.com/
8. Valmont Industries Inc. (NYSE: VMI) 3.86%	NO/ Sustainability and CSR reports	Public, worldwide, Central pivot and linear irrigation equipment	US, Nebraska, http://www.valmont.com

(continued)

Table 12.1 (continued)

Name of company and % from ZWI	<IR>/other nonfinancial reports	Description	Headquarter/ Web site
9. Calgon Carbon Corporation (NYSE: CCC) 3.79% (filtration)	NO/sustainability report	Public, worldwide Manufactures and markets products that remove contaminants and odors from liquids and gases, both for industrial, municipal, and consumer market, carbon recycling	US, Pittsburgh, Pennsylvania http://www.calgoncarbon.com/
10. Agilent Technologies (NYSE: A) 3.69%	No, Sustainability, environmental, corporate citizenship reports	Public, worldwide, Food, environmental and forensics, pharmaceutical, diagnostics, chemical and energy, and research	US, Santa Clara, California, http://www.agilent.com/
11. Lindsay Manufacturing Co. (NYSE: LNN) 3.65%	NO/sustainability reports	Public, Center pivot irrigation systems, infrastructure	US, Omaha, Nebraska http://www.lindsay.com/
12. Lindsay Manufacturing Co. (NYSE: MWA) 3.56%	NO/sustainability reports	Public, North America, Water infrastructure	US, Atlanta, Georgia, http://www.muellerwaterproducts.com/
13. Pentair, Inc. (NYSE: PNR) 3.50% (pumps, motors, filtration, water tanks)	NO/sustainability and CSR reports	Public, worldwide, Water & Fluid Solutions Valves & Controls, technical solutions	UK Worsley, greater Manchester, Incorporated in Ireland http://www.pentair.com/
14. ITT Industries (NYSE: ITT) 3.38%	NO/sustainability report	Public, Conglomerate, Worldwide, Pumps	US, New York, http://www.itt.com/
15. Watts Water Technologies (NYSE: WTS) 3.37%	NO/sustainability reports	Public, worldwide, Valves Global provider of plumbing, heating, and water quality solutions for residential, industrial, municipal, and commercial settings	US http://www.wattswater.com US: North Andover, Europe: Amsterdam
16. Pall Corporation (NYSE: PLL) 3.25%	NO/sustainability reports	Subsidiary of Danaher corporation Filtration, fluid management, electronics, municipal and industrial water purification, aerospace... R&D	US, New York, http://www.pall.com/
17. Pall Corporation. (NYSE: NLC) 3.20%	NO/sustainability reports	Owner: Ecolab (conglomerate)-public, Wholly subsidiaries, Chemicals and water treatment	US, Naperville, Illinois, http://nalco.ecolab.com/

(continued)

Table 12.1 (continued)

Name of company and % from ZWI	<IR>/other nonfinancial reports	Description	Headquarter/ Web site
18. Insituform Technologies Inc. LLC (NASDAQ: INSU) 3.15%	NO	Public, worldwide, Subsidiary of Aegion corporation, Pipeline installation and repair	US, St. Louis http://www.insituform.com http://www.aegion.com/
19. NOV Ameron International (NYSE: AMN) 3.10%	NO/sustainability report	Public, worldwide, Water transmission supplier of highly-engineered concrete and steel pipe systems	US Pasadena, California, http://www.nov.com/ameron.aspx
20. General Electric (NYSE: GE) 3.01% (GE Energy)	**YES** (integrated "summary" report)	Public, worldwide, Conglomerate, Power & Water sector including GE Energy (water and process technologies)—a division of General Electric with headquartered in Atlanta, Georgia, United States-http://www.ge-energy.com/ Water & Process Technologies Water treatment technologies	US, Boston, Massachusetts, http://www.ge.com/ http://www.ge-energy.com/
21. Flowserve Corp (NYSE: FLS) 2.95%	NO	Public, Pumps, valves Water resources industries Water supply	US Irving, Texas, http://www.flowserve.com/
22. FEP Holding Company LLC Franklin Electric (NASDAQ: FELE) 2.85%	NO	Public, worldwide, Residential water systems, Pumps and motors	US, Fort Wayne Indiana, http://franklinwater.com US Nearby Fort Wayne, Indiana http://www.franklin.com franklin-electric.com/
23. Siemens AG Ads (NYSE: SI) 2.82%	NO Sustainability reports	Public limited company (Aktiengesellschaft) Worldwide	Germany Berlin, Munich http://www.siemens.com/
24. IDEX Corporation (NYSE: IEX) 2.79%	NO Sustainability report	Public, Fluidics systems, hydraulic rescue tools	US Lake Forest http://www.idexcorp.com/
25. Layne Christensen Co. (NASDAQ: LAYN) 2.79%	**YES**	Public Global water management, construction and drilling company	US Houston, Texas, http://www.layne.com/en/

(continued)

Table 12.1 (continued)

Name of company and % from ZWI	<IR>/other nonfinancial reports	Description	Headquarter/ Web site
26. Gorman-Rupp (NYSE MKT: GRC) 2.65%	NO	Public Worldwide, Manufacturers pumps for municipal, water, wastewater, sewage, industrial, construction	US Ohio http://gormanrupp.com
27. Roper Industries (NYSE: ROP) 2.51%)	NO	Public Conglomerate Water and fluid handling, pumps, appliances, pumps, industrial controls	US, Florida, http://www.ropertech.com/
28. Consolidated Water Co. Ltd. (NASDAQ: CWCO) 1.56% (water utility)	NO	Private water utility company, CWCO Water utility Retail water operations, bulk water operations and services operations.	Cayman Islands http://www.cwco.com
29. Southwest Water Co. (NASDAQ: SWWC) 1.33%	NO	Water utility, Sewer services and connections	UK http://www.southwestwater.co.uk
30. American States Water (NYSE: AWR) 1.29%	NO	Public, Water utilities	US San Dimas, California, United States http://www.aswater.com/
31. Aqua America (NYSE: WTR) 1.24%	NO	Public Water and wastewater utility company	US Bryn Mawr, Pennsylvania, http://www.aquaamerica.com/
32. Companhia de Saneamento Basico do Estado de Sao Paulo (Sabesp); (NYSE: SBS) 1.03% (water and wastewater utility)	NO	Public limited company Water Waste services	Brazil, São Paulo, Brazil http://www.sabesp.com.br/

intellectual and structural capital and have high value man-made assets. The leaders in the water utilities business are high-tech companies that create and deliver technology and equipment to the water sector, for the detection and extraction of water and its delivery to consumers. There is a growing tendency for water companies to use the integrated reporting model for management purposes and for the purposes of their business communication.

The various WUCs have ~~a~~ varied use of incoming raw water: license, purchases, concessions, with various agreements. This makes it difficult to account and report water as a natural capital the integrated company report in a comparable way.

It also makes it difficult to assess the value created by their "property rights". Some WUCs have their own water collection dams built with their own investments; others are using high-tech methods incurring large capital expenses. Other WUCs use dams built and maintained by the state or with municipal funds and buy from them raw water at different prices.

For example, the Veolia-owned "Sofyiska Voda" Water Supply Company in Bulgaria buys at 0.02 Euro 1 cubic meter of water from the state-owned "Iskar" Dam, built with state funds, financed by state funds, owned by the National State Energy Company, whereas it sells water for household needs at the price of 1.24 Euro. In this way, it has paid 1.5 million Euros for incoming water and has sold water for 75 million Euros, with 50% loss of incoming water in the transmission system. Nevertheless, the price of water has increased from 2000 to 2016 by 337,93% (Protocol from the Public Discussion for approval of prices of water supply and sewerage services of Sofyiska Voda AD 2016).

Such a company cannot be compared to companies that have made their own investment in dams and in modern methods of extracting raw water or purifying mud water with their own installations, these being the result of their own technological and scientific developments and investments.

Given these two circumstances, it is possible to make use of integrated reporting and the rental approach in relation to the value assessment of their property rights and approximation to the calculation of the full value of the product they produce—extracted, purified and/or water delivered to the full value (and value added), created by the companies.

"Economic Rent" is a long time developed economic category, starting with Adam Smith and David Ricardo. For Ricardo, it is an amount that is paid to the owners due to the limitation of an important productive resource (Ricardo 1817, p. 38). Ricardo explores the rent of land and mines. According to him, rent from mines does not differ from that of soil, as a scarce resource. Ricardo specifically states that rent is not paid for air and water, as they are abundant "*... no rent can be paid for ... nothing is given for the use of air and water or for any other for the gifts of nature who exists in a bounder quantity.*" (Ricardo 1817, pp. 40–41). Ricardo's classical view for the emergence and existence of a rent is linked to two criteria: ownership (or "property rights") and quantitative resource constraints. Two other definitions of the neoclassical theory of economic rent are in circulation: "Marshallian rent", and the so called "Rent of Pareto". The neoclassical treatment of economic rent practically abandons the *property attribute* and stresses Ricardo's resource limitation. The reason for this is historical. Historically, property rights were established on the land, whereas this was not done with water either at Marshall's or Pareto's time. This historical circumstance continues to affect the perception of water resources as free, the same way air is free. The measurement of the various types of economic rent (Ricardian, Marshallian and Paretian) gives different end results when measuring the wealth created for the owner and the producer by the limited resource (Brar 1977). When there are insufficiently specified property rights to water resources, and when not WUCs, but others are involved in their collection and preservation, in the construction and maintenance of dams, the water balance, etc., in terms of satisfying the *mainstream*, the most suitable for use is the so-called Paretian rent. Thus the Paretian rent is a surplus of earnings over the amount necessary to keep the factor in its present use (Brar 1977). The use of Paretian rent also gives us the opportunity to keep track of the sustainability of the water system.

Various tools exist for calculating value added. Especially popular is the Economic Value Added (EVA). Essentially, EVA is the economic profit of the company. It compares the operating profit and the cost of capital. Incorporating the value of water as a non-man-made capital into the EVA calculations would yield different results for the value created/degraded by WUC.

Economic rent, in turn, shows the return on factors of production. In the case of water, the *Paretian rent* shows the return on water as the only and limited resource for WUC's business and the profit from water treatment and supply. It also shows the wealth of the owner through his ownership of the factor of production. The <IR> could include this issue in water company accountability in relation to their business model and their social impact through the distribution of the rent between them and the owner of the water resources—the society represented by its organizations (The creation of tools in this respect is the subject of different scientific research).

Contemporary economics usually neglects rent as an analytical category. In recent times, the rental approach has been revived in works such as "*Rents, Rent-Seeking and Economic Development: Theory and Evidence in Asia*" by, Khan and Jomo (2000), "*Rent-Seeking, Institutions and Reforms In Africa: Theory and Empirical Evidence for Tanzania*" by Pius Fischer (2006), "*Skin in the Game, Hidden Asymmetries in Daily Life*" by Nassim Nicolas Taleb (Taleb 2018); Michael Hudson's and Dirk Bezemer's "*Incorporating the Rentier Sectors into a Financial Model*",(Hudson and Bezemer 2012); Era-Dabla Norris and Paul Wade, "*Rent Seeking and endogenous Income Inequality*", (Norris and Wade 2001).

Another difficulty in introducing the rental approach in assessing the value added of WUC is the treatment of the National Income and Product Accounts of the annuity recipients as "providing a service, an economic contribution equal to what the rentiers receive as 'earnings'" (Hudson 2012, p. 5).

The question of "seeking for rent" or "seeking for profit" is relevant to <IR> of water companies and its disclosure in the Business Model for several major reasons: (1). The large profits that private water companies receive; (2). The large loans they receive from financial institutions, the state and the municipalities; (3) Regulated prices of water and the formation of water tariffs by regulators; (4). WUC operate under the conditions of the so-called "natural monopoly". As long as there is "trade with water rights" in some countries and regions, it is complementary on a national scale and does not change their local monopoly situation. Rent is one of the major forms of the manifestation of property rights.

The WUC's position of monopolists in this field also raises the need to highlight the question of whether their profits are rent seeking or the result of investments and effective work.

The proof of "rent-seeking" or normal business "seeking for profit" in the case of monopoly status directly affects the WUC's business model and has a direct relevance to the issue of value added. According to Trucost's research, 25% of unpriced natural capital costs are from water (Roberts 2013).

Gordon Tullock's original "rent-seeking" concept dates back to 1967 (Tullock 2005, p. 35). It is further developed by himself and other scientists such as Arye L. Hillman, Eliakim Katz, William J. Corcoran, Gordon V. Karels, Richard S. Higgins, William F. Shughart, Robert D. Tollison, Richard S. Higgins, Fred S. McChesney and others dealing with the problem. The establishment of rent—seeking is well developed (Rowley et al. 1988)."*Rent seeking is defined as the study of how individuals compete for artificially contrived transfers*" (Tollison 1982, p. 601).

When rent seeking is a result of free demand and supply, it is equivalent to profit seeking and is a normal business activity. When we have a case of government intervention, corruption or regulation in favor of collector groups and in monopoly status, clarifying whether profits are the result of normal profit or rent seeking is necessary.

Striving for rent is in itself a positive phenomenon in economy because it is a search for profit that is normal for every business. In this case, he seeks a profit, accumulated in the usual economic way, an attempt to obtain a surplus over the normal return on resources in a decent manner. In this sense, "green/ecological rent" and "natural/water rent" are positive economic phenomena. Such WUC rents are the result of the use of more productive technologies, more efficient deliveries without losses in the water transmission system.

However, rents can be obtained through non-economic means by the abuse of power-influence or use of non-economic means such as lobbying and hidden government decisions. This leads to the usurpation of the distribution of state or public resources. Such is the monopoly rent. In this case, as Tollison argues, the profit is transferred from customers and the ultimate owner of property rights to the monopolist (Tollison 1982, p. 576).

The theory of rent has evolved not only in the argumentation of "rent- seeking". "Water rent" and "environmental rent" can also be defined and calculated.

Water rent within the sphere of water supply is a relationship between the "common pool resources—the owner of the resource—private, public or a state WUC and the customers".

In the different cases water rent can be:

- A monopoly rent- when there is one supplier, as with water suppliers who use the exclusive properties of the created by nature water resource or natural resource—water. This rent is expressed in the imposition of monopolistically high prices on delivered water. The reasons may be different: a monopoly on technology or a variety of "rent-seeking" in monopoly and regulation conditions such as the described case-low efficiency, large losses in the system, but as a result—monopolistically high prices.

- Differential (Ricardian rent)—when different water companies compared to one another use sources of water supply acquired and/or extracted at a different degree of difficulty.

The rental approach allows assessing the quality and availability of the water resource for water companies; their investment needs, and further exposes the cost of their supply. The chartered rental approach can also provide information on fairness/price justification when water companies buy water from owners of water repositories.

At the same time, it can tell us if their business is related to "rent seeking" or their profits are the result of their fair efforts.

Rent, including water rent, may be equal to zero if there is poor management, inefficient technology and other factors that prevent profitability. There are WUCs that claim to have zero profits, insignificant profits or are at a loss due to their low water supply prices set by regulators. This is a statement for which there is no objective evidence. Rather, it should be assumed that they do not acquire water rent because of poor management, outdated technology, unskilled staff and, as a result, poor performance.

Water rent is a kind of natural rent. In many cases, private companies largely privatize it. The distribution of the natural water rent is a matter that can also be solved in the Integrated-reporting model. It is logical to distribute it between the supreme owner—the population, the state and the water companies. The equitable distribution of water rent is also a matter of ethical nature and there is reason for it to be considered (Yakovets 2003, p. 8).

Natural rent can be seen as a profit above the industry average, realized at the expense of using more efficient technologies, better management, higher qualification of staff, own scientific and technological developments. Natural anti-rent occurs also from non-compliance with established environmental standards for quality, efficiency and limitations. It creates national and international expenses related to the protection of the environment and especially the waters that would be avoided in compliance with the standards and limitations. Anti-rent damages are both short-term and long-term and are anti-sustainable.

Along with positive annuity, anti-rentals can also be defined (Yakovets 2003, pp. 76–107).

The anti-rent is an overpayment received "not by the rules" and can be seen as a rent seeking option.

12.5 Conclusions

Water as a resource that belongs to the planet and the whole of humanity differs from the value of the man-made capital that WUC use in treatment and supplying water resources. Water as a natural capital, according to the definition of <IR>, it is a gift of nature, a non-man-made natural material capital. The water should be treated as a capital asset, different from that created by man in the business of the water supply companies. Economic science, in spite of its special attention to value theory, seems to have refused to treat natural capital as a separate capital asset in companies' work. <IR> gives the opportunity for water to be defined and valued as a separate material and intellectual capital from man-made capital assets in water business because of the particular importance and status of water as a gift of nature and the most important resource for life in economic use, and as a human right. The rental approach in <IR> makes it possible to specify the property rights to water resources. Applying different types of rents in calculating the WUC's value added makes it possible to assess the latter more accurately. The rental approach sheds more light on the WUC's business model, clarifies the return on the productivity factor—water and the investments made. This approach can clarify to a greater extent the actual return on the water companies' investments. Along with this, the rental approach provides greater insight into the source of the profits of these companies and what they are due to: rent seeking, monopoly, unfair pricing, or technological advances and good governance. Applying this rental approach can be of benefit to society, bearing in mind its property rights, and nature, the creator of water.

References

AECOM. http://www.aecom.com, https://www.aecom.com/about-aecom/sustainability/sustainability-report/

Agilent Technologies. http://www.agilent.com/, https://www.agilent.com/environment/esr/home.html

American States Water. http://www.aswater.com/, http://www.annualreports.com/Company/american-states-water-company

Amundsen, E., Andersen, C., & Sannarnes, J. (1992). Rent taxes on Norwegian hydropower generation. *The Energy Journal, 13*, 1.

Aqua America. http://www.aquaamerica.com, http://www.annualreports.com/Company/aqua-america-inc

Assessing the Environmental and Economic Value of Water: Review of Existing Approaches and Application to the Armenian Context. Final Report. Retrieved from https://www.oecd.org/env/outreach/AM%20Water%20Value.pdf

Badger Meter. https://www.badgermeter.com, https://badgermeterinc.gcs-web.com/financial-information/annual-reports

Brar, J. S. (1977). Estimation of economic rent as a measure of factor owners' welfare. *Western Journal of Agricultural Economics, 1*(1), 215–220.

Calgon Carbon. http://www.calgoncarbon.com/, http://www.sustainability.calgoncarbon.com

Companhia de Saneamento Basico do Estado de Sao Paulo. http://www.sabesp.com.br/, http://site.sabesp.com.br/site/interna/subHome.aspx?secaoId=95

Consolidated Water Co. Ltd. http://www.cwco.com, http://investors.ropertech.com/Annual-Reports

Danaher Corp. http://www.danaher.com/, http://www.danaher.com/how-we-work/corporate-social-responsibility

Dietz, T., Fitzgerald, A., & Shwom, R. (2005). Environmental values. *Annual Review of Environment and Resources, 30*(November), 336–365. https://doi.org/10.1146/annurev.energy.30.050504.144444.
FEP Holding Company LLC (Franklin Electric). http://www.franklin.com
Fischer, P. (2006). *Rent-seeking, institutions and reforms in Africa: Theory and empirical evidence for Tanzania*. New York: Springer.
Flowserve Corp. http://www.flowserve.com, https://www.flowserve.com/en/more/suppliers/anti-slavery-and-human-trafficking-statement
Fogel, R. (1994). *New findings on trends in life expectancy and chronic diseases: The implications for health costs and pensions* (p. 4). University of Chicago, Selected Papers Publication of the Selected Paper was supported by the Albert P. Weisman Endowment, N 76. Retrieved from https://chicagobooth.eu/~/media/3d8f716b52494584ab4154da79a1d5f1.pdf
Fogel, R. (1999, March). *Catching up with the economy*. The American Economic Review, Retrieved from https://chicagobooth.eu/~/media/3d8f716b52494584ab4154da79a1d5f1.pdf, https://www.die-gdi.de/fileadmin/user_upload/pdfs/Messner_WS_2016/Fogel_Catching_up_with_the_economy.pdf
General Electric (GE Energy). http://www.ge.com/, http://www.ge-energy.com/
Gorman-Rupp. http://gormanrupp.com, https://www.csrhub.com/CSR_and_sustainability_information/Gorman-Rupp-Industries
Green, C. (2003). *Handbook of water economics, principles and practice* (p. 21). Chichester: Wiley.
http://franklinwater.com, http://www.franklin-electric.com/
http://www.danaher.com/how-we-work/corporate-social-responsibility/
Hudson, M. (2012). *Veblen's institutionalist elaboration of rent theory* (Working Paper). Levy Economics Institute, No. 729.
Hudson, M., & Bezemer, D. (2012). Incorporating the rentier sectors into a financial model. *World Economic Review, 1*(6 September), 1.
IDEX Corporation. http://www.idexcorp.com/, http://phx.corporate-ir.net/phoenix.zhtml?c=83305&p=irol-reportsannual
Insituform Technologies Inc. LLC. http://www.insituform.com
International Integrated Reporting Council. (2013, March). *Business model background paper for <IR>* (p. 11).
International Integrated Reporting Council (IIRC). (2015). *The international framework for integrated reporting* (IR) (p. 12).
ItronInc. http://www.itron.com/, http://investors.itron.com/financial-information/annual-reports
ITT Industries. http://www.itt.com/, http://www2.lucidconnects.com/2017-sustainability-outlook-report?gclid=Cj0KCQjwzIzWBRDnARIsAAkc8hEq0194ScpYKVcMj3QfSv8PjQTHfjaKjXM9dTNA-opAfmR2Gf5BYCEaAlO5EALw_wcB
Khan, M. H., & Jomo, K. S. (2000). *Rents, rent-seeking and economic development: Theory and evidence in Asia*. Cambridge: Cambridge University Press.
Layne Christensen Co. http://www.layne.com/en/, http://www.annualreports.com/HostedData/AnnualReports/PDF/NASDAQ_LAYN_2016.pdf
Lindsay Manufacturing Co. http://www.lindsay.com/, http://www.lindsay.com/sustainability
Lindsay Manufacturing Co. http://www.muellerwaterproducts.com/, http://www.lindsay.com/sustainability
Market Watch. http://www.marketwatch.com/investing/index/zwi/charts?symb=XX%3AZWI&countrycode=XX&time=12&startdate=1%2F4%2F1999&enddate=3%2F4%2F2017&freq=2&compidx=SP500&compind=XX%3AWSJIXUSBNK&comptemptext=Enter+Symbol%28s%29&comp=none&uf=7168&ma=1&maval=50&lf=1&lf2=4&lf3=0&type=2&size=2&style=1013

Myrick Freeman III, A., Herriges, J. A., & Kling, C. L. (2014). *The measurement of environmental and resource values. Theory and methods* (3rd ed.). New York: Routledge.

Nash, R. (1989). *The rights of nature: A history of environmental ethics*. Madison, WI: University of Wisconsin Press.

Norris, E., & Wade, P. (2001, February). *Rent seeking and endogenous income inequality* (Working Paper). International Monetary Found.

NOV Ameron International. http://www.nov.com/ameron.aspx

Ostrom, E., Gardner, R., & Walker, J. (1994). *Rules, games and common-pool resources*. Ann Arbor, MI: University of Michigan Press.

Pall Corporation. http://nalco.ecolab.com/

Pall Corporation. http://www.pall.com/, https://www.pall.com/en/about-pall/corporate-sustainability.html

Pentair, Inc. http://www.pentair.com/, https://www.pentair.com/en/about-us/corporate-social-responsibility/sustainability

Plottu, E., & Plottu, B. (2007). The concept of total economic value of environment. *Ecological Economics, 61*, 52–61.

Protocol from the Public Discussion for approval of prices of water supply and sewerage services of Sofyiska Voda AD. (2016). Sofia, 13.04.2016, the draft decision on the submitted application with input. № B-17-44-2 / 31.03.2016, Retrieved from http://www.dker.bg/files/DOWNLOAD/prot-oo-sofvoda-13apr2016.pdf, http://sofia.dir.bg/news/-21578398, http://www.periscop.bg/sofiyska-voda-kupuva-kubik-voda-za-4-stotinki-a-prodava-a-sofiantsi-za-216-lv/

Ricardo, D. (1817). *On principals of political economy and taxation* (p. 39). 3rd Edition 1821, Batoche Books, 2001, Canada.

Roberts, D. (2013, April 17). *None of the world's top industries would be profitable if they paid for the natural capital they use*. Retrieved form https://naturalcapitalcoalition.org/protocol/sector-guides/food-and-beverage/

Rogers, P., Bhatia, R., & Huber, A. (1998). *Water as a social and economic good: Global water partnership* (p. 10). Stockholm: Development Cooperation Agency.

Roper Industries. http://www.ropertech.com/, http://investors.ropertech.com/Annual-Reports

Rothman, M. (2000, July). *Measuring and apportioning rents from hydroelectric power developments* (World Bank Discussion Papers, N 419).

Rowley, C. K., Tollison, R. D., & Tullock, G. (Eds.). (1988). *The political economy of rent-seeking*. Boston, MA: Springer.

Saint-Marc, Ph. (1977). *Socialization of nature*. Cited from Saint-Marc, Ph., Socializacii prirody, (2005), Moscow, p. 331; Sosnina, T. N., The Value. Historical and Methodological Research, Samara, p. 212.

Siemens AG Ads. http://www.siemens.com/, https://www.siemens.com/global/en/home/company/sustainability/corporatecitizenship.html

Southwest Water Co. http://www.southwestwater.co.uk, https://www.southwestwater.co.uk/about-us/documents/Annual-reports/

Sustainability Accounting Board (SASB). (2016). *Provisional standard*. Sustainability Accounting Standards Board. Retrieved from https://www.sasb.org

Taleb, N. N. (2018). *Skin in the game. Hidden asymmetries in daily life* (p. 129). New York: Random House Publishing Group.

Tetra Tech. http://www.tetratech.com/, http://www.tetratech.com/en/sustainability-report-card

The International Integrated Reporting Council (IIRC). (2013). *The international framework for integrated reporting (IR)* (p. 12).

The Law Dictionary. Retrieved from https://thelawdictionary.org/palisades-water-index/

Tollison, R. (1982). *Rent seeking. A survey, kyklos, international review for social sciences* (Vol. 35, p. 601). Oxford: Wiley-Blackwell.

Tullock, G. (2005). *Public goods, redistribution and rent seeking* (p. 35). Cheltenham: Edward Elgar.

United Nations. (1992a). *The Dublin statement on water and sustainable development*. UN Documents Gathering a body of global agreements.

United Nations. (1992b, June 3–4). *The Rio declaration on environment and development* (p. 1). The United Nations Conference on Environment and Development.

United Nations (UN). (1972, June 5–16). Report of the United Nations conference on the United Nations environment conference, Stockholm.

United Nations (UN). (2008). System of National Accounts (SNA).

United Nations (UN). (2012). *System of environmental-economic accounting*. Central Framework (SEEA).

URS Corporation. http://www.urs.com/

Valmont Industries Inc. http://www.valmont.com

Veolia Environment. https://www.veolia.com/en/annual-and-sustainability-report-2016-veolia

Watts Water Technologies. http://www.wattswater.com, https://investors.wattswater.com/Sustainability/default.aspx

Web Finance Inc. (2018). *Business dictionary*. Retreived from http://www.businessdictionary.com/definition/environmental-value-added.html

Yakovets, Y. (2003). *Rent, anti-rent, quasi-rent in the globally-dimensioned dimension, IKTS "Akademkniga"* (pp. 76–107). Academic Book. [Яковец Ю. В., (2003), Рента, антирента, квазирента в глобально-цивилизационном измерении].

Young, L. (2002, September). *New Zealand treasury* (Working Paper 02/21) (pp. 1, 4–6).

Chapter 13
Corporate Reporting Practices Concerning Non-financial Aspects: A Possible Prolix?

Adriana Tiron-Tudor, Diana-Lavinia Martin, and Teodora Viorica Farcas

13.1 Introductory Elements About Integrated Reporting and Sustainability Reporting

Reporting has taken a different road from traditional financial reporting, to new reporting trends that add non-financial information on the list of users' interests, along with the purely financial data (Radley 2012 cited by Fărcaş 2015b). The financial reporting became insufficient for satisfying the necessity of information for the users of corporate reporting. Corporate reporting refers to all the reports voluntarily or compulsorily drawn up by corporations. Along with financial reporting (which is compulsory), nowadays the majority of corporations are publishing non-financial reports. These types of reports evolved together with the modern financial reporting and are consequences of the industrialization process and advancement of technology (Fărcaş 2015a).

The evolution of corporate reporting practices distinguishes three stages of corporate reporting: (a) the initiatives regarding corporate reporting; (b) sustainability reporting era and (c) the evolution of integrated reporting (Dragu and Tiron-Tudor 2013). In the non-financial reporting (Jastrzebska 2016) are included four main categories: social reporting, environment reporting, sustainability reporting and integrated reporting.

Nowadays, more and more investors and stakeholders are asking for information regarding environmental, social and governance (ESG) aspects that affects business (Hughen et al. 2014). The Global Reporting Initiative (GRI) framework, is considered to be the most frequent type of reporting used to disclose these types of information (English and Schooley 2014). Disclosing information related to environment and social responsibility can determine the involvement of stakeholders and

---

A. Tiron-Tudor (✉) · D.-L. Martin · T. V. Farcas
Babes-Bolyai University, Cluj Napoca, Romania
e-mail: diana.martin@econ.ubbcluj.ro; teodora.farcas@econ.ubbcluj.ro

S. O. Idowu, M. Del Baldo (eds.), *Integrated Reporting*, CSR, Sustainability, Ethics & Governance, https://doi.org/10.1007/978-3-030-01719-4_13

identify material risks in terms of sustainability but also can increase the confidence of stakeholders and also represents a proof of accountability from organizations. GRI is considered to be a major supporter of integrated reporting and the objective mentioned by GRI when G4 was launched was to offer guidance on 'how to link the sustainability reporting process to the preparation of an integrated report aligned with the guidance to be developed by the International Integrated Reporting Council (IIRC)' and to fasten and to facilitate the adaptation of <IR> (English and Schooley 2014). Eccles et al. (2010) defines integrated reporting as a process that encompasses in the annual reports of the companies' information related to environment, society and governance (ESG).

Villiers et al. (2014) mention in their research that <IR> and GRI have equivalent features but withal the two types of reporting have dissimilarities. On this basis, in order to find out the similarities between different types of non-financial reporting, a comparative analysis of principles and content elements was carried out by Idowu et al. (2016). The study highlighted that <IR> contains definitions and principles underlying the sustainability reporting standard (GRI) and the social responsibility framework (ISO 26000). Also, Hughen et al. (2014) mentioned that <IR> is going to be the new trend in reporting practices and has links with sustainability reporting through their strategies. Through integrated reporting principles, corporate reporting can be improved by offering to the investors a long-term perspective on value creation to complete the information provided by financial statements (Pwc 2014). All this information in a brief report that would show the organization's material, social, environmental actions, also the risks and opportunities in order to reflect the way this elements are integrated within the organization (Morros 2016). There are also studies that present the <IR> as the report that brings together sustainability reporting and financial reporting with the purpose of simplifying and diminish the numbers of reports a corporation must issue (Fărcaş 2015a). Stacchezzini et al. (2016), revealed that integrated report is not intended to be a sustainability report but through its principles focuses on issues of sustainability and develop the integrated thinking. Related to sustainability, James (2015) conducted a study on the benefits of publishing sustainability reports and the largest share in the answers was referring to increasing corporate reputation with the publication of sustainability reports.

Also in the case of CSR reporting, Sierra-Garcia et al. (2015) pointed out that there is a connection between <IR> and the assurance of CSR reports. The sample for the research was based on 7144 observations from the GRI database during the period 2009–2011 and showed that 16.2% of the companies draw up reports integrating financial and non-financial information. From the total of 7144 observations, 20.5% of the companies were assuring both integrated report and CSR report.

Our research intends to contribute to the literature concerning nonfinancial information in corporate reporting with some insights about a possible prolix in this area. We aim to analyze also if the integrated report could be sufficient regarding the sustainability issues or there is still the need for preparing sustainability reports.

We start our research with a review of the sustainability reporting and integrated reporting in banking and insurance sector. Further we make a review of the sustainability reporting (GRI type) and integrated reporting (issued by IIRC). The research continuous with a presentation of Generali Group and the main reports issued for the year 2016. The main part is the analysis of the Generali Group report, namely the *Annual Integrated Report and Consolidated Financial Statements 2016* and the *Sustainability Report 2016*. First, we compare the reports with the frameworks to observe the appliance of the frameworks. Then we analyze the extent to which the information related to sustainability is available in both <IR> and <SR> issued by Generali Group.

13.2 <SR> and <IR> in Banking and Insurance Sector

Kolk (2010) made a research on 213 multinational companies (MNCs) in the Fortune Global 250, from different sectors (among which the financial sector with the niche banks and insurance) and analyzed the extent to which those companies issued sustainability reporting in the period 1998–2005. The results revealed that for the analyzed period a limited number of MNCs from the insurance sector published sustainability reporting, while the companies from banking sector used to publish more sustainability reports but few in comparison with other sectors. Speaking in general about the financial sector, it was noticed by Kolk (2005) that European companies issue more sustainability reports than the one's from US and Japan.

Regarding sustainability, Weber et al. (2014) revealed that European financial companies have real concerns regarding environmental and labor issues because of the regulations, in comparison with companies from North America and Asia-Pacific.

Perrault Crawford and Clark Williams (2010) made a research on three financial companies in France (BNP Paribas, Credit Agricole and Société Générale). In France in 2008 there were 23 companies from different sectors that issued GRI reports and only two from financial services, one of them was BNP Paribas. At the same time all the three companies from financial services (BNP Paribas, Credit Agricole and Société Générale) reported to the Nouvelles Regulations Economiques (NRE) in a manner that is in accordance to GRI.

Lodhia (2015) focused the research on a bank from Australia and its transition to integrated reporting. The research was based on document analysis and on semi-structured interviews with the key decision makers and operation personnel and revealed that till the bank disclosed both annual reporting and sustainability reporting, was very difficult to depict the real bank performance and to distinguish themselves from other banks. Because integrated reporting is a complex process, to draw up an efficient integrated reporting it is important to take into account more than economic, social and environmental issues.

At European level there are a couple of financial institutions that deliver integrated reports. According to Sofian and Dumitru (2017), there are nine companies registered in the IIRC Database but only eight companies (ABN AMRO, Achmea, Aegon, AXA, FMO, Generali Group, ING Group, UBS) have available reports.

Aegon (Netherlands) issued the first report available in the examples database in 2011, followed by Achmea in 2012.

A research related to Generali Group internal integrated report (internal application of IR principles) was made by Fasan et al. (2016). The study consisted of an analysis of how the internal integrated report conducted by Generali was able to develop the management control system related to the connection with strategy and organizational culture, the usage of non-financial indicators and better understanding of cause effect relationships.

We selected our case study from the banking and insurance sector because of the interested parties in this area. The companies reporting from this area have a great responsibility towards the general public. The reports from this area should be comprehensive and should contain detailed sustainability information. We choose the particular case of Generali Group because the company was one of the companies from the <IR> pilot program and was considered in literature to be one of the firsts examples in issuing integrated reports (Fasan et al. 2016).

13.3 Generali Group

Generali is one of the major insurance company in the world with more than 70 billion euros in premium income and 55 million clients. The company was set up in 1831 in Trieste under the name Assicurazioni Generali Austro-Italiche. The company offers a wide range of insurances: life insurance (71%), property and casualty insurance, insurance for companies and global business lines and asset management (29%). Generali succeeded to be a multinational Group with presence in more than 60 countries: in mature regions (Italy, Germany and France), in Eastern Europe, in EMEA region (Austria, Belgium, Greece, Guernsey, Ireland, Netherlands, Portugal, Spain, Switzerland, Tunisia, Turkey and Dubai), in Asia and Latin America. The Group has more than 73,000 employees (50.6% men and 49.4% women) all over the world and 500 billion euros in assets under management. For the year 2016, their operating result was over 4.8 billion euros and the net profit was 2.1 billion euros, with taxes paid totaling 1 billion euros. Generali Group was nominated in 2015 among the world's 50 smartest companies (MIT Technology Review 2015) and one of the world's 50 largest companies (Fortune Global 500 2015).

In Italy, Generali Italia is the market leader with over 10 million clients and a total premium income, in 2014, of 23.5 billion euros which mark off 33% of total premium income and 48% of the Group's operating result. Generali Deutschland is the second insurance company in Germany with more than 13.5 million customers, representing 24% of premium income and 17% of the total operating result. Then comes the third important market that is Generali France, with 7 million customers. In France, Generali takes the seventh place in life insurance segment and the sixth place in property and casualty insurance. Another important market is represented by Central and Eastern Europe where Generali is the leader regarding the Group worldwide profitability, on the strength of property and casualty insurance.

During the years, while they maintained their leadership in Italy, they have continuously consolidated their position in the rest of the world. As a result, in 2015, 65% of their premium income was from outside Italy. Generali prospects are to continue their growth by expanding in regions as Asia and South Africa and by becoming the retailer leader in Europe.

13.4 Methodology

Generali Group presented for the year 2016 an important number of reports for their stakeholders and shareholders. They started a tradition from 2014 in preparing different types of reports. Our attention is focused on two of their reports, namely the *Annual Integrated Report and Consolidated Financial Statements 2016* and the *Sustainability Report 2016*. We analyze two main frameworks for reporting: Integrated Reporting Framework issued by International Integrated Reporting Council and for the Sustainability Reporting, GRI G4, issued by the Global Reporting Initiative (Global Reporting Initiative 2017) considered to be the most well-known and the most used type of sustainability reporting (Perrault Crawford and Clark Williams 2010; Kolk 2010).

In the following, the research is grouped around the subsequent issues:

(a) General remarks about the Generali reports for the year 2016
(b) Generali Group Sustainability Report 2016 and the parallel with the GRI Framework
(c) Generali Group Annual Integrated Report and Consolidated Financial Statements 2016 and the parallel with the IIRC Framework
(d) Comparative analysis of Sustainability Report 2016 and the Annual Integrated Report and Consolidated Financial Statements 2016 issued by Generali regarding the social, environmental, human and natural information contained.

The main technique used in the research is the content analysis that helps the researcher to make deductions from text with the context of the use of information (Krippendorff 2013). We will examine the content of the two reports by highlighting were exactly in the reports are described, in order to see if the same information related to sustainability is contained by the two reports. The content elements of the reports (*Sustainability Report 2016* and the *Annual Integrated Report and Consolidated Financial Statements 2016*) will be analyzed and compared first with the frameworks and then between them.

The results of the research are presented as it follows:

- In case of the parallel between the reports (*Sustainability Report 2016* and the *Annual Integrated Report and Consolidated Financial Statements 2016*) and the Frameworks (GRI, <IR>), the results consist of comparative tables that highlight if the element is present in the report, where can be found in Generali reports and the explanation (citation from the report).

- In case of the comparison between the reports (*Sustainability Report 2016* and the *Annual Integrated Report and Consolidated Financial Statements 2016*) were taken into account the social and human aspects and the environmental and natural aspects. The results draw the similarities and dissimilarities between the aspect mentioned before related to the reports.

13.5 General Remarks About Generali Reports for 2016

Over time, Generali issued many reports and starting from 2008 the reports are available on their website. The Group took part in the IIRC pilot program (2011) and their first integrated report issued was in 2012; in the literature, as we mentioned previously, Generali Group is considered an initiator in applying integrated reporting (Fasan et al. 2016).

For the last year, 2016, as shown in the table below, Generali Group (2017) issued a number of self-contained reports regarding financial aspects but also regarding non-financial aspect like sustainability, corporate governance and management reports at different periods of year at individual or group level Table 13.1.

At the end of 2016, Generali Group prepared a total of 1441 pages of reports: Annual Integrated Report 2016, which presents the financial and non-financial performance of the Group in an integrated manner and Annual Integrated Report and Consolidated Financial Statements 2016, details the Group financial performance in accordance with the national and international regulation presenting the overall complete picture of the Group.

Table 13.1 Generali reports—year 2016

Year 2016	Name of report	Number of pages
First Quarter	Interim management statements as of 31 March 2016 (press release)	20
First Semester	Interim management statements as of 30 June 2016 (press release)	20
	Consolidated financial half-year report 2016	160
Nine Months	Interim financial information as of 30 September 2016 (press release)	4
Full year	Consolidated Results as of 31 December 2016 (press release)	21
	Sustainability Report 2016	102
	Annual Integrated Report and Consolidated Financial Statements 2016	338
	Annual Integrated 2016	98
	Solvency and financial condition report	146
	Management Report and Parent Company Financial Statements 2016	314
	Corporate Governance Share Ownership Report 2016	162
	Remuneration Report 2016	56
Total		1441

Source: http://www.generali.com/info/download-center/results#2016

Sustainability Report 2016, presents the way in which Generali Group creates value for all its stakeholders while Corporate Governance and Share Ownership Report 2016, puts forward the corporate governance system of Generali. Another type of report prepared by the Group is the Management Report and Parent Company Financial Statements 2016, which discloses the company performance related to the regulations. Last but not least is the Remuneration Report 2016, offers information about the remuneration policy adopted by the Group.

Together with those reports, on Generali website are available additional documents for the first quarter of the year, the first semester and 9 months—Presentation of the results, Supplementary financial information and the Presentation for Investor Day and some of the reports are available only as press release.

13.6 Results

13.6.1 Generali Sustainability Report <SR> and the Framework

The sustainability report prepared by Generali Group and called *Sustainability Report 2016*, according to the information found in the report is considered by the Group a GRI type report and complies with the content elements and principles of GRI G4. In the GRI Database (2017), the Sustainability Report of the Group is considered by the Global Reporting Initiative a GRI G4 report type and the adherence level of the report is *in accordance—core.*

According to Global Reporting Initiative (2015), there are two possibilities in which an 'in accordance' sustainability report can be prepared:

the core option—contains the essential elements of a sustainability report and.
the comprehensive option—which involves beside the core option information related to organization's strategy, governance, ethics and integrity.

The <SR>, GRI type through its G4 version of the guideline is based ten principles and nine content elements.

The ten guiding principles defined by GRI G4 guideline are:

1. **Stakeholder Inclusiveness**—the organization should recognize its stakeholders and should present the way in which their expectations and interests are satisfied;
2. **Sustainability Context**—the presentation of the organizations future performance and how it will contribute for the improvement/deterioration of sustainability issue;
3. **Materiality**—in the report should be covered aspects related to economic, environmental and social impacts or the ones that influence the decision of stakeholders;

4. **Completeness**—the report should contain sufficient material aspect and their boundaries to reflect relevant economic, environmental and social aspects;
5. **Balance**—the report should contain both positive and negative aspects related to organizations performance;
6. **Comparability**—the information should be presented in a way that enables the evaluation of the performance over time of the organization;
7. **Accuracy**—the information presented in the report should be enough accurate and all the stakeholders should be able to measure the performance;
8. **Timeliness**—the report should be made constantly, on a schedule, to help stakeholder to make informed decisions;
9. **Clarity**—the information should be delivered in an accessible and understandable manner, reachable for all the stakeholders;
10. **Reliability**—the information used for the report should be used in a responsible way because it can be subject of examination to establish the quality and the materiality of information.

In *Generali Sustainability Report 2016*, the principles are not that explicitly presented like the content elements, but by analyzing the entire document we can say that implicitly the report is based on the principles presented above.

The content elements of the GRI G4 guideline are divided in two parts, the first seven content elements are related to general information and the last two are related to specific information:

1. **Strategy and analysis**—provides a strategic overview of the organizations sustainability;
2. **Organizational profile**—refers to the providing of information on: the status of the organization, the sector in which the organization operates, the customers it has, the number of employees, net sales, the total number of employees, the type of employees;
3. **Identified Material Aspects and Boundaries**—makes an overview of the process the organization has followed to determine the content of the reporting and include information on: the list of entities that are included in the organization's financial reports, entities that are included in the financial statements but are not included in the GRI report;
4. **Stakeholder Engagement**—the implications of the stakeholders during the reporting period;
5. **Report Profile**—is provided information regarding the reporting period (fiscal or calendar year), the date of the previous report, the cycle (annual, biannual);
6. **Governance**—the transparency of the governance structure and organization is pursued to ensure accountability; information about the governance describes how management is set to pursue the goal of the organization from an economic, environmental and social point of view.
7. **Ethics and integrity**—provides an overview of the organization's values, principles and standards; it also provides an overview of internal and external mechanisms for applying ethical and legal behavior;

8. **Disclosures on management approach**—the organization has an opportunity to explain how the economic, social and environmental impacts are related to material aspects and how they are managed;
9. **Indicators**—refers to the information provided by the results of these indicators on the economic, environmental and performance of the organization.

The Sustainability Report is a report that is drafted annually (the report from 2008 is the first available) by Generali Group and enables the Group to communicate how they are able to create value in a responsible way for all its stakeholders over the long term. After comparing the elements of the GRI G4 Framework with the Sustainability Report 2016, it can be concluded that the report contains and expands the content elements that the G4 Guideline refers to. In Table 13.2 is available a synthesis of the content elements that can be found in the Sustainability Report 2016:

13.7 Generali Integrated Report <IR> and the Framework

According to Generali Group their Annual Integrated Report and Consolidated Financial Statements 2016, complies with regulations and principles of <IR> issued by International Integrated Reporting Council (IIRC). In the following we are going to establish what actually the IIRC guideline requires for a report to be integrated and to which extent Generali integrated report follows the guideline, the principles and the content elements.

As stated by IIRC, an integrated report should have at the base seven guiding principles, that should be applied individually and collectively for the scope of the preparation of the integrated report. The seven guiding principles of <IR> are (IIRC 2015):

1. **Strategic focus and future orientation**—the integrated report should present the organization's strategy and the way in which can create value on short, long, medium and long term, with reference to the capitals (financial, manufactured, intellectual, human, social and relationship, natural);
2. **Connectivity of information**—the integrated report should present the interrelations and the dependences between the factors that influence the capacity of the organization to create value over time;
3. **Stakeholder relationships**—the integrated report should supply information regarding the organizations relationship with the stakeholders and how the organization takes into consideration their needs and interests;
4. **Materiality**—the integrated report should present information related to the relevant and essential matters than help the organization to create value on long, short and medium term;
5. **Conciseness**—the integrated report should be concise but at the same time should cover enough information to facilitate the understanding of organizations strategy, governance and performance;

Table 13.2 Elements of GRI guideline in Generali Group <SR>

Elements of GRI	Yes/No	Where?	Section	Explanation
Content elements—general information				
Strategy and analysis	✓	Pages 4–5	Letter from the chairman and the Group CEO	*"outlines our idea of corporate social responsibility and describes how it is a tangible part of the Group's everyday life"* *"In our new Charter of Sustainability Commitments, we establish a strong connection between the work we carry out and our commitment towards a healthy, resilient and sustainable society, where people can progress and flourish"* *"the ability to trigger ongoing dialogue with our stakeholders, in order to clearly understand, year after year, which issues are truly important for them, in the light of a complex society and of increasingly new and global risks"*
Organizational profile	✓	Pages 1, 6–7, 9, 65, 86	Name of the organization, location of the headquarters, number of countries were operating, scale of the organization, the workforce, the supply chain	*"Assicurazioni Generali S.p.A"* *"Company established in Trieste in 1831"* *"Registered office in Trieste, Piazza Duca degli Abruzzi, 2"* *"more than 60 countries"* *"gross written premiums € 70,513 mln"; "operating result € 4830 mln"; "net profit € 2.1 bln"; "our clients 55 mln"; "claims paid € 48,871 mln"* *"our people 73,727" from which "manager 1781, employees 53,498, sales force on payroll 18,238, other 210"* *"The process of identifying and selecting suppliers is based on principles of free competition and transparency, and is aimed at ensuring the high quality of our products and services, as set out in the Ethical Code for suppliers of the Generali Group"*
Identified Material Aspects and Boundaries	✓	Pages 9, 15, 17, 97	Entities included in the organizations consolidated financial statements, defying the report content, material aspect identified in the process for defying report content, boundary	*"The information refers to the Group's consolidation scope. Where this is unavailable, it refers to the insurance companies—with the exception of those of the Europ Assistance group—that operate in Austria, France, Germany, Italy, Czech Republic, Spain and Switzerland"* *"To guarantee the comparability of the most significant indicators over time and to enable readers to compare our results, the current values are compared against those of the previous period using graphics and tables."*

			within/outside the organization for each material aspect, significant changes from previous reporting periods	*"The Sustainability Report outlines the actions we take in response to our commitments and material issues."*
Stakeholder Engagement	✓	Page 16	The stakeholders engaged by the organization	*"employees", "clients", "agents and distributors", "contractual partners", "financial community", "community"*
Report Profile	✓	Pages 9, 90, 98	Reporting period, reporting cycle, external assurance, contact point regarding the report, GRI content index	*"drafted annually and approved by the Board of Directors"* *"the information and data mainly refer to the year that ended on 31 December 2016"* *"also complies with the G4 reporting guidelines of the Global Reporting Initiative (GRI), in accordance—core option"* *"has not undergone external assurance, the Group preferring to concentrate its available resources on improving its social and environmental performance and the efficiency of the reporting process, in terms of both quality and quantity"*
Governance	✓	Pages 18–19	Governance structure including committees responsible for decision-making on economic, environmental and social impacts	*"The Corporate Governance, Social and Environmental Sustainability Committee performs a consultative, recommendatory and preparatory role in favour of the Board regarding the decisions to be taken with respect to the structure of corporate governance rules and social and environmental sustainability issues"* *"Responsibility for sustainability at Group level is attributed to the Group CEO, who is charged with implementing the strategies and policies established by the Board of Directors."* *"EMS Review Committee Internal body that reviews the Group Environmental Management System, which respects the requisites of the ISO 14001 standard"* *"Country Sustainability Committees Bodies active at national level that make it possible to improve the organization and coordination of activities in the social and environmental sphere"*
Ethics and integrity	✓	Page 24	Rules for running business with integrity	*"The Ethical Code for suppliers highlights the general principles for the proper and profitable management of relations with contractual partners"* *"We condemn and combat all forms of corruption and financial crime."*

(continued)

Table 13.2 (continued)

Elements of GRI	Yes/ No	Where?	Section	Explanation
Content elements—specific information				
Disclosures on management approach	✓	Pages 78–79	Economic—economic performance	*"We take concrete actions to respond to climate change that has a serious global impact and significant effects on our business. In doing so we comply with our mission as an insurer by preventing risks and developing the best solutions to handle them."*
Indicators		Pages 79–80, 87	Environmental—materials, energy, emissions	*"Reduce our greenhouse gas emissions by 20% by 2020 (base year 2013)" "Incentivize our clients to adopt environmentally sustainable behavior", "Invest in a sustainable way", "Incentivize virtuous behavior in our supply chain" "Environmental Management System, which respects the requirements of the ISO 14001 standard", "Total emissions t 118,290 CO2e (−5.1% compared to 2013)" "Energy consumption, almost 60% of which consists of electricity, is the main cause of our greenhouse gas emissions... in 2016 we continued to develop new projects and launched further actions to improve the energy efficiency of our buildings, both at Group level and at the level of individual countries"*
		Pages 24–25, 33, 36–39, 45, 46–47, 50–51, 54, 56–57, 60, 61, 62–63, 65, 67, 70, 86, 87, 98	Social—labor practices and decent work *(employment, labor/management relations, training and education, diversity and equal opportunity, equal remuneration for women and men, local communities, anti-corruption)* – product responsibility *(product and service labeling, customer privacy, product portfolio, active ownership)*	*"In the event of company reorganizations, we safeguard our people by assigning, where necessary, new roles with targeted training or professional requalification measures, depending on each person's skills and experience" "The remuneration of our people is defined in line with the market and with what established by the National Collective Bargaining Agreements and Supplementary Company Agreements or local laws." "91.1% trained people; 37.3 average hours of training per capita; € 61.2 mln training costs" "We commit to valuing the uniqueness of people and diversity of thought, particularly in relation to generation, gender and geographical differences" "...we focus greatly on the issues of employment and integration..." "After successfully completing the largest integration plan in the insurance industry in Europe, Generali Italia has entered the second phase of development to become the market leader with the best customer experience."*

Source: Author's design based on Generali Group *Sustainability Report 2016*

6. **Reliability and completeness**—the integrated report should cover all material matters, positive or negative; the reliability of the information disclosed in the report can be increased by internal control, stakeholder engagement and internal audit; the completeness of the report is given by the extent of information disclosed and the peculiarity and the accuracy.
7. **Consistency and comparability**—the integrated report should contain information that is consistent over time and comparable with the information disclosed by organizations even if each organization creates value in its own way.

Similar to the case of the GRI Framework presented in the previous subchapter, we can say that the Generali *Annual Integrated Report and Consolidated Financial Statements 2016* complies with the seven guiding principles of the <IR> Framework and are presented implicitly in the report.

Beside the seven guiding principles presented before, an integrated report should pursue eight content elements posed as questions to which an integrated report should answer. The eight content elements are linked to each other and do not exclude one another. The enumeration of the content elements made by IIRC (2015) is not a standard structure, each content element can be disclosed in the order that the organization wishes and ensures the linking between the content elements. The eight components named content elements (IIRC 2015) and their linked questions are:

1. **Organizational overview and external environment**—the question to which an integrated report should answer is: ***What does the organization do and what are the circumstances under which it operates?***
2. **Governance**—an integrated report should answer the following question: ***How does the organization's governance structure support its ability to create value in the short, medium and long term?***
3. **Business model**—is considered the ability of transforming inputs through its activities into outputs that conducts to creating value over long, short and medium time; the question to which an integrated report should answer is: ***What is the organization's business model?***
4. **Risks and opportunities**—the integrated report establish and indicates the risks and opportunities that are specific for the organization; for this content element the specific question is: ***What are the specific risks and opportunities that affect the organization's ability to create value over the short, medium and long term, and how is the organization dealing with them?***
5. **Strategy and resource allocation**—with this content element the organization presents the strategy and how will be carried into effect in order to achieve its objectives; the question in the case of this content elements is: ***Where does the organization want to go and how does it intend to get there?***
6. **Performance**—in an integrated report there are qualitative and quantitative information related to the performance of the organization, combining financial measures with other components; the integrated report should answer the following answer: ***To what extent has the organization achieved its strategic objectives for the period and what are its outcomes in terms of effects on the capitals?***

7. **Outlook**—provides information related to future challenges that the organization can confront and how responds to uncertainties; the question provided by <IR> guideline is: ***What challenges and uncertainties is the organization likely to encounter in pursuing its strategy, and what are the potential implications for its business model and future performance?***
8. **Basis of preparation and presentation**—presents its basis of preparation and presentation including elements like: reporting boundary, organizations materiality determination, methods used to quantify and determine material matters; the question for this content element is: ***How does the organization determine what matters to include in the integrated report and how are such matters quantified or evaluated?***

Beside the principles and the content elements that are present in other frameworks as well, the capitals are a distinctive matter for integrated reporting (Fasan et al. 2016). In the IIRC are presented six capitals that are involved in the value creation process: *financial, human, intellectual, social and relationship, manufactured, natural.* According to the IIRC framework for <IR>, not all the capitals have the same relevance or the same applicability for the organizations. There are organizations that could consider that some capitals are not that important for their activity and won't be included in the <IR>. At the same time, according to IIRC an integrated report should disclose the interdependencies between the capitals and the components of capitals.

The *Annual Integrated Report and Consolidated Financial Statements 2016* drawn up by Generali Group, follows the principles and the content elements of the <IR> and the similarities can be observed in Table 13.3.

13.8 A Comparison of Social, Human, Environmental and Natural Aspects Disclosure in Integrated Report and Sustainability Report

The social, human, environmental and natural matters are important sustainability issues disclosed in both sustainability reports and integrated reports. In sustainability reporting (GRI type) the elements listed above are presented as indicators and divided into categories, while in <IR> (IIRC type) those elements are considered capitals which are defined in the <IR> guideline as items that all the organizations depend for their success.

A presentation and at the same time a comparison of the social, human, environmental and natural matters is going to outline the extent to which is disclosed this type of information in both types of reporting, in the guidelines and in the case of Generali Group. In order to make the comparison we are going to split the aspect in two groups: *social and human aspects* and *environmental and natural aspects*, because of the linkage between the aspects and the very thin line between them.

Table 13.3 Elements of <IR> guideline in Generali Group <IR>

Elements of <IR>	Yes/ No	Section	Where?	Explanation
Content elements				
Organizational overview and external environment	✓	Our history	Pages 2, 12, 14	*"Company established in Trieste in 1831"* *"Registered office in Trieste, piazza Duca degli Abruzzi, 2"* *"We are one of the largest global players in the insurance industry, a strategic and highly relevant sector for the growth, development and welfare of modern companies. In almost 200 years we have built a Group, that operates in over 60 countries through more than 420 companies and almost 74 thousand employees"* *"Gross written premium € 70,513 mln", "Operating result + 0.9%, € 4830 mln", "Net profit € 2.1 bln", "Dividend per share proposed + 11.1%, € 0.80", "Our clients 55 mln", "Our exclusive distributors 151 thousand"*
Governance	✓	Our Governance and remuneration policy	Pages 32–35	*"Approximately 233,000 shareholders in 2016"* *"38.4% female director, 57.5 average age"* *"The remuneration policy for non-executive directors establishes that remuneration consists of a fixed component as well as an attendance fee for each Board of Directors meeting in which they participate, in addition to the reimbursement of expenses incurred for participation in such meetings."*
Business model	✓	How we create value: our business model	Pages 20– 21	*"We develop simple, integrated, customized and competitive life and property and casualty insurance solutions for our clients. . ."* *"We distribute our products and we offer our services based on a multi-channel strategy, while also relying on new technologies: not only through a global network of agents and financial advisors, but also through brokers, banc assurance and direct channels which allow our clients to obtain information on alternative products. . ."*
Risks and opportunities	✓	Risks and opportunities of the external context	Page 36	*"New customer needs—In this currently uncertain economic environment, consumer attitudes to insurance products and services are changing in light of two global trends: digitalization. . .and economic uncertainty..."* *"Technological evolution—The unprecedented availability of customer data,*

(continued)

Table 13.3 (continued)

Elements of <IR>	Yes/ No	Section	Where?	Explanation
				combined with the technological capabilities of processing data quickly and efficient in terms of costs, allows the insurance business to create customized prices and identify potential fraud." *"Uncertain financial and macro-economic landscape—In 2016, numerous-political events have impacted the global economy. . .at macroeconomic level, Italy remains in difficulty, with expected growth of 0.8%..."* *"The global warming caused by greenhouse emission originating from human activity is triggering a rise in extreme weather events. . .Proper measures are therefore necessary to avoid higher losses and increased volatility that would impact on insurance policies' price. . ."*
Strategy and resource allocation	✓	Our strategy	Pages 24–25	*"The external context is increasingly challenging: on one hand, economic, financial and political uncertainty which results in greater volatility in interest rates, government and corporate bond spreads on the equity market and, on the other hand, different customer behavior, driven primarily by rapid technological evolution and more stringent regulation."*
Performance	✓	Our strategy	Pages 26	*"We aim to maintain a diversified global presence by focusing on markets in which we can achieve a significant position and excellent performance. We will therefore dispose of the companies that do not meet specific profitability and future growth requirements in order to invest in those activities that generate higher returns."—"we have already embarked upon a streamlining process. . .we expect to generate at least € 1 billion from the optimization of our international presence by the end of 2017. . ."* *"We are committed to constantly improving our operating machine to maximize the Group's potential, by implementing optimization initiatives such as streamlining the product portfolio, simplifying processes and integrating IT platforms. At the same time, we will invest in new competences enhance those activities that have higher value creation"—"in Germany we digitalized the entire claims management process, eliminating our use of paper. . .our clients are also able to use an app o view entire claims."*

Outlook	✓	Outlook	Page 116	*"Despite the challenging environmental and the high volatility of the financial markets, in 2017 the Group expects to increase shareholder remuneration, consistently with the strategic plan presented to the market"* *"In the life segment the Group will continue to deal with a range of dynamic restrictions deriving from market environment."* *In property and casualty, total premium income is expected to grow in the Generali Group's main geographical areas. . .this growth will take place within an environment of strong competitive pressure"*
Basis of preparation and presentation	✓	Basis of presentation	Page 144	*"The Generali Group's consolidated financial statements at 31 December 2016 were drawn up in accordance with the IAS/IFRS issued by IASB and endorsed by the European Union"*
Capitals				
Financial	✓	How we create value: our business model	Page 20–21	Generali Group presents in its own interpretation the creating value process with the involvement of the capitals
Manufactured	✓			
Intellectual	✓			
Human	✓			
Social and relationship	✓			
Natural	✓			

Source: Author's design

Table 13.4 References in Generali Group <IR> made to Generali Group <SR>

No.	<IR> Generali	Issue	<SR> Generali
1	Page 7	Materiality matrix process and representation	Page 24, 17
2	Page 17	History	Page 34
3	Page 23	Prevention of corruption	Page 24
4	Page 27	Long term value creation	Page 31
5	Page 28	Customer and distribution innovation	Page 54, 63
6	Page 31	Improving diversity and inclusion	Page 43
7	Page 37	Risk management	Page 57, 63
8	Page 39	Risk management	Page 78
9	Page 49	Group responsible investments	Page 36
10	Page 59	Market positioning and performance	Page 59
11	Page 61	Market positioning and performance	Page 30
12	Page 121	Principles—Strategic focus and future orientation; Materiality	Page 17
13	Page 122	Stakeholders and methods of dialogue	Page 16

Source: Author's design

During the content analysis of Generali Group *Annual Integrated Report and Consolidated Financial Statements 2016* we have observed that frequently for more information, there were made 13 references to the *Sustainability Report 2016*. In order to have a picture of the references made in the content of the *Annual Integrated Report and Consolidated Financial Statements 2016* we synthetized the information below in Table 13.4.

13.8.1 Social and Human Aspects

In GRI (2015) sustainability reporting framework, the social and human dimension consists in the impact that the organization has on the social system in which it operates and it has four sub-categories: labor practices and decent work, human rights, society and product responsibility and each of the four sub-categories is divided into aspects that should be taken into consideration for a comprehensive report.

In <IR> framework issued by IIRC (2015) the disclosure of social aspects is present in the capital called social and relationship and deals with the organizations relationship into and between the communities, main stakeholders and other networks. This capital highlights the ability to share information and to increase the individual and the collective welfare and comprises of: stakeholders' relationship, social license to operate, shared norms, brand and reputation (intangibles). The human capital mentioned in the <IR> framework highlights people's competences and experiences and the reasons why they are inspired and determined to innovate by aligning with organizations governance and ethical values, by understanding, developing and implementing the organization's strategy.

Generali Group Sustainability Report 2016, complies with GRI G4 reporting guideline and is drafted with in accordance-core option which according to GRI G4 (2015) guideline means that it provides the background against which an organization communicates the impacts of its economic, environmental and social and governance performance. In Generali Sustainability Report 2016 in the category social are presented information related to employment, labor/management relations, training and education, diversity and equal opportunities, equal remuneration for women and men, anti-corruption.

The *Annual Integrated Report and Consolidated Financial Statements 2016* prepared by Generali Group is a complex report that puts together both financial and non-financial information. Even the Generali Integrated Report has triple number of pages in comparison with the sustainability report, it doesn't disclose the same information regarding the social issues presented in the Sustainability Report 2016. Frequently, along their integrated report, there are references made to sustainability report were additional information can be found.

In the *Annual Integrated Report and Consolidated Financial Statements 2016* at page 23, Generali Group presents their rules of running business with integrity. At the bottom of the page 23 is made a reference to *Sustainability Report 2016,* regarding other information to the prevention of corruption topic. The *Sustainability Report 2016* present the topic of anti-corruption and integrity at page 24 in a more detailed manner, mentioning the Code of Conduct of the Group, the initiatives and the employees' engagement regarding integrity and anti-corruption.

Another reference made in the *Annual Integrated Report and Consolidated Financial Statements 2016* (page 28) to the *Sustainability Report 2016* is related to customer and distribution innovation. Comparing the two reports regarding customer and distribution innovation, the *Sustainability Report 2016* (page 54 and 63) contains more detailed information related to this topic. While in the *Annual Integrated Report and Consolidated Financial Statements 2016* the efforts of the Group about client satisfaction are presented in brief, in the *Sustainability Report 2016* this issue is presented in a detailed manner with examples from different countries were the Group carries out its activity and the initiatives made to create value by listening to the customer needs.

In the *Annual Integrated Report and Consolidated Financial Statements 2016* is made another reference to *Sustainability Report 2016* as respect to the issue of improving diversity and inclusion. The *Annual Integrated Report and Consolidated Financial Statements 2016* (page 31) presents in briefly the efforts and the results of the Group make to achieve diversity and inclusion (e.g. trained people, money spend for training). The *Sustainability Report 2016* (page 43) presents the issue of improving diversity and inclusion by adding information about future objectives like designing of a management school for new group managers, the opportunity of the employees to dialogue with their managers about their individual performance.

13.8.2 Environmental and Natural Aspects

According to GRI (2015) sustainability reporting framework, the environmental aspect of <SR> is given by the impact that an organization has on the natural system: land, air, water and ecosystem. On one hand <SR> follows the inputs (e.g. energy and water) and on the other hand also the outputs (emissions, effluents and waste) making a linkage between the resources and the outcomes.

The environmental category (GRI type) is divided into a number of aspects that should be detailed in the <SR> and they refer to: material aspect (materials used to produce and package the organizations products), energy (energy consumption in and outside the organization), water (water sources affected by withdrawal of water and the volume recycled and reused), biodiversity (protected areas), emissions, effluents and waste (the quality of the water discharged, total weight of waste by type), products and services, compliance (total number of fines and non-monetary sanctions for non-compliance with the rules), transport (the impact that the transporting of good and employees), overall (organizations costs with the environmental protection), supplier environmental assessment (new suppliers screened by the environmental criteria), environmental grievance mechanisms (number of grievances about environmental impacts).

Even if Generali Group provides insurance services and it is not a production organization, going through the Sustainability Report 2016—in accordance, core option—shows that the environmental matters are considered a real concern for the Group. In their Sustainability Report 2016 are presented aspects like climate change, initiatives made for preserving the artistic and natural heritage, the commitment for reducing their greenhouse gas emissions (GHC) with 20% by 2020, the investment in research and innovation, the reduction of the emission generated by the corporate fleet, energy, paper and water consumption.

From the IIRC (2015) point of view, the environmental matters are linked to the natural capital and consist of all renewable and nonrenewable resources of the environment in order to provide goods and services supporting the past, present and the future welfare of an organization. The natural capital includes preoccupations for: air, water, land, minerals and forest, biodiversity and eco-system health.

In the Annual Integrated Report 2016, prepared by Generali Group are presented some risks related to environment. The global warming caused by greenhouse gas emissions is a major concern for the Group because it is the cause of extreme weather conditions and have an impact on the insurance needs and can make the insurance too expensive. The Group supports researches and studies on environmental risks, in order to be closer to the consumer and to provide proper insurance services for the consumer needs. In order to deal with the environmental sustainability issues, Generali Group has a Governance and Sustainability Committee with a consultative, recommendatory and preparatory role in favor of the Board for taking the decisions with respect to the structure of corporate governance rules and social and environmental sustainability matters.

Along the report there are made references to the Sustainability Report 2016 and other reports were additional information can be found. It is mentioned that Generali has its own policy for environment and climate, named The Group Policy for Environment and Climate, and comprises of guiding principles of reference for the environmental management strategies and objective. In 2010, in France was launched La Carte du Sport Responsible, which supports and promotes six core principles for sporting with respect for others and for the environment.

13.9 Conclusion

Integrated reporting is the new challenge to which the academic and business world needs to answer. Integrated report should be a better communication channel with the stakeholders, but should not double the other forms of reporting. The IIRC framework issued in December 2013 was changed. The regulatory body has made changes to the initial draft from 2011 so that now the framework allows the existence in parallel of other forms of report. This is for sure not something that the corporations need, and we also consider that the value of the framework is diminished by this position that the regulatory body has taken.

The comparison made by the study revealed the fact that the sustainability report is more detailed and complex related to matters about social, human, environmental and natural aspects. It seems that if we are looking for sustainability information is recommended to read the <SR> not only the <IR>. In the case of Generali Group, even if the *Annual Integrated Report and Consolidated Financial Statements 2016* is longer than the *Sustainability Report 2016,* the information related to sustainability is not equivalent. This idea was debated by Stacchezzini et al. (2016) and mentioned that the integrated report is not a sustainability report, but focuses on sustainability and integrated thinking. Nevertheless, each type of report respects the recommendations made by the guidelines after which they were prepared.

Our research is a basis for practice because it punctuates the overlays that occur between frameworks, particularly between the reports, and comes to helps practitioners. With the theoretic aspects at the basis it is easier for practitioners to know exactly were different aspect related to sustainability and integrated reporting are presented.

A possible limitation of our current research is that we have taken into consideration only one sector (financial—banking and insurance) and one company with the available reports from 2016. In the future, the present research can be improved by referring to other companies from the same sector in order to have comparability or by extending our research with reports issued by Generali Group for other years.

Our study represents a little piece in the puzzle of knowledge related to corporate reporting. The intention of <IR> is to apply "principles and concepts that are focused on bringing greater cohesion and efficiency to the reporting process, and adopting 'integrated thinking' as a way of breaking down internal silos and reducing duplication." (<IR> Framework consulted at https://integratedreporting.org/

resource/international-ir-framework/). By using our study we demonstrate that <IR> reports and sustainability reports still go ahead together because some sustainability information is missing from IR. A further step in the research area would be to determine if that information is vital or needed for the users of corporate reporting.

References

Dragu, I., & Tiron-Tudor, A. (2013). New corporate reporting trends. Analysis on the evolution of integrated reporting. *Annals of the University of Oradea, Economic Science Series, 22*(1), 1221–1228.

Eccles, R., Krzus, M. P., & Tapscott, D. (2010). *One report: Integrated reporting for a sustainable strategy* (1st ed.). Hoboken, NJ: Wiley.

English, D. M., & Schooley, D. K. (2014). The evolution of sustainability reporting, accounting and auditing financial reporting. *The CPA Journal*, 26–35.

Fărcaş, T. (2015a). The development of corporate reporting over time: From a traditional system to an integrated system. *Revista Audit Financiar, 124*(4), 106–113.

Fărcaş, T. (2015b). Users' needs: A premise for corporate reporting change, RePec, Doaj, EBSCO and CABELLS publishing services. *The annals of the University of Oradea: Economic sciences* (pp. 939–943). http://anale.steconomiceuoradea.ro/volume/2015/AUOES-1-2015.pdf

Fasan, M., Mio, C., & Pauluzzo, R. (2016). Internal application of IR principles: Generali's internal integrated reporting. *Journal of Cleaner Production, 139*, 204–218.

Fortune Global 500. (2015). http://fortune.com/global500/2015/assicurazioni-generali-48/

Generali Group. (2017). http://www.generali.com/info/download-center/results#2016

Global Reporting Initiative. (2015). https://www.globalreporting.org/information/g4/Pages/default.aspx

Global Reporting Initiative. (2017, February 15). https://www.globalreporting.org/information/current-priorities/integrated-reporting/Pages/default.aspx

GRI Database. (2017). http://database.globalreporting.org/reports/46290/

Hughen, L., Lulseged, A., & Upton, D. R. (2014). Improving stakeholder value through sustainability and integrated reporting. *CPA Journal, 84*(3), 57–61.

Idowu, S. O., Dragu, I. M., Tiron-Tudor, A., & Fărcaş, T. V. (2016). From CSR and sustainability to integrated reporting. *International Journal of Entrepreneurship and Innovation, 4*(2), 134–151.

IIRC. (2015). http://integratedreporting.org/wp-content/uploads/2015/03/13-12-08-THE-INTERNATIONAL-IR-FRAMEWORK-2-1.pdf

James, M. L. (2015). Accounting majors' perceptions of the advantages and disadvantages of sustainability and integrated reporting. *Journal of Legal, Ethical & Regulatory Issues, 18*(2), 107–123.

Jastrzebska, E. (2016). Reporting of non-financial information as a stakeholder engagement method. *Research Paper of Wroclaw University of Economics, Nr. 423* (pp. 61–78). ISSN 1899-3192, e-ISSN 2392-0041.

Kolk, A. (2005). Sustainability reporting. *VBA Journal, 21*(3), 34–42.

Kolk, A. (2010). Trajectories of sustainability reporting by MNCs. *Journal of World Business, 45*, 367–374.

Krippendorff, K. (2013). *Content analysis: An introduction to its methodology* (3rd ed.). California, CA: Sage ISBN 978-1-4129-8315-0.

Lodhia, S. (2015). Exploring the transition to integrated reporting through a practice lens: An Australian customer owned bank perspective. *Journal of Business Ethics, 129*, 585–598.

MIT Technology Review. (2015). https://www.technologyreview.com/lists/companies/2015/

Morros, J. (2016). The integrated reporting: A presentation of the current state of art and aspects of integrated reporting that need further development. *Intangible Capital, 12*(1), 336–356.

Perrault Crawford, E., & Clark Williams, C. (2010). Should corporate social reporting be voluntary or mandatory? Evidence from the banking sector in France and the United States. *Corporate Governance: The international journal of business in society, 10*(4), 512–526.

Pwc. (2014). *B20 Panel of six international accounting networks*. https://www.pwc.com/gx/en/capital-projects-infrastructure/assets/pwc-b20-investment-infrastructure.pdf

Sierra-Garcia, L., Zorio-Grima, A., & Garcia-Benau, M. A. (2015). Stakeholder engagement, corporate social responsibility and integrated reporting: An exploratory study. *Corporate Social Responsibility and Environmental Management, 22*, 286–304.

Sofian, I., & Dumitru, M. (2017). The compliance of the integrated reports issued by European financial companies with the international integrated reporting framework. *Sustainability, 9*, 1319.

Stacchezzini, R., Melloni, G., & Lai, A. (2016). Sustainability management and reporting: The role of integrated reporting for communicating corporate sustainability. *Journal of Cleaner Production, 136*, 102–110.

Villiers, C., Rinaldi, L., & Unerman, J. (2014). Integrated reporting: Insights, gaps and agenda for future research. *Accounting, Auditing & Accountability Journal, 27*(7), 1042–1067.

Weber, O., Diaz, M., & Schwegler, R. (2014). Corporate social responsibility of financial sector – Strengths, weaknesses and the impact on sustainable development. *Sustainable Development, 22*(5), 321–335.

Part III
Implementing Integrated Reporting in Different Countries and Organisations: Issues, Benefits and Challenges

Chapter 14
Integrated Reporting and Materiality Process Disclosure in European Sustainability Oriented Companies

Tiziana De Cristofaro and Carmela Gulluscio

14.1 Introduction

Given the ever-evolving stakeholder expectations about information provided by companies, these latter are continuously rethinking their reporting. Integrated Reporting (hereinafter IR) stands out among recent reporting foci because its emphasis on value creation is generating a wealth of new non-financial information and reports (Eccles and Krzus 2015: 61–62). In this context, materiality becomes a critical principle, especially for integrated reports (Busco et al. 2013: 15).

However, no unique definition of the materiality reporting principle exists. This is because several definitions of materiality were coined to reflect its implications on both financial and non-financial reporting (for a selection, see Chong 2015 and Adams 2013).

According to the integrated reporting framework of the International Integrated Reporting Council (IIRC), the concept of materiality is understood as follows: "a matter is material if it could substantively affect the organization's ability to create value over the short, medium or long term" (IIRC 2013b: 18). Companies should therefore carry out a four-phase 'materiality determination process' (hereinafter MDP) which shows both commonalities (e.g. focus on potential influence of material issues on the judgement of report users) and unique features (e.g. value creation criterion regarding relevant issue selection) when compared with other frameworks (see IIRC 2013a).

T. De Cristofaro
G. d'Annunzio University of Chieti-Pescara, Chieti, Italy
e-mail: tiziana.decristofaro@unich.it

C. Gulluscio (✉)
Unitelma Sapienza University of Rome, Rome, Italy
e-mail: carmela.gulluscio@unitelma.it

S. O. Idowu, M. Del Baldo (eds.), *Integrated Reporting*, CSR, Sustainability, Ethics & Governance, https://doi.org/10.1007/978-3-030-01719-4_14

Despite this increasingly crucial role, materiality within non-financial reporting contexts and for IR has only recently risen above the research horizon. This study aims to contribute to this research branch by proposing a statistical analysis to ascertain any association between companies' attitude to Integrated Reporting implementation and their attitude to providing information about the MDP adopted. More in detail, some predictors of the above attitudes are used to test our association hypothesis on a sample of European sustainability-oriented RobecoSAM award winners. Contrary to expectations, only weak connections between IR orientation (IRO) and MDP disclosure (MDPD) are identified.

The chapter starts by describing a background (Sect. 14.2) that frames materiality in a twofold perspective (standard setting for non-financial information and relevant research) and at the same time focuses on IR. Then the main research hypothesis (Sect. 14.3) and the methods applied (Sect. 14.4) are presented. Afterwards, results are outlined and discussed (Sect. 14.5). Concluding remarks, limitations, and suggestions for further research close the chapter (Sect. 14.6).

14.2 Background: Materiality Standard Setting and Research. A Focus on Integrated Reporting

Interest in the relevance of information developed first within solely local financial reporting contexts (in the 1970s, e.g. FASB 1975) and it then developed to cover non-financial reporting contexts as well (in the 2000s). This pattern derives from the late development of both international accounting convergence (e.g. the adoption of international accounting standards IAS/IFRS by countries the world over) and non-financial reporting (Mio 2013: 80). This evolution of reporting contexts ('local/financial→international/financial→local and international/non-financial') has determined materiality-related consequences at a twofold level: (a) standard setting on non-financial reporting (Sect. 14.2.1); (b) research on non-financial reporting (Sect. 14.2.2).

14.2.1 Materiality and Standard Setting for Non-financial Reporting

Since the early 2000s, both standard setters and organizations have provided guidelines for non-financial reporting. The concept of materiality often recurs and has been gradually gaining importance, as evidenced by Table 14.1.

Table 14.1 Some guidelines dealing with materiality for non-financial reporting

Year	Reference in this chapter	Title including (T) or Dealing with (D) ... Materiality	IR
2006	AccountAbility (2006)	T	D
	GRI (2006)	D	D
2008	AccountAbility (2008a)	D	D
	AccountAbility (2008b)	D	D
	UNCTD (2008)	D	D
2011	GRI (2011)	D	D
2013	IIRC (2013a)	T	T
	IIRC (2013b)	D	D
	GRI (2013a)	D	D
	GRI (2013b)	D	D
	IIRC (2013c)	D	T
2015	IFAC (2015)	T	T
2016	GRI (2016b)	D	D
2017	EC (2017)	D	D

Key: *EC* European Commission, *GRI* Global Reporting Initiative, *IFAC* International Federation of Accountants, *IIRC* International Integrated Reporting Council, *UNCTD* United Nations Conference on Trade and Development

Both the 2016 GRI Standards (superseding the 2013 G4 Guidelines, to be phased out starting 1 July 2018)[1] and legislative institutions (e.g. EC)[2] joining an otherwise voluntary-based list of standard setters are noteworthy (since Table 14.1 is not exhaustive, see Eccles et al. 2012b: 66; Eccles and Krzus 2015: 125). Multiple sources of guidance originated definitions of materiality that vary according to the main scope and features of the standards/frameworks issued.

Comparisons among frameworks return important insights: (a) while the AccountAbility (2008a) framework is 'process-based', the GRI (2013a) and IR frameworks are 'principle-based' (Mio 2013: 83–84; Busco et al. 2013: 12); (b) although the GRI and IR frameworks are to some extent aligned (principles, content elements and reporting boundaries), "there is no straight forward solution to directly reconcile the two materiality definitions" and "each organisation will need to develop their own approach to aligning" them (KPMG 2014, Sect. 14.6).

The evolution of individual frameworks is also a source of insight. For example, document-structure analysis reveals an increased operative orientation of the IIRC-based guidance. In fact, while the first document issued focuses on definition, importance and determination processes (IIRC 2013a), intermediate documents (IIRC 2013b, c) focus only on definition and process, and the later IFAC guidance

[1]These new Standards include materiality among the 'Reporting Principles for defining report content', distinguishing them from those 'for defining report quality'.

[2]EC (2017: 9–10) provides specific additional guidelines on materiality for non-financial statements that, according to Directive 2014/95/EU (article 19a), certain European entities have to include in the management report.

(2015) emphasises "exploring basic concepts" and "determining materiality", also offering "concepts in practice" boxes.

Both this evolution and the journey towards IR entailed a renewed role of materiality. On one hand, before IR there was no real debate on materiality, as the GRI had set a 'fixed' list of disclosures, thereby reducing the materiality problem (Mio and Fasan 2014: 7). On the contrary, IR exalted conciseness and materiality regained prominence (ibidem). On the other hand, in the IR context materiality is important also because IR encourages enhancing and increasing value creation information (Black Sun & the International Integrated Reporting Council 2014: 11). As new or enhanced value creation information needs to fit with previous financial and non-financial data, suitable materiality 'filters' need to be devised.

14.2.2 Research on Materiality in Non-financial Reporting Contexts

Many have researched materiality in non-financial reporting contexts. *Standard setters* have contributed to the debate in particular with research collaborations (see Table 14.2).

Table 14.2 shows a recent orientation towards materiality applied to specific sectors. Indeed, the process of determining materiality is also based on the industry factor (IFAC 2015: 4), and this operative problem needs to be addressed. Although some authors have supported sector-specific materiality standards (see Eccles et al. 2012b), only standard setter SASB (2013) looks at sector as its primary unit of analysis for materiality.

When *audit organizations* study IR, materiality is the core concept (e.g. Ernst and Young 2013; Hespenheide and Koehler 2013) and a common driver, as it "should be the overarching filter when defining report content" (KPMG 2014, Sect. 14.6).

Finally, while academic research on materiality initially focused on financial and auditing reporting (e.g. Messier et al. 2005; Edgley 2014 and their references), current studies on materiality often compare definitions from the two 'paradigms' using the evolutionary perspective from financial towards non-financial reporting

Table 14.2 Standard setters' studies on materiality for non-financial information: a selection

Year	Standard setters and organizations involved	Reference in this chapter
2003	AccountAbility	Zadek and Merme (2003)
2006	AccountAbility, BT Group and *Lloyd's Register Quality Assurance*	Forstater et al. (2006)
		Zadek et al. (2006)
2013	AccountAbility	Murninghan (2013)
2015	Global Reporting Initiative and RobecoSAM	RobecoSAM and GRI (2015)[a]
2016		GRI (2016a)[a]

[a]Materiality applied to specific sectors

(Mio and Fasan 2013). In both cases, the first guidelines were modelled on those for financial information (Eccles et al. 2012b: 66); therefore, these definitions share their underlying criterion of information usefulness for decision making (Mio 2013: 81) to the point that "international and US guidance on materiality for financial and nonfinancial information is substantively the same [. . .]" (Eccles et al. 2012a: 6). As for differences, emphasis is usually placed on thresholds for non-financial reporting (Mio 2013: 81); moreover, IFAC acknowledges that "the subject matter is more precisely prescribed in international and national reporting standards" than in IR standards (IFAC 2015: 10).

Current academic literature on materiality in non-financial reporting mainly focuses on sustainability reporting (Lydenberg 2012; Hsu et al. 2013; Unerman and Zappettini 2014; Calabrese et al. 2016), while often highlighting the link with assurance (Eccles et al. 2012a; Moroney and Trotman 2012; Mio 2013). A less common focus is stakeholder engagement (Zhou and Lamberton 2011).

Despite the increasing relevance of materiality in IR, authors have investigated links between IR adoption and other factors:

(a) general factors, such as sector concentration (Frías-Aceituno et al. 2014), country-level determinants (Jensen and Berg 2012), country 'performance' (Turcu 2015) and national systems (Frías-Aceituno et al. 2013a; García-Sánchez et al. 2013);
(b) reporting-related company features, such as Bloomberg scores (Lai et al. 2016) and CSR reports assurance (Sierra-García et al. 2015);
(c) not reporting-related company features, such as governance (Frías-Aceituno et al. 2013b), size (Frías-Aceituno et al. 2014; Sierra-García et al. 2015) and profitability (Frías-Aceituno et al. 2014).

Academic studies on IR have only recently begun to consider materiality. Narrative sections of these studies focus on rising challenges and questions (Eccles and Krzus 2015: 120; Gelmini et al. 2015: 145), both theoretical (e.g. relationships among principles; materiality as a firm-specific social construct; cultural context influence on materiality meaning) and operational (e.g. MDP; stakeholder engagement; materiality matrix).

Empirical sections of these academic studies focus on different operational aspects of materiality. For example, Mio and Fasan (2013) provide evidence of 5 companies tackling the issue of materiality out of the 83 participating to the IIRC pilot programme. Their research has found that companies with a corporate governance system more focused on sustainability "tackle the issue of materiality more effectively" (p. 14).

Mio and Fasan (2014) test what drives the way in which IIRC pilot program companies disclose materiality in their integrated reports. They found that "materiality disclosure is not company-specific but it is rather industry-induced".

Eccles and Krzus (2015) studied 2012 integrated reports from 124 companies. Results indicated that, although most companies discussed materiality as a risk factor, disclosure on their definition of materiality or on "if and how it was making progress in managing these risks" was very poor (ivi: 202).

Gelmini et al. (2015) analysed 2014 GRI-G4-compliant integrated reports from 19 European companies; results show a range of methods used for materiality identification, vague narrative about the timing of issues related with materiality assessment, and a strong link between stakeholder dialogue and materiality.

Finally, Wee et al. (2016) explored materiality (and conciseness) in 252 integrated reports from 28 countries. In order to discover factors affecting preparers' and auditors' judgements on materiality in IR, they first investigated companies' determination of material issues and the related disclosure, finding that only 50% of integrated reports describe the MDP. The main focus is on MDP 'border' phases (stakeholders and material items selected) while central phases (evaluation and prioritisation) are neglected.

As seen, a special focus on empirical research on links between MDP and IR is still lacking.

14.3 Research Hypothesis Development

Section 14.2 highlighted the link between IR orientation and MDP disclosure as an under-researched area, especially if considered with a specific geographical perspective. The literature therefore lends itself to support a research question about their association, which we aim to answer in this paper with the help of some literature excerpts useful to develop our basic research hypothesis. These excerpts help frame the company sample (a) and expectations regarding potential associations (b).

As for the company sample (a), previous research showed that there is a strong positive relation between sustainability rating (i.e.: environmental-social-governance scores) and IR adoption (Lai et al. 2016: 14). It is therefore more likely for sustainability-oriented companies to embark on IR. Moreover, both the remarkable success of IR in Europe (Eccles and Krzus 2015: 61, and Turcu 2015: 68) and likely "significant differences in the meaning of materiality between countries" (Eccles and Krzus 2015: 136) suggest focusing on a defined geographical area.

As for the expectations about materiality (b), relevant literature suggests that companies all around the world still show "relatively poor levels of disclosure about [. . .] materiality"; nevertheless, some authors note that "some integrated reports appear to be still in a transition phase from sustainability reports to integrated reports" (Wee et al. 2016: 13). All these considerations suggest how, in an increasingly IR-oriented context, materiality could also gain importance in reporting practices.

Moreover, some affirmed that "[. . .] the integration process "forces" companies to implement a materiality determination process [. . .]. This activity lowers the amount of non-material information disclosed, and this is central to avoiding information overload or misunderstanding in investors" (Druckman 2016: xiv). This means that the integrated reporting process, if adequately implemented, should imply a suitable materiality exercise.

The study by Mio and Fasan (2013) can offer insight on both (a) and (b). Its findings about the chain 'experience on stakeholder needs→sustainability-oriented governance→more effective tackling of materiality' suggest that companies with higher awareness of non-financial performance are likely to focus more on materiality (ivi: 14).

Based on the literature, we chose to study companies' "attitudes" towards both IR implementation and MDP disclosure rather than their outputs (e.g. the 'effective' level of information integration and the quality/quantity of issues reported on, respectively). We deemed that, in a context of sustainability-oriented companies (i.e. a context where IR awareness is common), the higher the IR orientation, the higher the potential MDP disclosure level. Given the context, we expected a positive association between IR orientation and MDP disclosure, and therefore tested the following basic research hypothesis:

In sustainability-oriented companies, a positive attitude to IR implementation is associated with a positive attitude to MDP disclosure.

Note that, since high and low IR orientation degrees could theoretically coexist with strong and weak MDP disclosure due to the variety of IR orientations and MDP practices, testing this association could return insightful results.

14.4 Methods

14.4.1 The Sample

To select our sustainability-oriented company sample, we used the RobecoSAM Yearbook 2016. Our choice was consistent with Bocken et al. (2014: 47), who suggest the use of sustainability rankings to identify companies likely to adopt sustainable business models. Therefore, we selected our sample based on the final results of the 2015 RobecoSAM annual Corporate Sustainability Assessment. RobecoSAM assessed 2126 companies belonging to 59 sectors, of which 464 qualified for the Sustainability Yearbook 2016.

Among the 464 eligible companies, 248 (i.e. the top 15% of companies from each industry) were classified into three award categories: RobecoSAM Gold, Silver and Bronze Class. Almost half of the 248 awards (i.e. 116) went to European companies. Among these, eight companies were excluded from the sample (one for lacking publicly available information, and another seven that have gone through mergers). The resulting 108 companies represent the subpopulation selected for this research (hereinafter 'sample'). Table 14.3 details the composition of the sample.

Table 14.4 shows the 108 selected companies by country.

In Table 14.4 it is possible to observe that companies from 14 out of the 27 EU countries were awarded by RobecoSAM. Most of the sustainability-oriented

Table 14.3 From the RobecoSAM sustainability medal awards to the sample

Award	Description of awarding criteria	No. of companies awarded		
		Overall	Europe	Sample[a]
Gold class	Companies whose score is within 1% of the industry leader's score	77	36	34
Silver class	All companies receiving a score within a range of 1–5% from the score of the industry leader	74	32	29
Bronze class	Companies whose score is within a range of 5–10% from the score of the industry leader	97	48	45
	Totals	248	116	108

[a]European companies less companies excluded

Table 14.4 Sample coverage of the European territory

Country	Companies	
	No.	%
Belgium	1	0.93
Denmark	3	2.78
Finland	3	2.78
France	16	14.81
Germany	16	14.81
Italy	8	7.41
Ireland	1	0.93
Netherlands	11	10.19
Norwey	1	0.93
Portugal	1	0.93
Spain	12	11.11
Sweden	2	1.85
Switzerland	11	10.19
United Kingdom	22	20.37
Tot.	108	100.00

companies in the sample are from big western European countries, on which this study focuses indirectly due to sample composition.

14.4.2 *Data Collection and Analysis*

In our investigation of the publicly available online material from the 116 European companies in the initial sample (see Table 14.3) we found eight companies to be excluded, and therefore focused on the remainder (108). We then proceeded to (a) recording information on investor relations/finance and sustainability/responsibility areas; (b) gathering 2015 or 2015/16 reports and annexes forming the companies' financial and non-financial reporting.

Online data related to variables linked with IRO degree (IR1, IR2, etc.) and MDPD predictors (MDP1, MDP2, etc.) were collected for the subsequent Principal Component Analysis (PCA) and assigned a 'non-dichotomous' score (scales from 0–2 to 0–5). For both IRO and MDPD individual Pearson correlation matrices were elaborated. With this procedure, we discovered strong and weak links as well as a lack of correlation between variables, to then select the two sets of IR and MDP variables to use for the PCA carried out by means of the SPSS statistical program. The PCA returned one principal component for each set (i.e. PCir and PCmdp). After analysing these two principal components (by setting a correlation threshold between main component and other variables, focusing only on variables exceeding their threshold), we clustered their two z-score distributions into three groups in order to decrease their degrees of freedom. Then these two new distributions were associated with two tests: Chi-Square and Cramer's V. The resulting values were used to test our basic research hypothesis on positive correlation between IRO and MDPD.

Figure 14.1 summarises this process highlighting sections whose findings will be expounded.

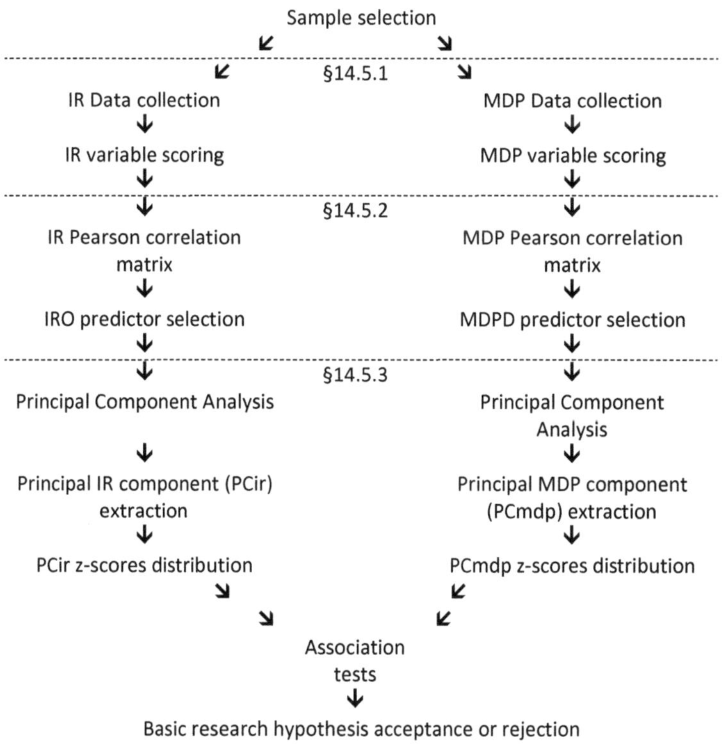

Fig. 14.1 Research phases by chapter section: from sample selection to data analysis

14.5 Results and Discussion

This section presents and discusses the main results we obtained.

The two subsections expound the results of the first two phases shown in Fig. 14.1 (until predictor selection, excluding sample selection), respectively for IR and MDP, to which they apply descriptive statistics. Note that variable content and related scoring criteria are presented here rather than in Sect. 14.4 because score criteria were set after a preliminary observation of IR and MDP disclosure phenomena. The aim was to have a realistic coverage of occurrences and avoid zero frequencies. Essentials notes about the PCA and association test are presented in Sect. 14.5.3.

14.5.1 Results and Discussion: IR and MDP Variable Collection and Scoring

14.5.1.1 IR Orientation Variables

In order to search for IR orientation predictors, we collected data about and scored the following variables[3]:

- IR1 (Website integration)—Investigates the degree of joint accessibility of financial and non-financial reporting on corporate websites. Reports lacking easily accessible cross-references score lower.
- IR2 (Core report number)—Evaluates the number of reports deemed 'central' in each company reporting, assigning higher scores to lower numbers. Other reports and annexes were not counted.
- IR3 (Core report length)—Counts pages dealing with core reporting. Lower numbers of pages score higher.
- IR4 (IIRC-compliant reporting)—Measures the degree of IIRC framework adoption. Reports entirely compliant score higher.
- IR5 (Six-capital model)—Searches for mentions of IIRC's six capitals in core reports. Capital-based reports score higher than partial uses or mentions.
- IR6 (Value chain)—Counts mentions of value chain, as sensitivity to this chain is expected in reporting of IR-oriented companies. Scoring thresholds are set by citation number.
- IR7 (IR aims)—Analyses the purpose of embracing the IIRC framework, scoring implementation under way higher than prospective implementation.
- IR8 (Reporting standards)—Rewards the use of reporting guidelines, regardless of the IIRC framework.
- IR9 (IIRC adoption)—Rewards the explicitly stated adoption of the IIRC framework, with or without other guidelines.

[3]Further information about some IR variables is provided further in the chapter.

- IR10 (IR communication)—Considers how the company's orientation toward IR is communicated in reports (coversheet, section titles, narration). High-impact solutions score higher.
- IR11 (IR mentions)—Counts mentions of IR in core reports, with more frequent mentions scoring higher.
- IR12 (Website non-financial focus)—Considers the visibility of sustainability/ responsibility sections compared to investor relations/finance sections on corporate websites.

Table 14.5 presents a selection of the results from some of the above variables.

Table 14.5 highlights differing numbers (Panel A) and types of reporting documents, with a prevalence of annual reports, followed by sustainability and responsibility documents (Panel B). Although about 90% of companies adopt at least one set of guidelines (Panel E), about 67% is not IIRC-compliant (Panel C) and, more in detail, only 23% mentions its six-capital model (Panel D). Regardless of IIRC compliance, IR is not mentioned by 57% (Panel G) and only 12% communicate IR aims through report coversheets (Panel F).

Jointly considered, IR2 and IR4 depict the current landscape of reporting patterns. Table 14.6 groups reporting styles under the "Integrated" label (i.e. IR accompanies financial and non-financial reports) and compares them with 'financial' (i.e. traditional report-based) and 'sustainability' (i.e. including sustainability or responsibility reports) styles.

Table 14.6 shows that about half of the companies adopting the integrated style do not opt for a stand-alone IR. This confirms that "One Report doesn't mean Only One report" (Eccles and Krzus 2010: 10). More generally, for the main differences among annual, sustainability and integrated reporting (e.g. audience, mandatory degree, regulation or guidelines, comparability) see Fasan (2013: 50) and IFAC (2015: 11).

14.5.1.2 MDP Disclosure Variables

In order to search for predictors of MDPD attitude, we collected data on the following 14 variables[4]:

- MDP1 (MDP website pages)—Investigates the existence and number of MDP pages in corporate websites, with higher scores awarded to pages fully dedicated to MDP.
- MDP2 (MDP document-based reporting section)—Investigates the existence of information on MDP in document-based reports, with special regard to dedicated sections. Higher scores are awarded for reports including separate MDP sections (regardless of their importance).
- MDP3 (Non-financial materiality pages)—Assesses the space occupied by issues of non-financial materiality (% of dedicated pages in document-based reports).

[4]Further information about some MDP variables is provided further in the chapter.

Table 14.5 IR variable results: a selection

Panel A (IR2)—Number of corporate document-based core reports	**Core report number**	**No.**	**%**
	Four or more	4	3.70
	Three	13	12.04
	Two	61	56.48
	One	30	27.78
	Tot.	108	100.00
Panel B (IR2)—Titles of corporate pdf reports	**Reports**	**No.**	**%**
	Annual report	75	30.74
	Sustainability report	39	15.98
	Responsibility report	25	10.25
	Financial statements/report	16	6.56
	Registration/legal documents	15	6.15
	Integrated report	12	4.92
	Corporate and management reports	11	4.51
	Annual review	5	2.05
	Activities and business reports	4	1.64
	Mix titles	14	5.74
	Other (not core) reports	28	11.48
	Tot.	244	100.00
Panel C (IR4)—IIRC-compliant reporting	**Compliance degree**	**No.**	**%**
	The whole reporting or at least one report is compliant	29	26.9
	Some reporting elements are compliant	6	5.6
	The company will adopt IIRC guidelines	1	0.9
	Not compliant at all	72	66.7
	Tot.	108	100.00
Panel D (IR5)—Six-capital model	**Compliance degree**	**No.**	**%**
	The model is mentioned (regardless their current or future application)	25	23
	Statement of imminent use of the model	2	2
	One or some capitals are mentioned	34	31
	No capital is mentioned at all	47	44
	Tot.	108	100.00
Panel E (IR8)—Reporting standards	**Groups of guidelines**	**No.**	**%**
	Three sets	16	15
	Two sets	52	48
	One set	30	28
	No one	10	9
	Tot.	108	100.00

(continued)

Table 14.5 (continued)

Panel F (IR10)—IR communication	**Style of communication**	**No.**	**%**
	Both within coversheet and core document text	13	12
	Within core documents' text only	18	17
	Some doubts hints within core documents	11	10
	No communication	66	61
	Tot.	108	100.00
Panel G (IR12)—IR mentions	**Groups**	**No.**	**%**
	Several mentions (5 or more)	25	23
	Some mentions (1–4)	21	19
	No mention	62	57
	Tot.	108	100.00

Table 14.6 Main reporting styles found

Style	Sub-style	No.	Subt.	%
Financial	Stand-alone financial report	11	19	17.59
	Several financial reports	8		
Sustainability	Several reports (financial and non-financial), including sustainability/responsibility reports	57	57	52.78
Integrated	Stand-alone IR or IIRC-compliant report	17	32	29.63
	Several financial and non-financial reports including an IR	9		
	Several financial and non-financial reports, including a self-declared IR or IIRC-compliant report	6		
	Tot.	108	108	100.00

- MDP4 (Document-based pages citing materiality)—Counts percentages of core reporting pages citing both financial and non-financial materiality.
- MDP5 (MDP phase description)—Indicates whether MDP phases are detailed in reports. Any dedicated sections are scored higher.
- MDP6 (Introduction to material issues)—Searches for preliminary information on selected MDP outputs. Detailed presentation of information scores higher.
- MDP7 (Location of introduction to material issues)—Searches for indications about the location of preliminary information on material issues. Clear indications facilitating access score higher.
- MDP8/MDP9/MDP10 (MDP orientation)—Scores webpages and documents separately and jointly, investigating the aims associated to MDP by companies. Multiple aims (including reporting) score higher, followed by reporting aims only.
- MDP11 (Materiality matrix)—Considers in how many locations (website, reports and annexes) a materiality matrix is proposed. Multiple locations score higher.

- MDP12 (Issue list)—Considers issue selection and prioritisation, comparing the list of issues against the initial list for prioritisation. Lists including only issues considered of high or medium relevance score lower than full lists.
- MDP13 (Issue clustering)—Examines how material issues are disclosed: not clustered (highest score), clustered or not indicated at all.
- MDP14 (Table of contents including MDP)—Investigates if tables of contents include MDP. Higher scores are assigned to multiple citations in several tables of contents.

Table 14.7 presents a selection of the results from some of the above MDP variables. It shows that about a third of companies does not even hint at MDP (Panels A and B) or emphasise it within the table of contents (Panel G). As for materiality issues, about 13% of companies list only MDP final outputs (i.e. highly relevant issues, Panel F) but about half of them do not offer readers a preliminary introduction to the outputs themselves (Panel C). Moreover, circa 70% use the materiality matrix (Panel E) while only about 10% does not state aims attributed to MDP (Panel D).

14.5.2 Results and Discussion: IR and MDP Variable Selection for PCA

14.5.2.1 IR Variable Selection

When IR variables were associated two by two with Pearson correlation analysis, 7 out of 12 resulted as positively correlated (we considered both medium—i.e. more than 0.3—and high correlations). For purposes of conciseness, Table 14.8 shows only the coefficients these seven variables returned.

The seven correlated variables were selected for the subsequent PCA. Uncorrelated variables are nevertheless worthy of comment, with specific regard to IR2–IR3 and IR1–IR12.

As for the IR2–IR3 set (report number and length), these variables were selected as IR predictors because:

- according to Paternostro (2013), the application of the connectivity principle in IR can originate different report aggregation degrees involving varying report numbers. Two approaches (i.e. weak and strong aggregation) out of the three proposed are "mere aggregation-based" (p. 69). In fact, intermediate IR-oriented steps based on issuing simple 'combined' financial/non-financial documents are reasonably expected, regardless of integration perspective (see Paternostro and Quarchioni 2013). Since an integrated report is often an additional document, we expect IR to encourage an increase in companies' reporting rationality and openness to revising their reporting structure as well as a decrease in document number (as demonstrated in Havlováa 2015: 235). The supposed predictive power of IR2 is strong in a developing IR context like the current scenario. At

Table 14.7 MDP variable results: a selection

Panel A (MDP1)—MDP website pages	**Existence and location**	**No.**	%
	A clearly dedicated section exists (relevant title)	51	43.30
	An unclearly dedicated section exists (irrelevant or partly relevant title)	12	10.34
	Only a hint/phrase can be found	11	9.48
	No information at all	42	36.21
	Tot.	108	100.00
Panel B (MDP2)—MDP reporting section	**Existence and location**	**No.**	%
	A clearly dedicated section exists (relevant title)	50	46.30
	An unclearly dedicated section exists (irrelevant or partly relevant title)	12	11.11
	Only a hint/phrase can be found	11	10.19
	No information at all	35	32.41
	Tot.	108	100.00
Panel C (MDP6)—Introduction to material issues	**Type of introduction**	**No.**	%
	An introduction about all material issues exists	25	23.15
	An introduction about some material issues exists	34	31.48
	No introduction at all	49	45.37
	Tot.	108	100.00
Panel D (MDP9)—MDP orientation	**Aim assigned to MDP**	**No.**	%
	Both reporting and non-reporting	41	37.96
	Only reporting	22	20.37
	Only non-reporting	34	31.48
	No aims declared	11	10.19
	Tot.	108	100.00
Panel E (MDP11)—Materiality matrix	**Location of matrix(es) proposed**	**No.**	%
	Three	1	0.93
	Two	33	30.56
	One	41	37.96
	No matrix at all	33	30.56
	Tot.	108	100.00
Panel F (MDP12)—Issue list	**Type of material issues disclosed**	**No.**	%
	All	38	35.19
	Medium and highly relevant only	50	46.30
	Highly relevant only	6	12.96
	No issues at all	14	12.96
	Tot.	108	100.00
Panel G (MDP14)—Table of contents including MDP	**Number and location of mentions**	**No.**	%
	Double mention within one (some report) or more (various documents) main tables	10	9.26
	One mention within a main table	23	21.30
	Mention within internal tables	1	0.93
	No mention	74	68.52
	Tot.	108	100.00

Table 14.8 Pearson correlation matrix excerpt: seven IR positively correlated variables

	IR4	IR5	IR7	IR8	IR9	IR10	IR11
IR4		0.509	0.760	0.509	0.856	0.836	0.845
IR5			0.771	0.393	0.532	0.458	0.473
IR7				0.391	0.532	0.866	0.849
IR8					0.714	0.393	0.459
IR9						0.702	0.759
IR10							0.875
IR11							

a more 'mature' IR stage, a 'narrow integration' could be observed (ibidem) and it would become more likely for additional reports to co-exists with main reports;

- since the IIRC Consultation Draft (2013b), conciseness was associated to materiality. An integrated approach must avoid redundancy (Busco et al.: 15), and therefore a holistic approach to quantity disclosure aiming to contain information should be adopted. Length of information (e.g. page numbers) might appear more strictly associated to the conciseness principle than to the overarching IR one. However, our aim is not to measure IR effectiveness, but rather it is to represent predictors of integration attitude. We take the view that the issue of conciseness inevitably accompanies the move towards IR (as suggested by Havlováa 2015: 235), regardless of its status as an IIRC guiding principle; a recent study has also highlighted how reducing the reporting size represents both an opportunity and a problem (Black Sun and the International Integrated Reporting Council 2014: 19).

Although IR2 and IR3 are not strongly correlated with other variables, they are with each other; a higher number of reports seems therefore connected with an increase in page numbers. However, medium coefficients (0.3–0.5, depending on scoring) reveal that this happens only to an extent, testifying to a potential trend reversal in the practice of disclosing large reporting documents.

Another interesting set of variables is IR1–IR12 (website integration and website non-financial focus). These variables were selected as IR predictors because, in our opinion, respectively:

- the integrated thinking entailed by IR implies new coherent and flexible website structures aiming to be increasingly user-friendly for IR-oriented users. New structures must account for integrated availability of reporting documents published online or easy accessibility to both financial and non-financial reporting, regardless of the website area visitors are exploring. Hence, as for IR1, we assumed that companies moving towards IR realize that website structure can support this move (Krzus 2011: 271);
- where an IR orientation exists, a focus on communicating/highlighting non-financial aspects (e.g. sustainability) is expected, and this should be reflected in the reporting areas of corporate websites.

Unlike the previous set, IR1–IR12 are not correlated (the coefficient being close to zero). Although both variables are website-related, lack of correlations seems to suggest that awareness of websites as tools to integrate financial and non-financial information does not mean visibility of sustainability/responsibility website areas.

The weak link between document-based and website-based variables suggests a weak corporate attitude to shaping websites in line with their reporting style. In this regard scholars support a central use of the Internet in the IR context (Krzus 2011: 276), among other reasons in order to overcome the severe limitations of paper-based reporting.

Although IR1 and IR12 are generally uncorrelated with other IR document-based variables, sometimes IR1 is borderline correlated to variables such as IR4, IR10 and IR11 (coefficients are slightly higher than 0.3). Oddly, an orientation to support joint online availability of financial and non-financial reports is somewhat associated with choices like the degree of IIRC-compliant reporting (IR4), the way IR attitudes are communicated (IR10) and explicit mentions of IR (IR11).

All this foretells a difficult journey towards a corporate perspective on web resources harmonised both with stakeholder abilities (Krzus 2011: 271) and with an appropriate exploitation of the potential of the web for IR aims (about the IR-Internet relationship, see Eccles and Krzus 2010, Chap. 8).

14.5.2.2 MDP Variable Selection

When MDP variables were associated two by two with Pearson correlation analysis, 5 out of 14 resulted positively correlated with 0.3+ coefficients. For the following PCA we also added another two variables correlated with medium values with some of the above five variables. Hence, although not all the seven variables selected resulted strongly correlated, they were in any case selected, as Table 14.9 shows.

As for variables included despite some low coefficients, it can be supposed that results about MDP11 (Matrix) could derive from the prominent GRI-G4 origin of the matrix itself (GRI 2013b). Since this does not exclude that IR might adopt this GRI tool, and some coefficients nearing 0.3 were found, we included this variable in PCA. Similarly, even though MDP14 (Table of content) was expected to play a more

Table 14.9 Pearson correlation matrix excerpt: seven positively correlated MDP variables

	MDP2	MDP3	MDP4	MDP9	MDP11	MDP13	MDP14
MDP2		0.505	0.512	0.531	0.278	0.487	0.292
MDP3			0.400	0.512	0.214	0.257	0.251
MDP4				0.438	0.311	0.397	0.368
MDP9					0.246	0.305	0.202
MDP11						0.342	0.228
MDP13							0.100
MDP14							

significant role in reporting the basic role of MDP to stakeholders, some correlation coefficients with variables not selected pointed to its inclusion in PCA.

As for the remaining variables, two sets (MDP6–MDP7 and MDP1–MDP8) are worthy of comment. Although not correlated to other MDP variables, MDP6–MDP7 (introduction to material issues and its location) show an expectedly strong correlation. The absence of links with remaining MDP variables indicates how companies do not adequately introduce readers to material issues selected for their IR, regardless of the model chosen (IR, sustainability, etc.).

MDP1–MDP8 (MDP website pages and MDP orientation) are also, as expected, strongly correlated. In terms of MDP disclosure this indicates on one hand a certain homogeneity among companies' web choices, and on the other hand a full separation between document-based reporting and web-based communicative choices.

Finally, as for MDP12–MDP13 (issue list and clustering), although the variables are both issue-based, they were originally chosen as MDP variables because they are not only a measure of number of material issues listed. Indeed, the number of issues alone cannot and should not be used to measure MDP disclosure orientation: a company might well report on its materiality process regardless of the number of issues reported on, which anyway represent only the MDP output.

14.5.3 Results and Discussion: From PCA to Association Between IR and MDP Principal Components

When PCA was implemented on selected IR and MDP variables, one principal component (or factor) was extracted for each. Explained percentages of variance were 70.029% for the first (PCir) and 45.088% for the second (PCmdp) factor.

For both PCir and PCmdp, SPSS returned a pattern matrix that groups correlation coefficients with the variables each factor is composed of. The matrixes (not shown here)[5] help to extract the nucleus of the factorial analysis solution.

More in detail, since thresholds of 0.8 (for PCir) and 0.5 for (PCmdp) were chosen, the following variables resulted as exceeding such thresholds:

- IR4 (IIRC-compliant reporting), IR9 (IIRC adoption), IR10 (IR communication), and IR11 (IR mentions), for PCir;
- MDP2 (MDP document reporting section), MDP5 (MDP phase description), and MDP9 (MDP orientation), for PCmdp.

The final step of the interpretation process of the factorial model labels the principal components obtained. Since in this research PCA represents only an

[5]The matrixes will be sent upon request.

intermediate step to obtain two distributions for association, to our aims it is sufficient to highlight the following commonalities among the two variable sets:

- PCir variables are linked both by standard adoption and by choices regarding document preparation; since both aspects recall formal issues (compliance and wording) this factor is labelled 'form dimension';
- PCmdp variables are linked by aspects related to the information provided within reporting documents; this factor is labelled 'document-based reporting dimension'.

Finally, both PCir and PCmdp distribution of z-scores (whose ranges are shown in Fig. 14.2) were considered and grouped in classes to prepare for the statistical tests used (Chi-Square and Cramer's V). As shown in Fig. 14.2, when both association tests were applied to the two z-score distributions, the process returned surprisingly low values (11.521 and 0.231 respectively). Although both are significant at 0.05 alpha level, such very low statistical results suggest no relevant association between companies' IR orientation and their MDP disclosure.

This striking result could derive from the subjective nature of scoring. For example, when the adoption of the IIRC framework was scored, a joint application was scored higher than a stand-alone application. A simple change in scoring schemes could return higher correlation coefficients. Similarly, the low association

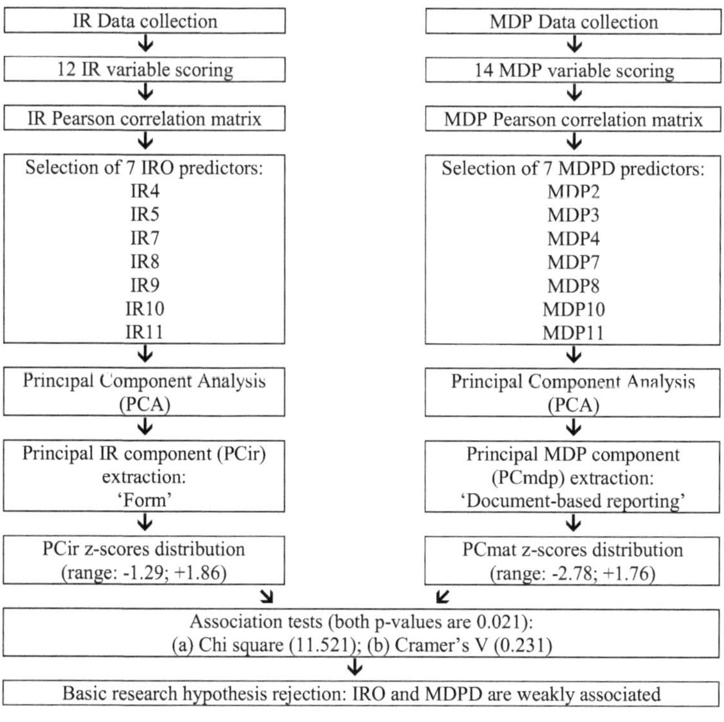

Fig. 14.2 Data analysis results within the research phases involved

could derive from the variables selected for factor extraction. As the comparison between Tables 14.5 and 14.8 highlights, correlations of IR matrix excerpts are higher than their MDP counterparts. This means potential bugs in the principal components extraction, as the comparison between PCmdp and PCir explained percentages of variance seems to confirm.

The weak correlation between IR orientation and MDP disclosure does not exclude a link between the phenomena, but neither does it consider that an attitude toward MDP disclosure could be expressed in sustainability reporting. Repeating the analysis with this focus, sustainability-reporting oriented companies might be found to be equally/more oriented to MDPD than those embracing IR.

Figure 14.2 summarizes the results of this section following the development of this study and completing with key figures and findings the phase scheme presented in Sect. 14.4.

14.6 Conclusions

Since IR implies special emphasis on non-financial information, materiality assumes a critical role, as the "sudden shake-up" (Thurm and De Ruiter 2014) of both standard setters and relevant research confirms. Materiality becomes important because its binary nature (i.e. an issue is either material or it is not) prevents intermediate solutions between reporting and not reporting (Eccles and Krzus: 119). Preparers are therefore more inclined to include (rather than cut off) borderline information so as to contain the risk of stakeholder disagreement about materiality judgements (ivi: 132).

This chapter aimed to verify if companies more IR-oriented show a corresponding higher attitude to MDP disclosure. Contrary to expectations, a weak association was found between companies' IR orientation and their MDP disclosure attitude. Therefore, in the current reporting landscape, a certain degree of IR does not necessarily correspond to a certain degree of transparency about MDP phases and selected issues. However, these findings confirm that, at least, the analytical direction suggested by the positive sign resulting from the Pearson test is correct. According to our findings, and consistent with previous research (Gelmini et al. 2015: 157), the materiality landscape in IR is still 'under construction'. When Wee et al. (2016) detected specific MDP phases, the few disclosures about the more 'competitive' phases (evaluation and prioritisation) were deemed useful only for novice IR users and less so for experts (p. 15). In this context, IR is not linked with a single MDP disclosure attitude and, at least until now, this seems not to provide specific operational solutions for companies undertaking the IR journey.

Descriptive statistics on IR and MDP is also useful for some final remarks. Companies' reporting patterns show a prominence of sustainability-based reporting, with IR being the second most common choice. Both the wide-ranging report number and the increasing IR value-oriented information seem to push towards a revision of the reporting structure to ensure disclosure effectiveness. The timing of such a revision can only be predicted to happen depending on the individual communication style of companies (i.e. a reporting style including fewer or more reports). Our analysis has detected a relatively wide adoption of the materiality matrix. However, in spite of the 'current' matrix version being not fully adopted yet, a newer version (the Sustainability Value Matrix) has recently been coined by prominent scholars (Eccles and Krzus 2015: 147).

Comparisons between IR and MDP results also point to weak links between website communication and document-based reporting. In our opinion, the integrated thinking underpinning IR will inevitably transform websites to cater for contemporary reporting practices, which in the future are likely to include stand-alone integrated reports. The question of the appropriate location for such reports on corporate websites will need to be answered, as companies have so far identified a variety of solutions (e.g. mutual links within finance and sustainability areas or separate reporting areas). In increasingly IR-oriented contexts, online tools take on a crucial role for both companies and stakeholders with their ability to renegotiate or remove boundaries between financial and non-financial reporting.

Some limitations to this study need to be acknowledged. In particular: (a) no normalization of core report number was introduced, despite companies being required to provide non-financial information in some European countries (e.g. Denmark, France and Sweden; see Krzus 2011: 273–274); (b) sample size prevented sectorial analysis; (c) variables selected as IR/MDP predictors can be both differently scored and enlarged. These limitations might be overcome in further research, for example by: (a) introducing a data normalization mechanism operating by country based on the national law regarding any mandatory non-financial information; (b) using bigger samples conducive to sectorial analysis; (c) attributing different scores to the variables used or selecting new variables.

The progressive consolidation of IR in terms of its shape and identity means an expansion of research centred on materiality, in particular in the following fields: (a) country-specific (Eccles and Krzus 2015) and sector-specific analyses (Mio and Fasan 2014); (b) link with stakeholder engagement, given its emerging critical role in defining materiality (Adams 2013; Gelmini et al. 2015); (c) development of material analysis models similar to those for sustainability contexts (Hsu et al. 2013).

Paradoxically, the materiality exercise itself originates new information if the following question is answered affirmatively: is disclosing materiality a material issue? The recent proposal of a 'Statement of Significant Audiences and Materiality' as a part of IR (Eccles and Krzus 2015: 132–133) seems to imply a positive answer. In Eccles and Krzus's view, this statement can accelerate IR adoption, avoid 'greenwashing' charges and "strengthen the social construction attribute of institutional symbolism" (ivi: 133). Therefore, despite the above paradox generated by MDP disclosure, this is well tolerated and overcome, probably because materiality is considered a compass for report readability.

Materiality therefore remains the ultimate rampart against the risk of non-financial information proliferating without control, to the point that companies' insiders believe that, "from an investor point of view, the winning reporting model should and will focus on materiality" (Eccles and Krzus 2010: 217).

Acknowledgement Tiziana De Cristofaro wrote Sects. 14.1, 14.2.2, 14.3, 14.4.2, 14.5.2, 14.5.3 and 14.6 and Carmela Gulluscio wrote Sects. 14.2.1, 14.4.1, 14.5.1 and the introduction to Sects. 14.2 and 14.5. The authors are grateful to Eugenia Nissi, associate professor of statistics at the University of Chieti-Pescara (Italy), for her support with the statistical analysis.

References

AccountAbility. (2006). *Guidance note on the principles of materiality, completeness and responsiveness as they relate to the AA1000 assurance standard*. Retrieved from http://www.envirotrain.co.uk/wp-content/uploads/2013/05/C3.4.7-AA1000-Guidance-Note.pdf

AccountAbility. (2008a). AA1000 Accountability principles standard 2008. Retrived from https://drive.google.com/file/d/0BwU82A_-zhEFMjdmLUlDdFo0dHc/view

AccountAbility. (2008b, October 24). *Introduction to the revised AA1000 assurance standard and the AA1000 AccountAbility principles standard 2008*. London: AccountAbility.

Adams, C. A. (2013, March 25). *Materiality: Financial reporting, sustainability reporting and integrated reporting*. https://drcaroladams.net/materiality-financial-reporting-sustainability-reporting-and-integrated-reporting/

Black Sun & the International Integrated Reporting Council. (2014). *Realizing the benefits: The impact of integrated reporting*. Retrieved from http://integratedreporting.org

Bocken, N. M. P., Short, S. W., Rana, P., & Evans, S. (2014). A literature and practice review to develop sustainable business model archetypes. *Journal of Cleaner Production, 65*, 42–56.

Busco, C., Frigo, M., Quattrone, P., & Riccaboni, A. (2013). Towards integrated reporting: Concepts, elements and principles. In C. Busco, M. L. Frigo, A. Riccaboni, & P. Quattrone (Eds.), *Integrated reporting. Concepts and cases that redefine corporate accountability* (pp. 3–18). Cham: Springer.

Calabrese, A., Costa, R., Levialdi, N., & Menichini, T. (2016). A fuzzy analytic hierarchy process method to support materiality assessment in sustainability reporting. *Journal of Cleaner Production, 121*, 248–264.

Chong, H. G. (2015). A review on the evolution of the definitions of materiality. *International Journal of Economics and Accounting, 6*(1), 15–32.

Druckman, P. (2016). Foreword. In C. Mio (Ed.), *Integrated reporting: A new accounting disclosure*. London: Palgrave Macmillan.

Eccles, R. G., & Krzus, M. P. (2010). *One report: Integrated reporting for a sustainable strategy*. New York: John Wiley & Sons.

Eccles, R. G., & Krzus, M. P. with Ribot, S. (2015). *The integrated reporting movement. Meaning, momentum, motives, and materiality*. John Wiley & Sons: Hoboken, NJ.

Eccles, R. G., Krzus, M. P., & Watson, L. A. (2012a). *Integrated reporting requires integrated assurance*. Q Finance. Retrived from www.qfinance.com

Eccles, R. G., Krzus, M. P., Rogers, J., & Serafeim, G. (2012b). The need for sector-specific materiality and sustainability reporting standards. *Journal of Applied Corporate Finance, 24*(2), 65–71.

EC-European Commission. (2017, June 26). *Communication from the commission. Guidelines on non-financial reporting (methodology for reporting non-financial information)*. Available at https://ec.europa.eu/info/publications/170626-non-financial-reporting-guidelines_en

Edgley, C. (2014). A genealogy of accounting materiality. *Critical Perspectives on Accounting, 25*(3), 255–271.

Ernst & Young. (2013). *The concept of 'materiality' in integrated reporting*. Integrated reporting update. Retrived from http://www.ey.com

Fasan, M. (2013). Annual reports, sustainability reports and integrated reports: Trends in corporate disclosure. In C. Busco, M. L. Frigo, A. Riccaboni, & P. Quattrone (Eds.), *Integrated reporting. Concepts and cases that redefine corporate accountability* (pp. 41–57). Cham: Springer.

FASB. (1975, March). *Discussion memorandum: An analysis of issues related to criteria for determining materiality*. Stamford, CT: FASB.

Forstater, M., Zadek, S., Evans, D., Knight, A., Sillanpää, M., Tuppen, C., & Warris, A. M. (2006). *The materiality report. Aligning strategy, performance and reporting*. London: AccountAbility.

Frías-Aceituno, J. V., Rodríguez-Ariza, L., & García-Sánchez, I. M. (2013a). Is integrated reporting determined by a country's legal system? An exploratory study. *Journal of Cleaner Production, 44*, 45–55.

Frías-Aceituno, J. V., Rodríguez-Ariza, L., & Garcia-Sánchez, I. M. (2013b). The role of the board in the dissemination of integrated corporate social reporting. *Corporate Social Responsibility and Environmental Management, 20*(4), 219–233.

Frías-Aceituno, J. V., Rodríguez-Ariza, L., & Garcia-Sánchez, I. M. (2014). Explanatory factors of integrated sustainability and financial reporting. *Business Strategy and the Environment, 23*(1), 56–72.

García-Sánchez, I. M., Rodríguez-Ariza, L., & Frías-Aceituno, J. V. (2013). The cultural system and integrated reporting. *International Business Review, 22*(5), 828–838.

Gelmini, L., Bavagnoli, F., Comoli, M., & Riva, P. (2015). Waiting for materiality in the context of integrated reporting: Theoretical challenges and preliminary empirical findings. In L. Songini & A. Pistoni (Eds.), *Sustainability disclosure: State of the art and new directions* (pp. 135–163). Bingley: Emerald Group Publishing Limited.

GRI. (2006). *G3 sustainability reporting guidelines*. Retrived from https://www.globalreporting.org

GRI. (2011). *G3.1 sustainability reporting guidelines*. Retrived from https://www.globalreporting.org

GRI. (2013a). *G4 sustainability reporting guidelines—Reporting principles and standard disclosures*. Retrived from https://www.globalreporting.org

GRI. (2013b). *G4 sustainability reporting guidelines—Implementation manual*. Retrived from https://www.globalreporting.org

GRI. (2016a). *Defining what matters do companies and investors agree on what is material*? Retrived from https://www.globalreporting.org

GRI. (2016b). *GRI sustainability reporting standards*. Retrived from https://www.globalreporting.org

Havlováa, K. (2015). What integrated reporting changed: The case study of early adopters. *Procedia Economics and Finance, 34*, 231–237.

Hespenheide, E. J., & Koehler, D. A. (2013, July 18). *Disclosure of long-term business value. What matters*? Deloitte University Press. Retrived from https://dupress.deloitte.com

Hsu, C. W., Lee, W. H., & Chao, W. C. (2013). Materiality analysis model in sustainability reporting: A case study at Lite-On technology corporation. *Journal of Cleaner Production, 57*, 142–151.

IFAC. (2015, November). *Materiality in <IR>. Guidance for the preparation of integrated reports*. New York, IIRC.

IIRC. (2013a, March). *Materiality. Background paper for <IR>*. New York: IIRC. Retrived from https://integratedreporting.org

IIRC. (2013b, April). *Consultation draft of the international framework*. New York: IIRC. Retrived from http://integratedreporting.org

IIRC. (2013c, December). *The international <IR> framework*. New York: IIRC. Retrived from https://integratedreporting.org

Jensen, J., & Berg, N. (2012). Determinants of traditional sustainability reporting versus integrated reporting. An institutionalist approach. *Business Strategy and the Environment, 21*(5), 299–316.

KPMG. (2014, September). *Bridging the gap between integrated and GRI G4 reporting*. Retrived from https://kpmg.com

Krzus, M. P. (2011). Integrated reporting: If not now, when? *Zeitschrift für Internationale Rechnungslegung, 6*, 271–276.

Lai, A., Melloni, G., & Stacchezzini, R. (2016). Corporate sustainable development: Is 'integrated reporting' a legitimation strategy? *Business Strategy and the Environment, 25*(3), 166–177.

Lydenberg, S. (2012). *On materiality and sustainability: The value of disclosure in the capital markets*. Initiative for Responsible Investment Hauser Center for Non-profit Organizations at Harvard University. Retrived from http://iri.hks.harvard.edu

Messier Jr., W. F., Martinov-Bennie, N., & Eilifsen, A. (2005). A review and integration of empirical research on materiality: Two decades later. *Auditing: A Journal of Practice & Theory, 24*(2), 153–187.

Mio, C. (2013). Materiality and assurance: Building the link. In C. Busco, M. L. Frigo, A. Riccaboni, & P. Quattrone (Eds.), *Integrated reporting: Concepts and cases that redefine corporate accountability* (pp. 79–94). Cham: Springer.

Mio, C., & Fasan, M. (2013). *Materiality from financial towards non-financial reporting* (Working paper series no. 19). Venice: Università Ca' Foscari.

Mio, C., & Fasan, M. (2014). *The determinants of materiality disclosure in integrated corporate reporting* (Working paper no. 9/2014). Venice: Ca' Foscari University of Venice. http://virgo.unive.it/wpideas/storage/2014wp9.pdf

Moroney, R., & Trotman, K. T. (2012). *Differences in auditors' materiality assessments when auditing financial and nonfinancial reports* (Working paper). Monash University.

Murninghan, M. (2013). In T. Grant (Ed.), *Redefining materiality II: Why it matters, who's involved, and what it means for corporate leaders and boards*. London: AccountAbility.

Paternostro, S. (2013). The connectivity of information for the integrated reporting. In C. Busco, M. L. Frigo, A. Riccaboni, & P. Quattrone (Eds.), *Integrated reporting. Concepts and cases that redefine corporate accountability* (pp. 59–77). Cham: Springer.

Paternostro, S., & Quarchioni, S. (2013). *Il percorso verso il report integrato. Tre prospettive a confronto*. Rome: Aracne.

RobecoSAM & Global Reporting Initiative. (2015). *Defining materiality: What matters to reporters and investors. Do investors and reporters agree on what's material in the technology hardware & equipment and banks & diverse financials sectors*? Retrived from https://www.globalreporting.org/resourcelibrary/Defining-Materiality-What-Matters-to-Reporters-and-Investors.pdf

SASB-Sustainability Accounting Standards Board. (2013). Conceptual framework. Retrived from https://www.sasb.org

Sierra-García, L., Zorio-Grima, A., & García-Benau, M. A. (2015). Stakeholder engagement, corporate social responsibility and integrated reporting: An exploratory study. *Corporate Social Responsibility and Environmental Management, 22*(5), 286–304.

Thurm, R., & De Ruiter, N. (2014). *The sudden materiality shock*. Retrieved from www.sustainablebrands.com

Turcu, R. D. (2015). Integrated reporting: The next step ahead for a sustainable society. *European Journal of Business Science and Technology, 1*(1), 63–75.

UNCTD-United Nations Conference on Trade and Development. Guidance on Corporate Responsibility Indicators in Annual Reports. (2008). Retrived from http://www.unctad.org/en

Unerman, J., & Zappettini, F. (2014). Incorporating materiality considerations into analyses of absence from sustainability reporting. *Social and Environmental Accountability Journal, 34*(3), 172–186.

Wee, M., Tarca, A., Krug, L., Aerts, W., Pink, P., & Tilling, M. (2016). *Factors affecting preparers' and auditors' judgements about materiality and conciseness in integrated reporting*. Retrieved from http://www.accaglobal.com

Zadek, S., & Merme, M. (2003). *Redefining materiality*. London: AccountAbility. Retrieved from www.accountability.org

Zadek, S., Tuppen, C., & Evans, D. (2006). *The materiality report. Aligning strategy, performance and reporting. A briefing*. London: AccountAbility.

Zhou, Y., & Lamberton, G. (2011, November 1–30). Stakeholder diversity versus stakeholder general views: A theoretical gap. In *Sustainability materiality conception, 1st world sustainability forum proceedings*. Basel: MDPI.

Chapter 15
Are Integrated Reports Really Integrated in Spain?

Belen Fernandez-Feijoo, Silvia Romero, and Silvia Ruiz Blanco

15.1 Introduction

The integrated report (IR) is the latest reporting approach developed by the International Integrated Reporting Council (IIRC). The IIRC is a global coalition of regulators, investors, companies, standard setters, the accounting profession and NGOs. The coalition is promoting communication about value creation as the next step in the evolution of corporate reporting. The IR is the combination of financial and narrative information (from the annual report (AR)) and non-financial and narrative information (from the sustainability report (SR)) (Eccles and Krzus 2010). Hence, it is more than putting together two different documents with financial and non-financial information. According to the IIRC, integrated reporting "aims to give a holistic view of the organization by putting its performance, business model and strategy in the context of its material social and environmental issues" (IIRC 2011). Similarly, Ernst and Young (2012) defines integrated reporting as giving context to financial and non-financial information and goals, connecting strategies with commitment to environmental, social and economic issues. It was developed as a consequence of the global financial crisis, when the need for improved reporting mechanisms seemed evident. Eccles and Krzus (2010) posit the urgent need of IR because "it will make more apparent, to both the company and its many stakeholders, the relationship between financial and nonfinancial performance and the extent to which financial performance for shareholders imposes externalities on other stakeholders" (p. 23). Although the IIRC highlights the shareholders as main

B. Fernandez-Feijoo (✉) · S. Ruiz Blanco
Universidade de Vigo, Vigo, Spain
e-mail: belen@uvigo.es; sblanco@uvigo.es

S. Romero
Montclair State University, Montclair, NJ, USA
e-mail: romeros@mail.montclair.edu

S. O. Idowu, M. Del Baldo (eds.), *Integrated Reporting*, CSR, Sustainability, Ethics & Governance, https://doi.org/10.1007/978-3-030-01719-4_15

addressees of the IR, this report also offers useful information on the organization's impact to other stakeholders. In fact, the stakeholder engagement is one of the fundamentals of the integrated reporting process (Sheridan 2014).

Given the novelty of the subject, there is little research on it. Lizcano et al. (2011) conducted an empirical study in Spain with data from 2010. They found that more than an integrated report, companies presented a collection of different independent documents such as financial statements, sustainability reports and corporate governance reports. To the best of our knowledge, there are no studies on the subsequent years about what companies present as an IR. The purpose of the chapter is three fold. First, to understand the evolution of integrated reporting in Spain. Second, to assess the integrated characteristics of the IR issued by Spanish companies, through the analysis of the integrated thinking in the CEO presentation pages of the IR; and third, to study to what extent are integrated reports committed to sustainability and really disclose integrated information. These assessments will uncover issues and benefits from IR and its implementation process. We choose Spain because of the high number of companies reporting, and the high quality of their CSR reports (KPMG 2013). Our findings indicate that the CEO/Chairman plays an essential role in transmitting integrated thinking to all members of the company. It seems that they need to adequate their message to achieve an effective involvement in sustainability of the organizational culture. We also find disparities between the underlying integrated thinking of the CEO/Chairman's presentations and the reference to the capitals proposed by the IIRC.

Our findings contribute to academic knowledge on the role of the IR as a result of integrated thinking within the organization. If the IR is a tool to send signals to specific stakeholders, employees, as well as to financial capital providers, it is necessary to establish the connection between the IIRC framework and the language used by companies in their reports. Finally, we contribute to literature by developing an analytic procedure for text analysis, combining both qualitative and quantitative characteristics of text.

The development of the chapter is as follows. After this introduction, we present the conceptual framework of the IIRC followed by a reference to the Integrated Reporting evolution in Spain. The empirical analysis of the CEO presentation pages and the content of the IRs is showed in Sect. 15.4. Section 15.5 presents the conclusions.

15.2 Background: The International Integrated Reporting Council (IIRC) and the International Integrated Reporting Framework (IIRF)

The IIRC is a global coalition of regulators, investors, companies, standard setters, the accounting profession and NGOs. Its mission is "to establish integrated reporting and integrated thinking within mainstream business practice as the norm in the public and private sectors" (http://integratedreporting.org/). Its vision is "to align capital allocation and corporate behavior to wider goals of financial stability and sustainable development through the cycle of integrated reporting and thinking" (http://integratedreporting.org/). Hence, integrated thinking is central to the IIRC framework.

Integrated thinking is defined as an "active consideration by an organization of the relationships between its various operating and functional units and the capitals that the organization uses or affects. Integrated thinking leads to integrated decision-making and actions that consider the creation of value over the short, medium and long term" (IIRC 2013, p. 2). The governance body has the responsibility of starting the change in this direction and transmitting it to the different organizational levels; however, it concerns to everyone in the organization the contribution to the value creation (IFAC 2015). Thus, the presentation letter of the IR should reflect the drivers of the value creation in the organization.

IIRC considers Integrated reporting as a natural evolution of corporate reporting, focusing on conciseness, strategic relevance and future orientation. The IIRC developed and issued the IIRF in 2013. The IIRF establishes a set of Guiding Principles and Content Elements that aim to define the content, as well as guide on the preparation of the IRs, which include financial and non-financial information.

The main purpose of an IR is to explain to financial capital providers how an organization creates value over time (IIRC 2013). Despite its approach towards financial capital providers, the IIRF highlights that an IR will also benefit all stakeholders. For this purpose, the IIRF proposes the disclosure of quantitative and qualitative, financial and non-financial information. IIRF approaches the reporting process not only searching for information quality, but also looking for the understandability of those key elements that might create value over time. This approach, as IIRC suggests, "can lead to behavioral changes and improvement in performance throughout an organization."(http://integratedreporting.org/what-the-tool-for-better-reporting/).

The IIRF establishes Guiding Principles and Content Elements to define the required disclosures and bases of an IR. The seven principles are: Strategic focus and future orientation; Connectivity of information; Stakeholder relationships; Materiality; Conciseness; Reliability and completeness; and Consistency and comparability. Regarding the content, the IIRF refers to eight Content Elements stated as questions, that the information disclosed in the IR should address. Content items are linked to each other and not mutually exclusive. The Content Elements are: Organizational overview and external environment; Governance; Business model; Risks and opportunities; Strategy and resource allocation; Performance; Outlook; and Basis of presentation. To be considered an IR prepared in accordance with the IIRF, the IIRC requires firms to report on at least 19 specific paragraphs out of the 168 that conform the total content requirements. These 19 items are collected in the IIRF appendix (IIRC 2013).

Integrated thinking aims to enhance accountability and foster stewardship. For this purpose, firms should take into account the effect of their decisions on the creation of value for their shareholders, over the short, medium and long term. Both, the effect of the decisions and the creation of value, are achieved through the capitals that the firms use. They are considered as "stocks of value" for the firms' activities. For this reason, the company's operations can affect or transform the capitals, and this should be reported. The six capitals are (IIRC 2013, pp. 11–12):

1. Financial Capital: the pools of fund available to an organization to produce goods or services.
2. Manufactured Capital: the physical resource available to an organization to produce goods or services.
3. Intellectual Capital: organizational and knowledge based intangibles available in an organization.
4. Human Capital: people's ability, competencies and experience in their contribution to the production of goods and services in an organization.
5. Social and Relationship Capital: the relationship established between individuals, organizations and the various stakeholders.
6. Natural Capital: all the renewable and non-renewable resources available for the production of goods and services.

Even though the framework does not require the report to be structured along those capitals, it intends to act as a theoretical underpinning for value creation (section 2B), and to make companies aware of all forms of capital.

The objective of the IIRC in the beginning was to promote sustainability accounting, but the current approach has abandoned this objective (Flower 2015). This author indicates as evidence for the change along the evolution of the framework, that the IIRC currently does not require companies to report harm inflicted outside the company (e.g. environment) unless it significantly affects the company. He also states that the reason for abandoning sustainability accounting is due to the composition of the governing council, the accounting profession and multinational companies. Other studies also highlight the scarce presence of sustainability disclosures in the IR (Alexander and Blum 2016; Thomson 2015). It seems that the IR aims to provide relevant information for investors rather than information about the firms' impact. Furthermore Thomson (2015) states that the IR may silence the radical elements of sustainability and reframe unsustainable practices as sustainable. Alexander and Blum (2016) posit that the framework has little relevance to sustainability and ecology.

During 2011 the IIRC invited a select group of companies to participate in a pilot program, with the purpose of testing the principles and practicalities of integrated reporting. The pilot program stated, "The aim of Integrated Reporting is to demonstrate the linkages between an organization's strategy, governance and financial performance and the social, environmental and economic context within which it operates." (http://integratedreporting.org/wp-content/uploads/2011/06/BriefingIntegratedReportingPilotProgramme.pdf).

Dragu and Tiron-Tudor (2013) studied the determinants for companies to participate in the program. They claimed that based on institutional theory those factors would be political, economic, and cultural. However, their findings suggest that these emergent factors have a small influence on the decision. Although with scattered and inconclusive results, these factors have been analyzed in several studies on the decision to issue an integrated report (Frías-Aceituno et al. 2013, 2014; García-Sánchez et al. 2013; Jensen and Berg 2012; Sierra-García et al. 2013; Vaz et al. 2016).

Wild and van Staden (2013) analyzed IRs submitted during the pilot period and evaluated their compliance with the IR guiding principles, content elements, and the capitals model. They found that although the reports were lengthy, they did not adhere to all the principles. Furthermore, the reports included general measures like strategy, operating context and organizational overview instead of more specific measures such as performance. They also found that the reports mostly covered four of the six capitals suggested in the guidelines, neglecting Manufactured and Intellectual capitals. Similarly, but considering companies in South Africa, Haji and Hossain (2016) find that although the adoption of IR was expected to change the reporting practices, the content of the reports is generic, biased and not relevant. Based on these previous results, we find a gap we aim to fill in this paper.

15.3 Evolution of Integrated Reporting in Spain

In order to provide a meaningful evolution of integrated reporting, we collected data from Spain. This decision was driven by a couple of facts. First, the high number of companies reporting, and the quality of the CSR reports (KPMG 2013). A study conducted by KPMG (2015) reveals that 27 out of the 100 largest companies in Spain produced IRs in 2013–2015. This high level of reporting is shared with The Netherlands. South Africa stands out with 91 of their 100 largest companies because firms are encouraged to apply the King III Code of Governance Principles, or disclose why they do not. And second, Spain is a pioneer in the evolution of corporate information. This is evidenced by the development of several projects directed to recognize intangible assets (intellectual capital, relational capital and human capital). Among these initiatives, Saitua et al. (2014) mention the Intelect Project presented in 1997, funded by the European Social Fund and the Meritum Project, presented in 1998 and funding by the European Union. Other initiatives were proposed by the Spanish Association of Accounting and Business Administration (AECA), a non-for-profit private entity created in 1979, with the purpose of promoting studies in the field of Business Sciences. This organization supports the reporting project developed by the IIRC. AECA aims to improve management techniques and information disclosure of Spanish companies. It has been developing and applying research in the standardization of non-financial information, stressing the use of the Integrated Scoreboard (Financial, Environmental, Social and Corporate Governance), since 2006. The role AECA plays in the development of the IR has been recognized by the IIRC. Furthermore, in 2017 AECA was responsible for the organization of one of the 10 meetings planned by the IIRC, within the program "Global call for feedback on business implementation of Integrated Reporting". This meeting was set to analyze the levels of implementation of Integrated Reporting.

When the IIRC issued its first draft in 2011, Lizcano et al. (2011) analyzed the differences between the conceptual proposals of the draft and the content of the 2010 IRs published by 16 Spanish firms. The authors found that the reports were generated by accumulating contents in a single document. This is far from an embedded structure that clearly evidenced linkages between financial and non-financial information, according to the concept of IR (Eccles and Krzus 2010). However, in their conclusions, they emphasize that, in general, the IRs of the sample included the contents proposed by the Discussion Paper of the IIRC. It is evident that there was a lack of compliance with the rules established in the IIRF, at least from a formal point of view.

In 2012, the IIRC launched an international pilot program, involving relevant companies all over the world. The objective was to receive a useful feedback from these companies on key aspects of the development and practical application of Integrated Reporting. Some Spanish listed companies participated in the IIRC 2012 pilot program (BBVA, Enagas, Inditex, Indra and Telefónica). The pilot program, involving well-known international companies, acted as a mimetic mechanism of institutionalization, creating a mirror effect among other companies.

Using data from the 2011 and 2013, PWC (2013, 2014) published two reports on the IR among IBEX-35 companies. The reports analyze all documentation published by these companies regardless of whether the information is published under the specific name of IR. In both reports, PWC studies areas such as the general description of the organization, strategy, business model, risks and opportunities, future prospects, corporate governance and performance. PWC (2014) emphasizes that the two areas of effective communication are the general description and the business model. The report also points out that corporate governance and performance indicators are the furthest from effective communication. The report highlights an improvement in the communication of integrated information with respect to previous years, albeit at a lower levels than the UK companies. The analysis reveals no large differences by sector.

In order to assess the evolution of the integrated reporting in Spain, we analyzed data of companies listed on the Madrid Stock Exchange (MSE) during the period 2013–2015. For each of the companies, we reviewed whether they published an IR or not. We classified a report as an IR when the reporting entity refers to it as an IR, or when the company explicitly states they follow the IIRC conceptual framework. The final figures show 66 reports from 30 companies. Table 15.1 presents the distribution of the IRs by year.

There is a sustained growth across the analyzed period. All the 14 companies that issued an IR in 2013, continued to do so in the subsequent years. Similarly, the companies that issued the IR report for the first time in 2014, continued in 2015.

Table 15.1 Distribution per year

Year	Frequency	Percentage
2013	14	21.2
2014	22	33.3
2015	30	45.5
Total (reports)	66	100.0

Table 15.2 Distribution by aggregated industry

Industry	Frequency	Environmental sensitive industry
Basic materials, manufacturing and construction	8	8
Consumer goods	4	
Consumer services	4	
Financial services and real estate	6	
Oil and energy	4	4
Technology and telecommunication	4	
Total number	30	12
Percentage over total	100%	40%

The most represented industry is Basic materials, Manufacturing and Construction followed by Financial Services and Real estate (Table 15.2). The remainder industries have the same number of companies in the sample. According to Fernandez-Feijoo et al. (2014), Basic materials, Manufacturing and Construction as well Oil and Energy are environmental sensitive industries. Companies from these industries represent the 40% of the sample.

To summarize, we posit that a significant number of MSE firms have been involved in the development of integrated reporting. Since the first drafts of this reporting model, well-known firms and relevant associations have paid attention and participated in its development. Hence, Spain is an interesting setting to carry out further research on the topic.

15.4 Empirical Analysis of 2015's Disclosures

In the empirical part of our research, we first study the integrated thinking in the CEO presentation pages, to find out if there is an underlying integrated thinking in the initial message. To achieve this aim, we include in our analysis the presentation letters/messages of the CEO and/or chairman of the 30 MSE companies that presented IR in 2015. Except for Bayer, all the companies are Spanish; hence, our results illustrate the reporting characteristics of the companies located in Spain. In the second step, we assess to what extent the IRs are committed to sustainability and really disclose integrated information. For that purpose, we analyze the content of the IRs submitted in 2015 by the same companies.

Table 15.3 presents the list of companies included in the study, their industry and their total assets.

The sample includes quoted companies, that is to say large companies, although there is a wide range that goes from total assets of 264.35 (Iberpapel) to 750,078 million euros (BBVA), being the average 68,066 million euros.

Table 15.3 Companies evaluated in the study

Company	Industry	Total Assets (thousands of euros)
Abengoa	Basic materials, manufacturing and construction	16,627,199
Acciona	Basic materials, manufacturing and construction	15,777,617
Aena	Consumer services	17,023,380
Amadeus	Technology and telecommunication	7,004,144
Atresmedia	Consumer services	1,1194,288
Banco Popular	Financial services and real estate	158,649,873
Bankia	Financial services and real estate	206,969,633
Bankinter	Financial services and real estate	65,140,000
Bayer	Consumer goods	73,917,000
BBVA	Financial services and real estate	750,078,000
Caixabank	Financial services and real estate	344,255,475
CCIP	Consumer goods	2,641,007
Cellnex	Technology and telecommunication	2,141,102
CIE automotive	Basic materials, manufacturing and construction	2,026,590
Ebrofood	Consumer goods	3,403,676
Enagás	Oil and energy	7,751,918
Ferrovial	Basic materials, manufacturing and construction	25,384,000
Fluidra	Basic materials, manufacturing and construction	746,920
Gas Natural	Oil and energy	48,132,000
Grupo Catalana Occidente	Financial services and real estate	13,295,985
Iberdrola	Oil and energy	104,664,182
Iberpapel	Basic materials, manufacturing and construction	264,350
Inditex	Consumer goods	17,357,148
Indra	Technology and telecommunication	3,064,299
Meliá	Consumer services	3,163,029
Prosegur	Consumer services	2,775,142
Sacyr	Basic materials, manufacturing and construction	10,457,474
Saeta Yied	Oil and energy	1,651,751
Técnicas Reunidas	Basic materials, manufacturing and construction	3,613,239
Telefónica	Technology and telecommunication	122,974,000

15.4.1 Analysis of the Integrated Thinking in the CEO Presentation Letters/Messages

15.4.1.1 Methodology

Integrated thinking is a process that affects internal management and connects to external reporting. Empirical research is needed in order to understand the process of integrated thinking (Feng et al. 2017). The CEO and/or the Chairman of the Board's presentation letters/messages allow the analysis of the commitment firms have to integrated reporting, as well as their integrated thinking. However, there are implicit references that have to be read between lines to find out the level of association the firm has with the integrated philosophy. For that purpose, we use interpretative textual analysis (Laine 2009), to methodically read the phrases and infer the underlying substance of the integrated thinking.

The interpretative textual analysis methodology is one of the several ways qualitative data is examined (Strauss and Corbin 1990). Its application engages researchers in a repetitive process of reading and discussing, to deal with the interpretation of the text (Laine 2005). One of the main limitations of this qualitative methodology is language and its interpretation (Willig 2008). To overcome this limitation, the process of analysis has to be carefully developed, following a rigid procedure to guarantee that relevant and material aspects of the text arise (Hood 2016). The procedure we follow began with a careful reading of the presentation letters/messages by each researcher, with the purpose of identifying evidence of the integrated thinking of the firms. To define the scores to be used, we ranked several aspects, inspired on the IIRF guiding principles: the focus of the letters/messages on shareholders or on other stakeholders; the vision on the creation value process from only an economic approach to considering others; and the involvement of a sustainability culture within the organizational culture. The result of this task is presented in Table 15.4. Low scores are assigned to companies that, despite preparing an IR, do not have a substantively approach to the holistic vision, and integrative idea that IIRC promotes. As an outcome of this phase, the researchers prepared a draft with their assessments based on the scores of Table 15.4. In a second phase, authors exchanged their drafts and scores, discussing their evaluations. When there was no coincidence in the assessment, the researchers reread the documents together and debated until they came to an agreement. The outcome of this process is a file that collects the final score. It seems necessary to mention that when the IR had presentation letters/messages from the CEO and the Chairman of the Board, we analyzed both documents and considered the one with the highest score.

Table 15.4 Scores and its definition

Score	Definition
1	Mainly focused on the shareholders. Description of the business strategy under an economic approach. Isolated reference to CSR policies, without link to the business activity
2	Mainly focused on the shareholders but referring to other stakeholders. Soft reference to the value creation process. Isolated reference to CSR policies, without link to the business activity
3	Focused on stakeholders. Isolated links among operational activity, value creation and CSR. Emerging dialogue with stakeholders
4	Focused on stakeholders. Links among operational activity, value creation and CSR. Open dialogue with stakeholders
5	Stakeholders as essential part of the business model. Clear relationship between sustainability and value chain. Organizational culture embraces sustainability

15.4.1.2 Empirical Analysis

The results of our empirical assessment of the presentation letters/messages are summarized in Table 15.5. The mean value is 2.4, the median is 2, and the mode is 2, meaning that on average, the IR is mainly focused on the shareholders but referring also to other stakeholders, with soft mention of the value creation process and isolated indications to CSR policies, also without linking them to the business activity.

Table 15.6 presents the scores of the presentation letters according to the researchers' assessment. The highest scores are those of Bayer (4), Inditex (5) and Sacyr (5). In these three companies, the integrated thinking is evidently underlying the text in the message of the CEO/Chairman. In these cases there is not only a declaration of intentions, but also a description of the actions carried out to achieve the intended objectives. For instance, Bayer's letter shows their intention to improve their performance based on CSR measures, when saying:

> In order to leverage our focus, we have given our organization a clear, integrated and more operations-based structure. Since January 2016, the heads of our new divisions – Pharmaceuticals, Consumer Health and Crop Science – have been members of the Board of Management of Bayer AG. This enhances operational accountability and accelerates decision-making, which are important for an innovation company. In this connection, I'm proud of the extensive measures we have taken to broaden Bayer's diversity in terms of gender and culture. This is now also reflected in the membership of the Board of Management (Bayer 2015, AR, p. 2).

Bayer links its activity (organizational structure, decision-making process, innovation) with non-financial CSR issues (accountability, board diversity, etc.):

> But it's not just through our innovations, but also through our activities, that we are helping people lead a better life. We seek to achieve economic growth in harmony with ecological and social responsibility. As evidence of this commitment, we adhere to the fundamentals of sustainable development and the ten principles of the United Nations Global Compact (Bayer 2015, AR, p. 6).

Furthermore, Bayer bases its business strategy on the dialogue with their stakeholders and shows its need to be legitimized by society for its activity:

Table 15.5 Number of firms by scores

Score	Number of firms
1	6
2	11
3	10
4	1
5	2
Total	30
Mean value	2.4

> In this context, it is essential to maintain an open and active dialogue with all our stakeholders. We act responsibly, have good arguments and need not shun controversy. Ultimately, no one can deny the huge benefit delivered by the innovations from our laboratories. In the life science industry, it's all too easy to overlook the significant downside of not deploying these innovations. We need the societal acceptance and appreciation to continue to contribute those benefits (Bayer 2015, AR, p. 6).

We ranked Inditex with the highest score because integrated thinking is underling the two pages of the CEO's letter. As evidence, we highlight the relationship Inditex establishes between growth and sustainability with their employees:

> Inditex sustained growth is driven by a team of over 150,000 people worldwide who are committed to the principle that all our activities at the Group, be it teamwork, innovation or management, are conducted hand-in-hand with social progress and environmental sustainability. This principle stems from the firm conviction that people, and respect for their dignity and integrity, should be our top priority (Inditex 2015, AR, p. 8).

The letter continues explaining the bases of its organizational culture on sustainability:

> This commitment to social progress and the endeavor for environmental sustainability forms the ethical principles held by all of us at Inditex. Based on these principles, we have championed and shared pioneering sustainable practices throughout our business operations, as well as with our partners and suppliers (Inditex 2015, AR, p. 8).

Social and environmental sustainability of its value chain is presented as an essential:

> Our collaboration since 2007 with the IndustriALL Global Union, which represents more than 50 million workers worldwide, is especially noteworthy. Recently, we significantly enhanced this collaboration through an agreement to establish union representatives in each of the 11 production areas or clusters Inditex uses to organize the sustainability of its production chain. The results of all of these initiatives underscore our belief that progress is best achieved by working together with expert institutions and organizations (Inditex 2015, AR, pp. 8–9).
>
> This year also saw the conclusion of our 2011-2015 environmental strategy. Notable milestones included the application of advanced environmental sustainability procedures for wet processes, an active recycling policy in our business operations, increased sourcing of sustainable raw materials, and the implementation of eco-efficient stores, offices and logistics centers (Inditex 2015, AR, p. 9).
>
> Building on these achievements, this year we launched our 2016-2020 environmental strategy, which incorporates new actions in our endeavour to close the loop at all stages of production – from the analysis and selection of raw materials to end-of-life recycling. A key project in this regard is further collaboration with social enterprises globally to support the reuse or recycling of our customers' unwanted clothes at the same time as adding social value (Inditex 2015, AR, p. 9).

Finally, we observe that Inditex communicates its business model taking into account its main stakeholders:

> I invite you to learn more about all of these topics in the pages that follow, which are a testament to the dedication and effort of the 150,000 strong team which forms Inditex. They are all proof of the company's enormous potential for growth and social progress. I would like to express my deep gratitude to these professionals for their hard work, and to our customers and shareholders, for continuing to place their trust in us (Inditex 2015, AR, p. 9).

Sacyr was also assigned 5 points, the highest score. Since the beginning of the CEO and Chairman's letter, the integrated thinking flows throughout the document. The first paragraph is devoted to the integrated strategy of the firm: "It was also a year that witnessed the Group's transformation, improved financial equilibrium, the consolidation of our business model and the achievement of sustainable growth." (Sacyr 2015, AR, p. 11). The letter continues presenting its business model following an integrated thinking approach:

> This report underlines the Sacyr Group's mission to develop complex infrastructure projects and services that help to improve people's quality of life and offer personal and professional opportunities to employees that generate value for our customers, partners, shareholders and society as a whole. This, coupled with our vision and values, is the driver of our business model. A more sustainable model that enables us to contribute positively to society as a whole (Sacyr 2015, AR, p. 11).

Interestingly enough is the link Sacyr shows among transparent management and communication, CSR, stakeholder dialogue, and sustainability:

> We firmly believe that transparency helps us to improve our business, and therefore we are committed to providing relevant, balanced and concise information for our stakeholders so that they can learn about the company's work in a transparent and open manner (Sacyr 2015, AR, p. 11).
>
> This openness to dialogue is reflected in our CSR policy, on which we base our work to develop the three dimensions of sustainability to build positive relationships with our stakeholders and uphold human rights. As a result of this commitment to stakeholders, we were included in the FTSE4GOOD IBEX sustainability index that recognises the efforts made by companies in the environmental, social and corporate governance areas (Sacyr 2015, AR, p. 11).

Sacyr's integrated thinking is also observed in the alignment of its activity/operational objectives and their global sustainability objectives:

> We are now faced with the exciting challenge of how to ensure the sustainable development of our planet, and the Sacyr group has joined the UN 17 sustainable development goals initiative to help achieve those goals that are aligned with the activities carried out by Group companies, thereby contributing to saving the planet by doing what we do best; our job. A sustainable company is a responsible company (Sacyr 2015, AR, p. 11).

Bayer's letter is addressed to shareholders and "friends" and it is 7 pages long. The two-page Inditex's letter is addressed to "friends". Finally, Sacyr has a three-page message without addressee. We conclude that neither the length nor the addressees seem to be important. The indicator is the real commitment expressing an integrated thinking, and the willingness to effectively communicate the firm's performance to its stakeholders in those integrated terms.

On the opposite side, we assign six companies (EbroFoods, Ferrovial, Gas Natural, Grupo Catalana Occidente, Indra, Prosegur) the lowest score (Table 15.6). In fact, five of the six letters/messages do not mention the fact that the report is an IR. Only Gas Natural mentions it, at the beginning and at the end of the letter. In these cases, integrated thinking is missing and there are only isolated references to CSR policies, mentioned in a very general way. Four out of six of these letters/messages are addressed to shareholders; one is addressed to shareholders, collaborators and clients; and other is a message without addressee. The number of pages ranks from 1 to 3.

The majority of the letters/messages are scored 2 (11) or 3 (10) points (Table 15.6). We assigned 2 points to those documents that, besides disclosing information about CSR policies, employees and other stakeholders, made reference to the value creation process as an antecedent of a future integrated thinking. Some examples of these references are as follows:

> At Atresmedia we are aware that our privileged position in the market carries with it a responsibility to society. With our multi-platform offer we reach almost 30 million people each day, and thanks to this effort we have a great ability to raise their awareness and mobilise them. For years, Atresmedia's commitment has been to use that potential to positively transform the environment in which the business operates and to enhance peoples' well-being (Atresmedia 2015, AR, p. 5).

> Without doubt, my main message is inspired by Enagás' mission: we will continue to develop and manage gas infrastructures efficiently and safely in order to create value for our stakeholders, always in a sustainable manner (Enagás 2015, AR, p. 7).

> Lastly, I would like to point out that at Telefónica we are committed to promoting a responsible business model capable of earning the trust of our customers, investors, employees and shareholders, and of the society at large. This relationship of trust is even more relevant in the new digital world. A world where traditional models no longer apply and new rules and values are needed. Values to respond to the demands and concerns of our stakeholders. Very clear values about what is right and what is not, what we believe in and how we behave (Telefónica 2015, AR, p. 6).

Letters/messages scoring 2 points rank form 1 to 5 pages and have several formats: one interview, three letters addressed to "friends", three letters addressed to shareholders, and four messages without addressee.

We assigned 3 points to those letters/messages that linked operational activity and effect on CSR, stakeholders, and sustainability. We labeled them as novel integrated thinkers. There is still room for improvement to spread integrated thinking all through the business. Although the set of letters/messages in this group refered to integrating sustainability in their operational activity, the flow of an assumed integrated thinking is not yet connected and underlying along the document. For instance, CIE Automotive initiates its way to integrated thinking by means of joining different elements that confirm it, but the perception of interactions and impacts is missing:

> Having reformulated our identity, we designed and approved CIE Automotive's CSR Policy. We also sought to make our commitment to a sustainable business model public by becoming a participant of the United Nations Global Compact. By means of this integrated annual report which I have the honour of introducing, we hope to present CIE Automotive in its various dimensions: economic, social, environmental and governance. We hope that by so doing we will strengthen our ties with society and all those we engage with day after day (CIE Automotive 2015, AR, p. 12).

Similar situations are found in the messages of Meliá and Iberdrola. The bases of integrated thinking seem to be established, but a further step is needed in order to get it off the ground:

This Annual Report attempts to reflect, in accordance with the Integrated Reporting principles, our strategic vision, how this is reflected in the performance, the Company's global figures as well as the impact that Meliá generates in the community and our relationship with stakeholders, without overlooking the "social cash flow" of Meliá, which reflects how our Company's revenue is redistributed in society as a whole (Meliá 2015, AR, p. 2).

Iberdrola will continue in this environment to develop its sustainable business model, which is focused on growth in regulated businesses and clean energy; geographic diversification in countries with high credit ratings; an increase in operational efficiency; the financial strength to exploit growth opportunities in the markets in which it does business; sustainable remuneration for shareholders; and a firm social commitment that seeks the creation of value for all of our Stakeholders (Iberdrola 2015, AR, p. 4).

Table 15.6 Letter/message scores

Company	Format	Number of pages	Score
Abengoa	Interview	1	3
Acciona	Interview	2	3
Aena	Letter addressed to "friends"	1	2
Amadeus	Message without addresser	2	2
Atresmedia	Message without addresser	5	2
Banco Popular	Message without addresser	1	3
Bankia	Message without addresser	2	3
Bankinter	Interview	5	3
Bayer	Letter addressed to shareholders and "friends"	7	4
BBVA	Letter addressed to shareholders	2	2
Caixabank	Message without addresser	1	2
CCIP (Coca-Cola Iberian Partners)	Message without addresser	1	3
Cellnex	Letter addressed to shareholders	3	2
CIE Automotive	Letter addressed to shareholders	2	3
EbroFoods	Letter addressed to stakeholders	3	1
Enagás	Interview	3	2
Ferrovial	Letter addressed to shareholders	1	1
Fluidra	Message without addresser	2	2
Gas Natural	Letter addressed to shareholders	2	1
Grupo Catalana Occidente	Letter addressed to shareholders, collaborators and clients	2	1
Iberdrola	Letter addressed to "friends"	2	3
Iberpapel	Letter addressed to "friends"	1	2
Inditex	Letter addressed to "friends"	2	5
Indra	Letter addressed to shareholders	2	1
Meliá	Message without addresser	2	3
Prosegur	Message without addresser	3	1
Sacyr	Message without addresser	3	5
Saeta Yield	Interview	4	3
Técnicas Reunidas	Letter addressed to "friends"	2	2
Telefónica	Letter addressed to shareholders	4	2

The ten letters/messages scoring 3 points rank form 1 to 5 pages and have several formats: four interviews, one letter addressed to "friends", one letter addressed to shareholders, and four messages without addressee. Interestingly, four of the five CEO/Chairman's letters evaluated in this group, were ranked with 3 points.

We also divide the sample based on different classification criteria, calculating the average points in each subsample. Table 15.7 shows the average points of the CEO/Chairman letter, grouped by industry classification, by format and by environmental sensitivity of the industry to which the company belongs. We observe that there is a positive relationship between number of pages and points for all observations but one. The companies with the highest score are the ones that present a letter/message with the largest number of pages. This means that companies oriented to their stakeholders, integrate in their business strategy the generation of value, and have a greater interest in presenting and explaining these circumstances in detail. The exception is for 3-page reports. This format may produce a symbolic message sent to the stakeholders. Table 15.7 shows also that companies in which the CEO/Chairman's letter is addressed to stakeholders or has no addressee, have lower average points. In relation to the industry, there is not a great dispersion in the scores. However, the companies with the highest scores are those belonging to the Consumer goods and the Financial sectors, being both well known by the general public. This behavior might be explained by the industry visibility that need the legitimation of society. According to the environmental sensitive classification used, highest scores are obtained by companies in non-environmental sensitive industries.

In our opinion, there is great variability of CEO/Chairman letters/messages in format and content. Except for 3 cases in which we find a clear orientation toward stakeholders and a well-defined relationship between sustainability and value chain, in the majority of the companies, a greater effort seems necessary to develop integrated thinking.

Table 15.7 Average points

By pages		By industry	
1 page	2.286	Basic materials, manufacturing and construction	2.375
2 page	2.384	Consumer goods	3.250
3 page	2.200	Consumer services	2.000
4 page	2.500	Financial services and real estate	2.667
5 page	2.500	Oil and energy	2.000
7 page	4.000	Technology and telecommunication	2.000
By format		By environmental sensitivity	
Interview	2.800	Environmental sensitive	2.250
Letter addressed to "friends"	2.800	Non environmental sensitive	2.500
Letter addressed to stakeholders	1.800		
Message without addresser	1.600		

15.4.2 Integrated Thinking Analysis of the IRs Issued by MSE Companies

As we previously mentioned, the IIRF defines integrated thinking as an "active consideration by an organization of the relationships between its various operating and functional units and the capitals that the organization uses or affects. Integrated thinking leads to integrated decision-making and actions that consider the creation of value over the short, medium and long term" (IIRC 2013, p. 2). Given the aforementioned definition, we expect a balanced disclosure of these capitals if the IR is really integrating the financial and non-financial information. However, we are aware that different industries might provide more information related to some forms of capitals than others. We conduct a content analysis of the reports to evaluate the level of disclosure on different capitals.

15.4.2.1 Methodology

In order to evaluate the integration of the reports, we follow the framework for the analysis of risk communication developed by Beretta and Bozzolan (2004). This methodology uses content analysis. "Qualitative content analysis is a research method for the subjective interpretation of the content of text data through the systematic classification process of coding and identifying themes or patterns." (Hsieh and Shannon 2005, p. 1278).

To build the sample we downloaded the IRs of the companies and separated the linguistic sentences using "RapidMiner". We then identified the target concept phrases based on the aforementioned capitals (Table 15.8). For each concept phrase we counted the number of sentences in which it was included. These values were used as elements to calculate the quality index as explained in the following sections. Given that the information on each of the concepts related to the six types of capitals might be expressed in different linguistic terms, to be sure that the communication complies with the IIRF framework we selected the terminology used in the IIRF. The IIRF does not require the use of the above mentioned terms, however, the guidelines indicates that the categorization of capital "serve as a guide to ensure that the organization does not overlook a capital that it uses or affects" (IIRC 2013, p. 12).

Table 15.8 Operating Capitals identified in the IIRC framework

English	Spanish
Financial Capital	Capital Financiero
Manufactured Capital	Capital Industrial
Intellectual Capital	Capital Intelectual
Human Capital	Capital Humano
Social and relationship Capital	Capital Social y relacional
Natural Capital	Capital Natural

All the IR's are available in Spanish except for Bayer's, for which we use the English version. Data were hand collected from companies listed in the Madrid Stock Exchange for the year 2015.

Different tools have been used to measure quality of reporting. One of them is the analysis of specific elements of a report (Beretta and Bozzolan 2004; Cohen et al. 2004; and Michelon et al. 2015). Beretta and Bozzolan (2004) point out that there are two aspects that have to be balanced when evaluating the quality of information: the absolute number of items disclosed and its weight in the overall information being provided. According to this rationale, Michelon et al. (2015) propose four dimensions to measure the CSR information quality, named density, accuracy, relative quantity and managerial orientation. When building our quality index we exclude the relative quantity measure because of the size of our sample. We believe that the density and accuracy measures include both the quantity and the weight of the selected phrases. We also exclude managerial orientation due to the high subjectivity that might be included in the analysis, given that in a previous run, different readers obtained different evaluations.

Our index includes the following constructs:

1. Density reflects the writing strategy and the number of phrases related to IR capitals. It is calculated as the ratio between the number of sentences in which a specific capital is included, and the total number of sentences in the report. Its value varies between 0 and 1, with values closer to 1 indicating more disclosure.
2. Accuracy reflects the type of disclosure. It is calculated by assigning 3 times more value to monetary sentences and 2 times more value to the quantitative sentences than the qualitative ones. Then the weight, considering the number of sentences in which a CSR concept phrase is included, ranges from 0 to 3, representing higher values more accurate information.

For each of these constructs we calculate one index per capital category.

The two variables have different scale. Thus, in order to avoid a scale effect, before calculating the Quality index variable, they have to be standardized. For this purpose, we apply Beretta and Bozzolan (2004)'s procedure, resulting on a variable that ranges from 0 to 1.

$$\mathrm{IND}_i^s = \frac{\mathrm{IND}_i - \min_i(\mathrm{IND}_i)}{\max_i(\mathrm{IND}_i) - \min_i(\mathrm{IND}_i)}$$

Where IND_i^s is the standardized index for observation i and IND_i is the value of observation i. After this standardization process, we calculate the Quality index for each company as the arithmetic media of the two variables already standardized.

15.4.2.2 Empirical Analysis

Table 15.9 presents the average quality index for each capital, and Table 15.10 presents the quality index for each company and capital.

Table 15.9 Quality index

Quality measures	MSE companies
Total	0.30189
Financial capital	0.19152
Industrial capital	0.11998
Intellectual capital	0.14836
Human capital	0.30365
Social capital	0.12384
Natural capital	0.14005

The average quality indexes for all categories of capital goes from 0.11998 (Industrial Capital) to 0.30365 (Human Capital), with the Financial Capital index (0.19152) being the second highest value of the capital set. Financial capital information is mainly oriented to shareholders, to whom the integrated report should be directed, according to the IIRF. Human capital information is mostly targeted to employees, who might be concerned on their contribution to the value creation (IFAC 2015). The fact that the information is clearly biased towards the financial capitals and human capital reveals the special importance given to shareholders and employees by MSE companies. The zero value of the quality index in a specific capital, means that the company does not report on that capital. Most of the reports covered, on average, two of the six capitals. This is less than the information reported by the companies in the Pilot Program (Wild and van Staden 2013). Similar to their results, we find that industrial capital has the lowest quality value.

MSE companies report on industrial, intellectual, social and relational, as well as natural capitals. This extended level of information disclosed might be explained by their reporting-on-sustainability tradition. It is widely recognized that Spain is a leading country in reporting about sustainability issues (KPMG 2015). In addition, the reference to all IIRF capitals might be related to the great interest that intangible capitals generated at the end of the last century. Evidence of this interest is the different projects on intangibles and intellectual capital developed in the 90's, as previously stated. In addition, this greater information breakdown might be a way to gain reputation and visibility.

The MSE companies' data (Table 15.10) distinguish three groups of companies. The first one includes 7 companies reporting on at least 5 of the six capitals. These companies are Abengoa, Banco Popular, BBVA, Ferrovial, Iberdrola, Sacyr and Saeta Yield. Five of the companies are classified as environmental sensitive and 2 are well known by the general public because they belong to the Financial sector. We label this group over concerned because it shows a clear interest in using the categorization stated in the IIRF, although it is not mandatory. In fact, Abengoa, the company with the highest total quality index, has been experimenting financial difficulties and, therefore, it needs to send signals to gain market confidence. In this group, only Sacyr has the highest score in the analysis of the CEO/Chairman's letters. The remainder companies scored between 1 and 3. The second group includes 7 companies not mentioning any of the capitals. These companies are Caixabank, CCIP, Cellinex, CIE Automotive, Gas Natural, Iberpapel and Prosegur.

Table 15.10 Quality index by Capitals

	Quality index							
Company name	Capitals reported	Total	Financial	Industrial	Intellectual	Human	Social and relational	Natural
Abengoa	6	0.83509	0.83333	0.83333	0.69434	0.72949	0.83929	0.84314
Acciona	1	0.33137	0.00000	0.00000	0.00000	0.43271	0.00000	0.00000
Aena	1	0.34376	0.00000	0.00000	0.41756	0.00000	0.00000	0.00000
Amadeus	2	0.34375	0.37138	0.00000	0.00000	0.00000	0.35100	0.00000
Atresmedia	1	0.34932	0.00000	0.00000	0.00000	0.39012	0.00000	0.00000
Banco Popular	5	0.46439	0.83534	0.69141	0.00000	0.71331	0.60477	0.64143
Bankia	2	0.37608	0.00000	0.00000	0.00000	0.39842	0.41622	0.00000
Bankinter	2	0.50470	0.00000	0.00000	0.00000	0.51668	0.51593	0.00000
Bayer	1	0.17167	0.00000	0.00000	0.00000	0.18445	0.00000	0.00000
BBVA	5	0.40916	0.47907	0.00000	0.48565	0.39400	0.37679	0.46439
Caixabank	0	0.00000	0.00000	0.00000	0.00000	0.00000	0.00000	0.00000
Catalana Occidente	1	0.36599	0.00000	0.00000	0.00000	0.44933	0.00000	0.00000
CCIP	0	0.00000	0.00000	0.00000	0.00000	0.00000	0.00000	0.00000
Cellinex	0	0.00000	0.00000	0.00000	0.00000	0.00000	0.00000	0.00000
CIE	0	0.00000	0.00000	0.00000	0.00000	0.00000	0.00000	0.00000
Ebrofood	1	0.37846	0.00000	0.00000	0.00000	0.49363	0.00000	0.00000
Enagás	1	0.43242	0.00000	0.00000	0.00000	0.79167	0.00000	0.00000
Ferrrovial	5	0.37734	0.41373	0.38559	0.42222	0.37241	0.00000	0.45629
Fluidra	1	0.17365	0.00000	0.00000	0.00000	0.19149	0.00000	0.00000
Gas Natural	0	0.00000	0.00000	0.00000	0.00000	0.00000	0.00000	0.00000
Iberdrola	6	0.55788	0.71022	0.65508	0.88889	0.61362	0.61126	0.73682
Iberpapel	0	0.00000	0.00000	0.00000	0.00000	0.00000	0.00000	0.00000
Inditex	1	0.42148	0.00000	0.00000	0.00000	0.43377	0.00000	0.00000
Indra	2	0.27954	0.44126	0.00000	0.00000	0.21913	0.00000	0.00000
Melia	1	0.34683	0.00000	0.00000	0.00000	0.38129	0.00000	0.00000
Prosegur	0	0.00000	0.00000	0.00000	0.00000	0.00000	0.00000	0.00000
Sacyr	5	0.36964	0.52245	0.47570	0.53021	0.50646	0.00000	0.49102
Saeta	5	0.51225	0.58949	0.55817	0.59893	0.54350	0.00000	0.56843
Técnicas Reunidas	2	0.37276	0.54940	0.00000	0.41296	0.00000	0.00000	0.00000
Telefónica	1	0.33915	0.00000	0.00000	0.00000	0.35398	0.00000	0.00000
Average	1.93333	0.30189	0.19152	0.11998	0.14836	0.30365	0.12384	0.14005

Two of them are classified as environmental sensitive, and the remainder belongs to different industries. 5 of the 7 companies are belong to the group of the smaller companies of our sample. It might be interpreted that they have fewer resources for the elaboration of this type of non-compulsory corporate disclosure. Gas Natural and Prosegur have the lowest values in the analysis of the CEO/Chairman's letters. Overall, the group scores 1 or 2, except for CCIP and CIE automotive that score 3, meaning that the letter is mainly focused on the shareholders, following an economic approach, or a soft reference to the value creation. They report their CSR policies, but in an isolated way, without link to the business activity.

The third group includes 16 companies reporting only one or two capitals. Seven of the companies are in industries classified as environmental sensitive. All of them disclose Human capital, except for Aena, Amadeus and Técnicas Reunidas, Two of them also report on Intellectual capital (Aena and Técnicas Reunidas). Bankia and Bankinter, both financial entities, report on Human as well as Social and Relational capitals and do not report on Financial capital. In this group, Inditex scores 5 in the CEO/Chairman's letter and EbroFood, Catalana Occidente and Indra score 1.

Table 15.11 presents the average quality score by industry. Our results are consistent with previous studies indicating that environmental sensitive industries report more and with higher quality than non-environmental sensitive industries, to differentiate themselves from the 'polluting companies' (Alali and Romero 2012; Campbell 2003; Deegan and Gordon 1996). We find that the environmental sensitive industries have higher scores than the non-sensitive ones. The quality scores in this group are higher for all types of capital, with the largest difference in industrial, intellectual and natural capitals.

15.5 Discussion and Conclusions

We conducted an analysis of the IR and its implementation process in order to understand the evolution of integrated reporting in Spain, to assess the integrated thinking in the CEO presentation pages of the IR, and also to study to what extent are integrated reports committed to sustainability and really disclose integrated information.

The implementation of IR in Spain has experienced an important growth among companies listed in the MSE for the period 2013–2015. The participation of 5 very visible Spanish companies in the pilot project of the IIRC, as opinion leaders, might have influenced other companies to issue IRs. Our results are consistent with those of PWC (2014), which highlights an improvement in the communication of integrated information in Spain with respect to previous years, and no large differences by sector.

Using the interpretative textual analysis methodology, we analyzed the integrated thinking in the CEO presentation letters/messages. Our analysis reveals that only 2 companies reach the highest values, meaning that stakeholders are an essential part of their business model. There is a clear relationship between sustainability and

Table 15.11 Average industry effect

	Quality index							
Industry	Capitals reported	Total	Financial	Industrial	Intellectual	Human	Social and relational	Natural
Basic materials, manufacturing and construction	2.5	0.29816	0.25279	0.24209	0.23525	0.31894	0.11990	0.25578
Consumer goods	0.75	0.24290	0.00000	0.00000	0.00000	0.27796	0.00000	0.00000
Consumer services	0.75	0.25998	0.00000	0.00000	0.10439	0.19285	0.00000	0.00000
Financial services and real estate	2.5	0.35339	0.21907	0.11524	0.08094	0.41196	0.31895	0.18430
Oil and energy	3	0.37564	0.32493	0.30331	0.37196	0.48720	0.15282	0.32631
Technology and telecommunication	1.25	0.24061	0.20316	0.00000	0.00000	0.14328	0.08775	0.00000
Sensitive industries	2.77	0.33690	0.28886	0.27270	0.30360	0.40307	0.13636	0.29105
Non-sensitive industries	1.44	0.27422	0.10556	0.02881	0.04633	0.25651	0.10168	0.04608

value chain, and organizational culture embraces sustainability. Most of the companies (11) were assigned a value of 2, meaning that the reports are mainly focused on the shareholders, with little reference to the value creation process and isolated reference to CSR policies. Ten companies are rated with 3 points, meaning that they are focused on stakeholders, with isolated links among operational activity, value creation and CSR, and with an emerging dialogue with stakeholders. Considering that CEO/Chairman play an essential role in transmitting integrated thinking to all members of the company, in our opinion, it is necessary to adequate their message to be in accordance with an effective involvement in the integrated thinking process.

Using content analysis and following the framework for the analysis of risk communication developed by Beretta and Bozzolan (2004), we explored the integrated thinking of the IRs issued by MSE companies. We calculated a quality index for each capital. Human capital and Financial capital have the highest average values of the quality index. The IR seems to be a tool to send signals to employees and involve them in the integrated thinking as well as financial capital providers, who, according to the IIRC, should be the addressees of the IR. Analyzing the number of capitals disclosed by companies, three groups were identified. First, the over-concerned companies, which report on 5 of the six capitals. As the quality indexes of each capital are similar in every company, it seems that all the capitals reported are equally relevant and applicable for them. These companies might use this report to send signals to the stakeholders, by adopting the language suggested by the IIRC. The CEO/Chairman presentation letter of a company in this group is focused on stakeholders being an essential part of the business model. The letter highlights a clear relationship between sustainability and value chain. In the presentation letter of the CEO/Chairman of another IR, the presentation is mainly focused on the shareholders. It includes a description of the business strategy under an economic approach and there are isolated references to CSR policies, without linkage to the business activity. The second group (7 companies) is labeled non-concerned companies, as they do not report on any of the capitals and the points of the CEO/Chairman letters do not exceed the value of 3. The third group includes concerned companies reporting on one or two capitals. In this group, all the possible scores of the CEO/Chairman's letters co-exist. The capitals disclosed by companies in this group are one or two of the following: Human, Intellectual as well as Social and Relational. The results on the second and third groups might be related to the wording used by the company, so that references to the capitals are stated using other terms. Future research is needed to better understand the disclosure of the six capitals identified by the IIRC. The capitals with the lowest quality scores are Industrial, Social, and Natural. Similarly, Wild and van Staden (2013) found a lack of disclosure on Industrial capital, but they also find the same deficiency in Intellectual capital.

In sum, we found disparities between the underlying integrated thinking of the CEO/Chairman's presentations and the reference to the capitals proposed by the IIRC. The Spanish language uses a lot of words. It avoids repetitions by means of an extensive use of synonymous. Therefore, similar concepts can be expressed by using different words and phrases. The disparities might be explained by firms with more integrated thinking, reporting on the six capitals in more flexible terms instead of following fixed structures. That is to say that if integrated thinking flows throughout firms, they will use a more natural verbiage in their reports, instead of sticking to the specific words suggested by IIRC. Further research is needed to develop and apply a similar methodology to analyze a set of synonym words that represent the capital that IIRC proposes.

We are aware that there are certain limitations in our study. First, the number of firms included in our sample. Although they are all the quoted companies that issued an IR during 2015, conclusions drawn from a population of 30 observations have to be cautiously analyzed. Further research should include companies from different countries to extend the sample size and also to identify cultural differences in integrated reporting. In addition, the IIRF does not require the use of the above-mentioned capital terms, but suggest them as a guide to report on them. The addition of different terms to be analyzed in the capitals as indicated above, would provide an opportunity to overcome this limitation.

References

Alali, F., & Romero, S. (2012). The use of the internet for corporate reporting in the Mercosur (southern common market) – The Argentina case. *Advances in Accounting, 28*(1), 157–167.

Alexander, D., & Blum, V. (2016). Ecological economics: A Luhmannian analysis of integrated reporting. *Ecological Economics, 129*, 241–251.

Beretta, S., & Bozzolan, S. (2004). A framework for the analysis of firm risk communication. *The International Journal of Accounting, 39*, 265–288.

Campbell, D. (2003). Intra- and inter-sectoral effects in environmental disclosures: Evidence for legitimacy theory? *Business Strategy and the Environment, 12*(6), 357–371.

Cohen, J., Krishnamoorthy, G., & Wright, A. (2004). The corporate governance mosaic and financial reporting quality. *Journal of Accounting Literature*, 87–152.

Deegan, C., & Gordon, B. (1996). *A study of the environmental disclosure practices of Australian corporations*. Kingston: Wolters Kluwer.

Dragu, I. M., & Tiron-Tudor, A. (2013). The integrated reporting initiative from an institutional perspective: Emergent factors. *Procedia-Social and Behavioral Sciences, 92*, 275–279.

Eccles, R. G., & Krzus, M. P. (2010). *One report. Integrated reporting for a sustainable strategy*. Hoboken, NJ: Wiley.

Ernst and Young. (2012). *Driving value by combining financial and non-financial information into a single, investor-grade document*. Retrieved from http://www.ey.com/Publication/vwLUAssets/Integrated_reporting:_driving_value/$FILE/Integrated_reporting-driving_value.pdf

Feng, T., Cummings, L., & Tweedie, D. (2017). Exploring integrated thinking in integrated reporting – An exploratory study in Australia. *Journal of Intellectual Capital, 18*(2), 330–353.

Fernandez-Feijoo, B., Romero, S., & Ruiz, S. (2014). Effect of stakeholders' pressure on transparency of sustainability reports within the GRI framework. *Journal of Business Ethics, 122*, 53–63.

Flower, J. (2015). The international integrated reporting council: A story of failure. *Critical Perspectives on Accounting, 27*, 1–17.

Frías-Aceituno, J. V., Rodríguez-Ariza, L., & García-Sánchez, I. M. (2013). Is integrated reporting determined by a country's legal system? An exploratory study. *Journal of Cleaner Production, 44*, 45–55.

Frías-Aceituno, J. V., Rodríguez-Ariza, L., & García-Sánchez, I. M. (2014). Explanatory factors of integrated sustainability and financial reporting. *Business Strategy and the Environment, 23*(1), 56–72.
García-Sánchez, I. M., Rodríguez-Ariza, L., & Frías-Aceituno, J. V. (2013). The cultural system and integrated reporting. *International Business Review, 22*(5), 828–838.
Haji, A. A., & Hossain, D. M. (2016). Exploring the implications of integrated reporting on organizational reporting practice: Evidence from highly regarded integrated reporters. *Qualitative Research in Accounting and Management, 13*(4), 415–444.
Hood, R. (2016). Combining phenomenological and critical methodologies in qualitative research. *Qualitative Social Work, 15*(2), 160–174.
Hsieh, H., & Shannon, S. E. (2005). Three approaches to qualitative content analysis. *Qualitative Health Research, 15*(9), 1277–1288. https://doi.org/10.1177/1049732305276687.
International Federation of Accountants, IFAC. (2015). *Creating value with integrated thinking. The role of professional accountants*. IFAC.
International Integrated Reporting Council (IIRC). (2011). *Towards integrated reporting. Communicating value in the 21st century*. Accessed June 10, 2012, from www.theiirc.org/the-integratedreporting-discussion-paper/
International Integrated Reporting Council (IIRC). (2013). *The international <IR> framework*. Retrieved from integratedreporting.org/wp-content/uploads/2013/12/13-12-08-THE-INTERNATIONAL-IR-FRAMEWORK-2-1.pdf
Jensen, J. C., & Berg, N. (2012). Determinants of traditional sustainability reporting versus integrated reporting. An institutionalist approach. *Business Strategy and the Environment, 21* (5), 299–316.
KPMG. (2013). *The KPMG survey of corporate responsibility reporting 2013*. Accessed March 7, 2017, from http://www.kpmg.com/Global/en/IssuesAndInsights/ArticlesPublications/corporate-responsibility/Documents/corporate-responsibility-reporting-survey-2013.pdf
KPMG. (2015). *Currents of change. The KPMG survey of corporate responsibility reporting 2015*. Retrieved from https://home.kpmg.com/content/dam/kpmg/pdf/2015/12/KPMG-survey-of-CR-reporting-2015.pdf
Laine, M. (2005). Meanings of the term 'sustainable development' in Finnish corporate disclosures. *Accounting Forum, 29*(4), 395–413.
Laine, M. (2009). Ensuring legitimacy through rhetorical changes? A longitudinal interpretation of the environmental disclosures of a leading Finnish chemical company. *Accounting, Auditing and Accountability Journal, 22*(7), 1029–1054.
Lizcano, J. L., Flores, F., & Rejón, M. (2011). In Retrieved from http://aeca.es/old/comisiones/rsc/estudio_empirico.pdf (Ed.), *El Informe Integrado: primeras aproximaciones de aplicación: El caso español*. Madrid: AECA.
Michelon, G., Pilonato, S., & Ricceri, F. (2015). CSR reporting practices and the quality of disclosure: An empirical analysis. *Critical Perspectives on Accounting, 33*, 59–78.
PwC. (2013). *Momento para la diferenciación Las empresas del IBEX 35 avanzan hacia el reporting integrado*. PWC. Retrieved from https://www.pwc.es/es/publicaciones/auditoria/assets/reporting-integrado-ibex-35.pdf
PwC. (2014). *En la senda de la integración Análisis del estado del reporting integrado en las empresas del IBEX 35*. PWC. Accesed from https://www.pwc.es/es/publicaciones/auditoria/assets/pwc-reporting-integrado-ibex-2014.pdf
Saitua, A., Andicoechea, L., & Albizu, E. (2014). El informe integrado: una propuesta de encaje para la información sobre capital humano. *Revista perspectiva empresarial, 1*(1), 33–47.
Sheridan, B. (2014). Integrated reporting more than a sum of parts. Deloitte corporate reporting. *Accountancy Ireland, 46*(1), 14–15.
Sierra-García, L., Zorio-Grima, A., & García-Benau, M. A. (2013). Stakeholder engagement, corporate social responsibility and integrated reporting: An exploratory study. *Corporate Social Responsibility and Environmental Management, 22*(5), 286–304. https://doi.org/10.1002/csr.1345.

Strauss, A., & Corbin, J. M. (1990). *Basics of qualitative research: Grounded theory procedures and techniques*. Thousand Oaks, CA: Sage Publications.

Thomson, I. H. (2015). But does sustainability need capitalism or an integrated report. A commentary on 'The international integrated reporting council: A story of failure' by Flower, J. *Critical Perspectives on Accounting, 271*, 8–22. https://doi.org/10.1016/j.cpa.2014.07.003.

Vaz, N., Fernandez-Feijoo, B., & Ruiz, S. (2016). Integrated reporting: An international overview. *Business Ethics: A European Review, 25*(4), 577–591.

Wild, S., & van Staden, C. (2013). Integrated reporting: Initial analysis of early reporters – An institutional theory approach. In *7th Asia Pacific interdisciplinary accounting research conference*, Kobe.

Willig, C. (2008). *Introducing qualitative research in psychology* (2nd ed.). Maidenhead: Open Univesity Press.

Integrated Reports (Accessed from February 2017 to July 2017)

Abengoa

http://www.abengoa.es/export/sites/abengoa_corp/resources/pdf/gobierno_corporativo/informes_anuales/2015/Tomo1/2015_Tomo1_IA.pdf

Acciona

http://memoria2015.acciona.com/d/informe-integrado-i.pdf

Aena

http://www.aena.es/InformeInteractivo_esp/files/assets/basic-html/page-1.html#

Amadeus

http://www.amadeus.com/msite/global-report/2015/es/index.html

Atresmedia

http://www.atresmediacorporacion.com/a3document/2016/04/15/DOCUMENTS/00905/00905.pdf

Banco Popular

http://www.grupobancopopular.com/ES/AccionistasInversores/GobiernoCorporativo/JuntaGeneraldeAccionistas/Documents/Junta%20Ordinaria%202016/Informe%20Integrado%202015_DEF.pdf

Bankia

https://www.bankia.com/recursos/doc/corporativo/20120927/anual/informe-anual-bankia-2015-ano-3.pdf

Bankinter

https://docs.bankinter.com/stf/web_corporativa/accionistas_e_inversores/info_financiera/memoria/2015/informe_anual_integrado.pdf

Bayer

http://www.annualreport2015.bayer.com/BBVA

BBVA

https://shareholdersandinvestors.bbva.com/wp-content/uploads/2017/09/BBVAen2015_tcm926-569028.pdf

Caixabank

https://www.caixabank.com/deployedfiles/caixabank/Estaticos/PDFs/responsabilidad_corporativa/Informe_Corporativo_Integrado_2015_CaixaBank.pdf

CCIP (Coca-Cola Iberian Partners)

http://www.ccepiberia.com/memoria-2015/es/downloads/CCIP-Informe-Anual-2015.pdf

Cellnex

https://www.cellnextelecom.com/content/uploads/2016/06/cellnex_IA_2015_ESP2-1.pdf

CIE Automotive

http://www.cieautomotive.com/documents/10182/152199/Informe+Anual+CIE+Automotive+2015%281%29.pdf/34b875eb-e1ce-4ec4-ae2b-390e178c0d5f

EbroFoods

http://www.ebrofoods.es/fileadmin/user_upload/informe_anual/InformeAnual2015/pdf/1-0-informe-anual.pdf

Enagás

http://www.enagas.es/stfls/ENAGAS/Relaci%C3%B3n%20con%20inversores/Informe%20Anual%20Interactivo_.pdf

Ferrovial

http://www.ferrovial.com/wp-content/uploads/2014/06/FERROVIAL-INFORME-ANUAL-INTEGRADO-CONSOLIDADO-2015.pdf

Fluidra

https://www.fluidra.com/wp-content/uploads/2017/05/Memoria-Fluidra-2015.pdf

Gas Natural

http://www.gasnaturalfenosa.com/servlet/ficheros/1297152555862/Informe_Integrado_cast.pdf

Grupo Catalana Occidente

https://www.grupocatalanaoccidente.com/~/media/Files/G/Grupo-Catalana-Occidente/information-economica/informe_anual_Interactivo_2016_v2_ESP.pdf

Iberdrola

https://www.iberdrola.com/wcorp/gc/prod/es_ES/inversores/docs/IA_InformeIntegrado16.pdf

Iberpapel

http://www.iberpapel.es/archivos/accionista_e_inversores/informacion_economica_financiera/memoria_anual/Informe-Anual-2015.pdf

Inditex

https://www.inditex.com/documents/10279/246747/Inditex+Memoria+Anual+2015+web.pdf/e94736f4-b0bd-4965-bea8-471fb1b8eb35

Indra

http://www.indracompany.com/sites/default/files/indra_informe_anual_rsc_2015_es_07062016_1.pdf

Meliá

http://annualreport.meliahotelsinternational.com/2015/themes/melia2015/assets/download/es/informe-anual-2015.pdf

Prosegur

http://www.prosegur.com/accionistas_inversores/informacion_financiera/informes_anuales

Sacyr

http://informesanuales.sacyr.com/2015/

Saeta Yield

http://www.saetayield.com/wp-content/uploads/2016/11/informe-integrado-saeta-yield-2015.pdf

Técnicas Reunidas

http://www.tecnicasreunidas.es/recursos/doc/accionistas-e-inversores/memoria-sostenibilidad/tr-informe-integrado-2015.pdf

Telefónica

https://www.telefonica.com/documents/153952/13347843/informe_anual_2015_es.pdf/1c6c42b3-0813-488e-995c-413229577fb6

Chapter 16
Early Adopters of Integrated Reporting: The Practical Evidence from Warsaw Stock Exchange Companies

Maria Aluchna and Maria Roszkowska-Menkes

16.1 Introduction

In a complex and turbulent business environment firm's ability to compete and grow largely depends on how well it manages its stakeholder relationships (Freeman 1999). As value creation process of a company is influenced both by tangible and intangible capitals, there is growing demand among shareholders and other stakeholders for transparent communication on company's performance—communication that could give a true and whole picture of business operations (Clayton et al. 2015; Dragu and Tiron-Tudor 2013; Eccles et al. 2011). In response to this demand non-financial reporting has proliferated, becoming one of the most dynamically developing themes in the literature on CSR and sustainability management. It originated from CSR and environmental disclosure and developed towards integrated reporting to provide a complete picture of multidimensional firm performance. Integrated reporting promises to balance short-, medium- and long-term perspectives of information disclosure and management orientation, to empower stakeholders incorporating their expectations into business strategy and, as a result, to move sustainability reporting from the business periphery to its mainstream. In contrast to traditional financial reporting focused more on short-term financial performance, integrated reporting offers more future-oriented value-based approach (Adams 2015; Ballou et al. 2012; Beattie and Smith 2013; Eccles and Krzus 2010; Geels 2011).

The idea of integrating sustainability and financial disclosure has gained interest among managers and researchers. More than 600 companies around the world (GRI Sustainability Disclosure Database) have already implemented this form of reporting. However, at the current development stage it is still neither institutionalized in

M. Aluchna (✉) · M. Roszkowska-Menkes
Warsaw School of Economics, Warszawa, Poland
e-mail: maria.aluchna@sgh.waw.pl

S. O. Idowu, M. Del Baldo (eds.), *Integrated Reporting*, CSR, Sustainability, Ethics & Governance, https://doi.org/10.1007/978-3-030-01719-4_16

corporate practice nor follows universal and widely recognized standards (Lodhia 2015). Academic research in this field is starting to expand and provide regulators and managers with valuable insights to foster further development of policy and practice (Villiers et al. 2014).

We would like to add to this literature examining the practice of 5 Polish listed companies who are the only firms quoted on the Warsaw Stock Exchange to currently adopt integrated reporting approach in their disclosure policy. We identify their organizational and sectorial characteristics, disclosure dynamics and evolution as well as the practice of integrated reporting. More specifically, using the IIRC framework we analyze the size, the scope, standards and the quality of non-financial information provided by our sample companies. Our goal is to identify challenges and success factors of early adopters of integrated reporting.

The remainder of the article is as follows. In the first section we explain the concept of integrated reporting as discussed in the existing literature and in the IIRC's framework. Further the research method and overview of sample companies is presented, followed by the results of the analysis of Polish early adopters' integrated reporting practice. Our findings are discussed in the discussion section. Final remarks are provided in conclusion.

16.2 Integrated Reporting

Integrated report is defined as "a concise communication about how an organization's strategy, governance, performance and prospects, in the context of its external environment, lead to the creation of value over the short, medium and long term" (IIRC 2013: 7). It demonstrates a reorientation of the focus of corporate reporting from short-term, backward-looking financial information to forward-looking, connected and strategic information that discusses an organization's ability to generate value over time (Adams and Simnett 2011). The primary purpose is to present a comprehensive picture of organization's performance and provide investors with information necessary for efficient capital allocation. However, it is also intended to help managers to create firm value (Demartini and Paolini 2013) by encouraging them to base their strategic decisions on criteria that include a full range of value drivers. In this way, it aligns goals of profit maximization and wider financial stability with sustainable development and stakeholder interests (Adams 2015).

The existing literature identifies various benefits of integrated reporting such as enhanced reputation, effective decisions and capital allocation, interest from shareholders, profit increases, future orientation, stakeholder engagement, retaining customers and employees, effective risk (including regulatory risk) management (Dragu and Tiron-Tudor 2014; Eccles and Krzus 2010; Frias-Aceituno et al. 2014; Higgins et al. 2014; IIRC 2013). This form of disclosure is perceived as a communications tool that enables to explain stakeholder company's business model, strategy, governance and operational performance (Lodhia 2015; Higgins et al. 2014). Studies show that early adopters of IR benefit from increased understanding of value creation in the long-term and improvements in decision-making due to changes in management information (IIRC and Black Sun 2015). Other researcher (Higgins et al. 2014; Stubbs and Higgins 2014), however, argue that integrated reporting stimulates merely incremental changes to processes and structures, since reporting managers do not have the agency or the responsibility to bring any fundamental ones.

The most recent significant global development in the area of integrated reporting is the formation of the International Integrated Reporting Council (IIRC), coalition of regulators, investors, companies, standard setters, accountants and NGOs, and the publication of International <IR> Framework in 2013 (IIRC 2013). The IIRC's concept is built around the notion of value that is generated from six categories of capital, not necessarily owned by the company: financial, manufactured, intellectual, human, social and relationship, and natural (IIRC 2013). The underlying assumption is that the dichotomy between social and business goals does not exist (Porter and Kramer 2011) and that CSR and competitiveness are linked in a synergic relationship (Perrini et al. 2006: 6). The satisfaction of various entities is interdependent and shareholders may enjoy higher firm value only when the expectations of other stakeholders are simultaneously met (Freeman 1999). However, since financial providers are the target audience of integrated reporting, the IIRC's framework focuses on value created for the organization. Thus, the materiality of a particular social or environmental issue and its inclusion in the report (as this depends on whether the issue is material or not) is determined by its impact on the organization's value.

The IIRC's framework (IIRC 2013: 16–23) delivers seven guiding principles for preparing and defining the content of an integrated report:

1. Strategic focus and future orientation—an integrated report should provide insight into the organization's strategy, how it creates value and how it effects particular capitals
2. Connectivity of information—an integrated report should present the holistic picture of the company's value creation process
3. Stakeholder relationships—an integrated report should provide information on how and to what extent the organization understands, takes into account and responds to their legitimate needs and interests
4. Materiality—an integrated report should disclose information about matters that substantively affect the organization's ability to create value over the short, medium and long term
5. Conciseness—an integrated report should be concise
6. Reliability and completeness—an integrated report should include all material matters, both positive and negative
7. Consistency and comparability—the information in an integrated report should be presented in a consistent way, enabling comparison with other organizations.

IIRC (2013) promotes integrated reporting as a voluntary disclosure practice. Since CSR strategies are highly organization-specific, the framework does not impose any specific obligations on companies. Being principle-based it includes a small number of general requirements and provides no indicators. The latter are provided, however, by Global Reporting Initiative (GRI) guidelines, that is internationally the most prominent and most widely used guidelines for non-financial data disclosure (Clayton et al. 2015). On the one hand principle-based approach ensure high applicability of the framework for organizations from different sectors. On the other hand, it hinders comparability (Zicari 2014) and creates space for abuse (Flower 2015). Thus, some argue that there is need for integrating reporting assurance as a fundament mechanism enhancing credibility and reliability (IIRC 2014).

16.3 The Practice of Integrated Reporting in Poland

16.3.1 Research Method and Sample Companies

The aim of the research was to draw upon the practice on integrated reporting by Polish companies listed on the Warsaw Stock Exchange (WSE). WSE is a successfully developing emerging stock market which reopened in 1991 and currently hosts 487 companies, including 54 foreign firms. The market capitalization estimated at €320 bn equals approximately 50% of Polish GDP. Since of 487 listed companies only five firms adopt the framework of integrated reporting we follow the case study approach. Table 16.1 presents the general overview of sample companies.

Table 16.1 reveals that sample companies differ in terms of the sector of operation and size however as referred to the general population of listed companies they belong to top 30 by market cap and assets and represent industries of significant negative impact on the natural environment. In addition, all sample companies are characterized by significant stakes in the hands of the government.

In our case study approach we studied corporate documents, websites and integrated annual reports. We analyzed the general characteristics of sample companies and identified the IR practice with respect to size of the report, adopted framework and sections covered. We also analyzed the procedure of the reporting and measures used. We provided an overview of integrated reporting of sample companies adopting IIRC framework to identify their practice. Specifically, we use the seven principles adopting the following qualitative measures:

Table 16.1 Overview of sample companies, data for 2016, in thousand PLN

	Sector of operation	Assets	Revenue	Net profit
LOTOS	Oil extraction and refining	19,326,309	20,931,090	1,015,247
KGHM	Copper mining	33,442,000	19,156,000	−4,371,000
TAURON	Energy generation	3,345,694	17,646,489	367,468
ORLEN	Oil extraction and refining	55,559,000	79,553,000	5,261,000
AZOTY	Chemical	10,777,169	10,024,405	609,499

Source: bankier.pl

- strategic focus—represents the case when reports contain information regarding future plans or goals and whether there is information regarding risks and opportunities related to environmental and social issues
- connectivity of information—represents the relations between core business strategy and sustainability strategy, information mutual interdependence of five capitals (non-financial and financial data); report structure (separating/integrating business and sustainability issues)
- materiality—provides information on the process of definition scope and range of the report
- stakeholder relations—addresses information on the process of defining the content of the report, the process need to be conducted within the cooperation with stakeholders (as opposed to situation when the dialog was replaced by expert opinion)
- conciseness—identification of size and/or tools (internet) increasing user-friendliness of the report
- reliability and completeness—information whether the report underwent the audit or verification by an external entity (auditor, consulting company)
- consistency and comparability—presentation of the data in comparison with previous years and use of GRI standards that allow comparability between organizations.

16.3.2 Grupa LOTOS SA

16.3.2.1 Company Characteristics[1]

LOTOS Group (Grupa LOTOS) is one of the Europe's major oil companies and the second largest producer of fuels in Poland. Its business consists of the extraction and processing of crude oil, as well as wholesale and retail sale of petroleum products. LOTOS Group is a producer and supplier of a number of products, including unleaded gasoline, diesel oil, diesel oil for heating purposes (light fuel oil), aviation fuel and heavy fuel oil. It also specializes in the production and sale of lubricating oils and bitumens. The LOTOS Group employs more than 5000 people in Poland and abroad.

LOTOS business group comprises of the parent company, which manages the refinery in Gdańsk and 15 affiliated firms which operate across the whole upstream and downstream value chain. Two of them are based outside Poland, in Lithuania and Norway. The refinery in Gdańsk is one of the newest and most advanced refineries in Europe in terms of applied technologies and environmental protection. Through LOTOS Petrobaltic S.A. and LOTOS Exploration and Production Norge AS, LOTOS is engaged in the exploration and production of crude oil from the

[1]Based on http://www.lotos.pl/en/866/this_is_lotos/activity/exploration_and_production

Baltic Sea and the Norwegian Continental Shelf. LOTOS Group extracts 28 barrels of crude oil each day. It is the sole company producing hydrocarbons in the Polish Exclusive Economic Zone of the Baltic Sea, and has almost exclusive rights to carry out exploration and production work in the Baltic Sea. The company also has access to onshore hydrocarbon deposits in Lithuania through its subsidiary AB LOTOS Geonafta. In the upstream segment: LOTOS focuses on enhancing the safety of supplies of oil processed by the refinery by obtaining access to hydrocarbon fields, and on raising hydrocarbon production by executing programs of increasing oil extraction from the Baltic Sea and the new projects outside Poland.

LOTOS has been listed on the Warsaw Stock Exchange since June 2005. The State Treasury remains the largest shareholder holding 53% of its shares. In November 2009 LOTOS joined the Warsaw Stock Exchange's RESPECT Index of socially responsible companies, which is the first such index in Central and Eastern Europe.[2]

16.3.2.2 Integrated Reporting

LOTOS Group is the national and sectorial leader of sustainability and integrated reporting. Already in 2007–2009, it published separate reports presenting its financial performance and its social and environmental performance. In 2010, as the first company in Poland, it adopted a new reporting model, meeting the requirements for both financial and sustainability reporting (triple bottom line reporting). The reporting was integrated following consultations held with key stakeholder groups in order to understand their expectations and to incorporate stakeholder interest in the reporting process. Over years the reports grew in terms of scope and size peaking in 2014 with 468 pages. Addressing the issues of functionality the size of the 2015 Report dropped to 264 pages. The general overview of 2014 and 2015 reports is presented in Table 16.2.

As presented in Table 16.2 the drop in the report size correlates with the change on its structure. While in both 2014 and 2015 LOTOS adheres to the framework of GRI, IIRC and Global Compact, it introduces the address to six capitals in 2015 and structures the reports in a different way.

In both cases LOTOS declares the rules of "accuracy, relevance, completeness, comparability, balance, and reliability" (LOTOS 2014: 21). In addition, its reports undergo independent assurance of CSR data which aims to ensure an appropriate quality of the reporting data collection process through the management systems in place at our organization. As formulated in the 2014 Report "all information presented in this Report has been gathered in a reliable and responsible manner and verified as true" (LOTOS 2014: 22). It uses majority of GRI G4 core and additional performance indicators covering economic, environmental and social aspects of our business and delivers indicators from the GRI Oil & Gas Sector

[2] See http://www.odpowiedzialni.gpw.pl/

Table 16.2 The general overview of LOTOS reports 2014–2015

Year	Size	Framework	Sections
2014	468 pages	Framework of GRI, Global Compact, EU Directive 2014/95/UE and frameworks of IIRC	Ethics and corporate governance Risk and opportunities Business strategy and model Results and prospects Financial information
2015	264 pages	Framework of six capitals, GRI and IIRC Addressing US sustainable Goals and Global Compact framework	2015 at the LOTOS group Strategy Innovation is key To lasting success Product, customer—commitment To quality, safety and brand image Efficient and stable organization Social impact Report

Source: own analysis based on annual reports, 2014 and 2015

Supplement. For the purpose of calculating the financial data the same methods as in the LOTOS Group's consolidated financial statements was adopted. While the consolidated financial statements of the LOTOS Group, we have included in this report have been audited by Ernst and Young Audyt Polska. In order to define key rules and procedures for assurance services, also those limited in scope the non-financial data, independent limited-scope external assurance was carried out in 2015 and 2016 by PwC in accordance with the International Standard on Assurance Engagements (ISAE) 3000, Assurance Engagements Other than Audits or Reviews of Historical Financial Information. ISAE 3000, issued by the International Federation of Accountants (IFAC), is based on the IFAC Code of Conduct and International Standards on Quality Control (ISQC-1).

Both reports emphasize the importance of stakeholder engagement as the company aims at "decision-making processes concerning economic, social and environmental matters to take into account the interests of various stakeholder groups, which are important to our business. It notes that "We also make sure that our stakeholders have access to timely information that meets their various needs" (LOTOS 2015a, b: 245). The 2015 Report as compared to its 2014 predecessor delivers a precise method of stakeholder identification and categorization and offers also a more detailed description of the process of defining the report content with the materiality analysis. It offers a clear procedure including three steps including (LOTOS 2015a, b):

- identification (key topics in corporate social responsibility and business based on the analysis of internal documents and materials as well as external reports and publications and media information)
- prioritization (dialogue with internal stakeholders with the identification of the materiality of key aspects for the organization)
- validation (workshops with management staff of 49 persons representing 11 companies to defined the final list of material aspects to be included in the 2015 report)

This procedure results in constructing a matrix which shows the significance of issues divided into aspects (economic, social and environmental) and simultaneously leads directly to the respective chapter (LOTOS 2015a, b). The size of the tile indicates whether a subject has been defined as an important or very important.

Data are disclosed in comparison to previous years and company presents risk and opportunities analysis (including social and environmental risks), as well as business and sustainability priorities for next years. However, the reader is not provided with any specific, measurable CSR goals, although the company declares that such operational goals have been set and reports on the level, in which they were met (LOTOS 2015a b: 65–68).

Both 2014 and 2015 reports are published in electronic form only for environmental reasons and assures also a number of functionalities so allow independent and flexible analysis and the interactive selection of data required. For instance readers can display selected numerical data by periods, GRI indicators, business segments, and other categories. The information presented in the publication is integrated with the contents of the organization's website, including earlier reports. The readers can also use LOTOS databook, glossary of industry and social terms as well as provide feedback. Additional functionalities have been implemented for the disabled, elderly and vision impaired, including a web reader, high-contrast option and adjustable font sizes. Economical printing option is also available to protect the environment.

16.3.3 KGHM Polska Miedź SA

16.3.3.1 Company Characteristics[3]

KGHM (KGHM Polska Miedź SA) specializes in technologically advanced exploration and smelting activities focusing on the extraction of ores, production of copper and other non-ferrous metals. It is one of the sixth producer of copper ore (ca. 550 thousand tons), the largest producer of refined silver (ca. 1300 tons) worldwide located in the south-western part of Poland. KGHM is organized as a business group with the parent company and 30 affiliated firms revealing product and geographical diversification. The company has located its facilities on three continents—Europe (Poland), the North (US, Canada) and South (Chile) America. Polish copper deposits—one of the biggest in the world—are exploited by three underground mines: "Lubin", "Polkowice-Sieroszowice" and "Rudna". The extracted material is enriched in the Concentrator Plant, while the production of copper, silver, gold, lead and other metals takes place in smelters: "Głogów", "Legnica" and "Cedynia". KGHM owns six mines: Robinson, Carlota (USA), McCreedy West, Morrison (Canada) and Franke and Sierra Gorda (Chile). Apart from copper, these

[3]Based on http://kghm.com/en

mines also produce molybdenum, nickel, gold, palladium and platinum. With its control over 22.7 million tons of copper ore resources worldwide KGHM has risen to the well-deserved position of a global mining industry leader. The company's portfolio also includes new metals like molybdenum, palladium or nickel which help KGHM join the international community of multi-resource companies. The company employs a total of 34,000 people.

Since July 1997 KGHM has been listed on the Warsaw Stock Exchange. The State Treasury is the largest shareholder holding nearly 32% of shares. In November 2009 KGHM joined the Warsaw Stock Exchange's RESPECT Index of socially responsible companies.[4]

16.3.3.2 Integrated Reporting

KGHM started its non-financial disclosure in 2011 with the CSR report addressing mostly the expectations of employees and the issues of working conditions, corporate governance and the relations with communities. In 2013 the company issued its first 300-page long integrated report alongside with the additional CSR disclosure continued throughout 2014. In 2015 company published integrated reporting including all aspects of its activity. Historically since 2013 reporting balances economic, social and environmental performance addressing the industry impact with respect to exploration and extraction operations. While the environmental section was added later, it now constitutes a section of the annual performance disclosing emission and waste production levels. The lesson learned from first version of integrated reporting resulted in a more coherent structure of 2014 and 2015 reports and led to the enrichment of the display on the interactive corporate website which at the moment of the analysis were not working. The general overview of 2014 and 2015 reports is presented in Table 16.3.

As presented in Table 16.3 KGHM integrated reporting follows relatively similar pattern every year (it was the same in 2013 which is not provided in Table 16.2). The structure indicates a greater emphasis on the market and financial results as well as the standards of corporate governance while social and environmental dimensions are viewed as sub-section of the annual performance (KGHM 2014, 2015).

Both reports were prepared in compliance with the guidelines of the International Integrated Reporting Council (IIRC) and the Global Reporting Initiative (GRI)—the G4 version, core level. None of the reports was verified externally. For both years the company followed the three steps of 1) identification of material sustainable development issues, 2) prioritization of material sustainable development issues and 3) validation workshop attended by senior management. However, no details on the adopted procedure in 2014 are provided. For 2015 KGHM discloses that during the identification and prioritization process, it analyzed documents, publications and press releases referring the Group and industry, including reports of competitors. "A

[4]See http://www.odpowiedzialni.gpw.pl/

Table 16.3 The general overview of KGHM reports 2014–2015

Year	Size	Framework	Sections
2014	158 pages	Framework of IIRC and GRI	Introduction About us KGHM in 2014 Our strategy and perspectives Corporate governance Risk management About the report Selected items from the standalone and consolidated financial statements
2015	88 pages	Framework of IIRC and GRI	Introduction Our business Market environment in 2015 Our results Corporate governance About this report Selected items from the separate and consolidated annual financial statements Additional GRI indicators

Source: own analysis based on annual reports, 2014 and 2015

chart of KGHM Polska Miedź S.A. and KGHM International Ltd. Stakeholders was created and their expectations within the scope of this Report were examined" (KGHM 2015: 68). Interviews were conducted with management and senior management of the Group. Key issues for KGHM included (KGHM 2015):

- resource levels of copper and silver, the outlook for mining operations, resource base development (Sects. 5.4, 5.5)
- improved production efficiency (Sect. 5.6)
- research and development (Sect. 5.8)
- ensuring the financial stability of the Company, reducing operational costs (C1) (Sect. 5.3)
- model for adding value (Sect. 3.3)
- KGHM vs. competitors (Sect. 5.1)
- the impact of KGHM on the environment (Sects. 3.9, 5.7, 5.9, 5.10)
- measures to protect the environment, reduction of energy and water consumption (Sect. 5.7)
- occupational health and safety, quality of human resources (Sect. 5.9)

Similar research was also conducted for KGHM International Ltd. stakeholders. Responses addressed indirect economic impact, improved management efficiency, anti-corruption and compliance, protecting human rights, diversity and equal opportunity, priorities of dividend policy and relationship with trade unions (KGHM 2015). Finally, both reports are viewed as a tool to identify stakeholder expectations however mostly limited to the content and structure of the report.

Analysis of risks and opportunities presented in the reports does not include any environmental or social issues. Although publications provide reader with some information regarding future business and CSR plans (CSR Strategy), no measurable goals are mentioned regarding the latter. Additionally, there is no information regarding the relation between business and sustainability strategy, no interdependencies between economic, social and environmental data are highlighted. However, the data are presented in a way that enables comparison with previous years.

While the pdf versions are concise and comprehensive, the company fails to offer presentation of selected aspects and interactive access to the data on its corporate website.

16.3.4 *TAURON Polska Energia S.A.*

16.3.4.1 Company Characteristics[5]

TAURON Polska Energia S.A. is also the second largest electricity generator and supplier in Poland and the largest supplier of heat in Upper Silesia. It is the corporate parent in the TAURON Group—one of the largest business entities in Poland with approximately PLN 16 billion equity and approximately 25,000 employees. The activity of TAURON Group comprises all components of the energy value chain—from coal extraction to the supply and sales of electricity to final customers. The holding controls approximately 29% of Polish hard coal resources and is responsible for 7% of domestic coal production operating in sectors of coal mining, generation, distribution and supply of electricity and heat. TAURON Group supplies over 49 TWh of electricity to more than 5.4 million customers per year which makes it the largest distributor of electricity in Poland. The Group's distribution grid covers an area of approximately 57,000 km^2 representing nearly 18.3% of Poland's area and consists of power lines measuring about 223,700 km.

Affiliated subsidiaries of TAURON Group include, among others:

- TAURON Wydobycie S.A. operating in coal mining
- TAURON Wytwarzanie S.A. specializing in electricity generation from conventional sources and biomas co-firing
- TAURON Ekoenergia sp. z o.o. specializing in electricity generation from renewable sources
- TAURON Dystrybucja S.A. dealing with electricity distribution
- TAURON Sprzedaż sp. z o.o. providing electricity supply to retail customers
- TAURON Obsługa Klienta sp. z o.o. dealing with customer service
- TAURON Ciepło sp. z o.o. specializing in heat distribution

[5]Based on http://en.tauron.pl/tauron/investor-relations/about-company/Pages/start.aspx

Since June 2010 TAURON has been listed on the Warsaw Stock Exchange. The State Treasury remains the largest shareholder holding 30% of its shares. In January 2013 TAURON joined the Warsaw Stock Exchange's RESPECT Index of socially responsible companies.[6]

16.3.4.2 Integrated Reporting

TAURON initiated its non-financial disclosure with the 2012 Sustainability Report. The CSR/Sustainability Reports were also published for 2013 and 2014 and then for 2015 the company published its first integrated report. While the report in available in the multimedia version, its highlights are presented in 24-pages brochure. 2015 Integrated Report covers Group highlights, investor relations in sections called TAURON on WSE and Financial results, stakeholder management in the section on Sustainable development and governance characteristics in the section on Corporate governance. The general overview of 2015 report is presented in Table 16.4.

The TAURON Group's 2015 Sustainability Report is its fourth non-financial report, similarly to prior editions published annually. The report was constructed according to Global Reporting Initiative standards—GRI G4 (compliance level basic). In addition, the content was subject to independent verification. As argued by TAURON "the extensive dialogue with the environment preceded the writing of this report, in accordance with the AA 1000SES standard. Consequently, the issues our stakeholders consider to be the most important are included. This year's report also contains the final summary of the TAURON Group's Sustainability Report for 2012–2015 with an outlook to 2020 and outlines the assumptions of the Group's updated strategy" (TAURON 2015: 3).

Issues covered in the report were selected via the consultation with key stakeholders conducted in cooperation with a CSR Consulting firm according to the AA1000 standard. The dialogue was divided into several stages allowing for the full picture of the expectations by internal and external constituencies including:

- individual interviews with experts—representatives of universities, CSR organizations and non-governmental environmental protection organizations

Table 16.4 The general overview of TAURON 2015

Year	Size	Framework	Sections
2015	Multimedia version only Two brochures added (key facts of 24 pages and sustainability Report 24 pages) 75 pages	Framework of IIRC and GRI	Group highlights TAURON on WSE Sustainable development Corporate governance Financial results

Source: own analysis based on annual report, 2015

[6] See http://www.odpowiedzialni.gpw.pl/

- stakeholder panel—representatives of trade organizations, the regulator, non-governmental organizations and trade unions attended this meeting
- online questionnaire—a record number of 5099 employees and clients filled out this electronic questionnaire
- workshops with TAURON Group managers—managers from strategy, sales, distribution, human resources management and PR attended this meeting.

Forty-four selected issues were divided into four topical areas covering market, environment, workplace and society. In the process of a dialog with the external stakeholders 11 topics were identified and ranked as high (weight of at least 4 points on a 5-point scale). In addition, seven issues were identified as being significant (weight of at least 4) by one of the constituencies. In sum, TAURON identified 68 GRI indicators, including 35 profile indicators, 10 sectorial indicators and 23 individual indicators.

The report presents both corporate as well as sustainability strategy. However, while the level of realization of particular sustainability goals for the strategy 2012–2015 is analyzed in a great detail, the report provides very general information related to the new 2016–2018 strategy. Similar observations can be made in regard to corporate strategy. The analysis of risks does not include social and environmental risks (except those related to CO_2 emission allowances). What is more, the publication shows no link between business and sustainability goals.

As argued by TAURON it attempts to improve the report in terms of the content and visually using graphic and technological tools and solutions to present non-business operations in the most accessible way. TAURON adopts GRI measures search and interactive tools. 2015 Sustainability Report is available in the version for disable people. The analysis revealed that while the integrated report is available in the multimedia version, there is also an option to download two separate reports in pdf format—the financial and CSR reports. Thus, as a matter of fact the company delivers two separate reports which are presented in the combined version via an internet application.

16.3.5 PKN ORLEN

16.3.5.1 Company Characteristics[7]

Being the largest company in Poland, PKN ORLEN is also one of the largest petrochemical companies in Central and Eastern Europe. It operates six refineries and the region's largest network of nearly 2700 service stations located in Poland, the Czech Republic, Germany and Lithuania. It processes crude oil into gasolines, diesel oil, fuel oil and aviation fuel. ORLEN is also a leading producer of petrochemicals, and its products are used as basic feedstocks by a large number of chemical

[7]Based on http://www.orlen.pl/

companies. In order to transform into an energy conglomerate, PKN ORLEN consistently develops its hydrocarbon exploration and production (upstream) and power segments. It holds ten licenses for oil and gas exploration throughout the country. It is involved in exploration for and extraction of natural gas from unconventional sources in Poland, and in hydrocarbon production from unconventional deposits in Canada through TriOil subsidiary. The company employs nearly 22,000 people.

The company has been listed on the Warsaw Stock Exchange for 15 years, and is currently included in the WIG20 and WIG30 indices, as well as the high-profile RESPECT Index. The general overview of integrated reporting in PKN Orlen is provided in Table 16.5.

16.3.5.2 Integrated Reporting

ORLEN has long history of non-financial disclosure dating back to 2002, when the company published its first CSR report. Starting from the 2008 Responsible Business Report non-financial data have been disclosed according to GRI standards. Simultaneously ORLEN was publishing detailed environmental reports. In 2014 the company presented its first integrated report.

Table 16.5 presents overview of the two integrated reports published so far. Their structure follows similar pattern, which indicates clear separation between business and social aspects of company's operations. CSR strategy and business strategy are presented separately. Although people and sustainable development are mentioned as the pillars of ORLEN's Strategy for 2014–2017, and CSR is declared to be "part and parcel of the Group's day-to-day operations" (PKN ORLEN 2015: 135), the

Table 16.5 The general overview of PKN ORLEN Reports 2014–2015

Year	Size	Framework	Sections
2014	Multimedia version only	Framework of IIRC and GRI	Our company Our strategy Our activities in 2014 Responsible company Corporate governance Risk and opportunities Financial results This report
2015	344 pages	Framework of IIRC and GRI	The Orlen group and its environment Our strategy Corporate governance Our operations in 2015 Responsible company Risk and opportunities Financial results Interactive KPI This report

Source: own analysis based on annual reports, 2014 and 2015

strategic goals do not reflect that (PKN ORLEN 2015: 71–72), neither do risk and opportunities analysis (PKN ORLEN 2015: 211–217).

Data are disclosed in comparison to previous years and in some areas, including those related to production capital (PKN ORLEN 2015: 56), safety of employees and subcontractors (PKN ORLEN 2015: 59) or environmental responsibility (PKN ORLEN 2015: 164) plans for the next year are presented. Detailed information about measurable objectives defined in CSR Strategy is presented in a separate document available on the ORLEN's website. Direct hyperlink is to found in the report (PKN ORLEN 2015: 139).

The company prepared its reports according to the IIRC framework and made effort to define the content through different kinds of capital—whether financial, manufactured, intellectual, human, social or natural. However, the interdependencies between the capitals and the company's value are not clearly illustrated. In non-financial reporting the company follows GRI guidelines. Additionally, limited third party assurance was provided to non-financial part of ORLEN's integrated reports.

While the 2014 Integrated Report provided only brief information on how relevant reporting aspects were defined, the next report explained the process in detailed. It included three steps:

- Analysis of perception studies, media coverage of the ORLEN Group and internal analyses that enabled to identify key financial, social and environmental aspects of the ORLEN Group's operations
- Preparation of the final list of relevant reporting aspects in consultation with members of PKN ORLEN's CSR Strategy Committee
- Consultations with various internal and external stakeholders, including employees representing various business areas, trading partners and members of local authorities. They were requested to fill in surveys designed to determine the relevance of the individual aspects to the ORLEN Group and its environment

Both of the reports are published in online multimedia version with interactive tools that allow readers to display particular information by GRI indicators, type of capitals or to compare data with those presented in the report from the previous year. Taking into account the extensive size of the reports (over 300 pages in 2015) the electronic version largely increases user-friendliness of both publications.

16.3.6 Grupa Azoty

16.3.6.1 Company Characteristics[8]

The Grupa Azoty Group is the largest Polish and one of the largest European chemical holding. The Group has concentrated complementary companies with

[8]Based on http://grupaazoty.com/

different expertise offers its customers a diversified product portfolio—from mineral fertilizers and engineering plastics through OXO alcohols and plasticizers to pigments. Its products are sold to over 50 countries all over the world. At present Grupa Azoty is formed by four largest chemical plants in Poland: from Tarnów, Puławy, Police and Kędzierzyn-Koźle, as well as over ten subsidiaries, and production companies from Germany and Senegal. The dominant entity in the Group is Grupa Azoty S.A. with main office located in Tarnów. The company employs nearly 14,000 people.

Grupa Azoty has been listed on the Warsaw Stock Exchange since 2008. The largest shareholder of the Group is the State Treasury that controls 33% of its shares. Since 2009, Grupa Azoty S.A. has been listed in the RESPECT index portfolio.[9]

16.3.6.2 Integrated Reporting

Grupa Azoty started to disclose non-financial information in 2008. Between 2008 and 2012 it published four environmental reports that also covered the Group's activities in the area of workers health and safety, and stakeholders dialogue. Starting from 2009 non-financial information was also disclosed in company's annual reports. The first truly integrated report was released in 2013, offering a systematic perspective on the company's general economic, social and environmental performance. This 164-page long publication was already prepared in accordance with GRI G4 core version and verified by a third-party external verification agency. While financial data were verified by KPMG Audyt Sp. z o.o., non-financial data limited assurance (in line with ISAE 3000) has been provided by Deloitte Advisory Sp. z o.o. (Grupa Azoty 2013). The general overview of integrated reports in Grupa Azoty is provided in Table 16.6.

As it has been indicated in the Table 16.6 the size of the reports has not changed over the consecutive 2 years. Neither has the structure of the reports. The company's reports comprise of five sections dedicated to information on the Group, its business operations, its social responsibility (including employee, community and environmental issues), its financial performance and the report itself. This structure indicates that the reports are rather combined than integrated. Indeed, financial and non-financial data are discussed in separation from each other and the company provides little information on the interdependencies between social, environmental and economic sources of value creation. No information is provided on the relation between the company's business strategy and sustainability strategy: whether the latter supports company's business goals or the former contributes to sustainable development. Although non-financial data are disclosed in comparison to previous reporting period, no plans or goals for future are presented. Additionally, the report does not include analysis of company's risks and opportunities related to

[9]See http://www.odpowiedzialni.gpw.pl/

Table 16.6 The general overview of Grupa Azoty reports 2014–2015

Year	Size	Framework	Sections
2014	152 pages	GRI	The group Our business Our responsibility Financial information Our report
2015	142 pages	Framework of IIRC and GRI	Our group Our business Our responsibility Our finances Our report

Source: own analysis based on annual reports, 2014 and 2015

environmental and social issues (Grupa Azoty 2014, 2015). Reporting aspects of average importance.

The scope and range of the Group's reports, like the applied measurement methods, have not changed since 2013. The impact areas and indicators were specified by the Expert Committee for CSR, which included employees responsible for different areas of the Group's business supported by third-party experts. To ensure that the CSR aspects identified as material meet stakeholders' expectations, at the stage of defining the scope of the reporting the Committee held meetings directly at the Group's companies and engaged in the process of collecting data around 70 people, who are responsible for maintaining dialogue with stakeholders on day-to-day basis. No additional forms of stakeholder engagement were used in this process (Grupa Azoty 2014: 138–139). The scope of the reports is defined around 17 issues of high relevance and eight issues of average relevance (Grupa Azoty 2015: 132).

All of the company's reports were published in pdf. Format and can be downloaded from the corporate website. No online, interactive version is provided, what highly decreases the reports functionality and comparability of the data.

16.4 Discussion

LOTOS, KGHM, TAURON, PKN Orlen and Grupa Azoty are the only five companies of 487 listed in the Warsaw Stock Exchange, the largest Central European stock market, which adopt the framework of integrated reporting in the analyzed period. While, these companies differ significantly in terms of integrated disclosure history and experience (ranging from the start in 2009 to the first integrated report published in 2015), they show significant similarities with respect to sector of operation burdened with negative impact on the natural environment and the control by the government. Our analysis draws upon the history and practice of integrated reporting. In addition, we summarize the disclosure according to IIRC principles as presented in Table 16.7.

Table 16.7 Comparison of integrated reports according to quality assessment criteria

Company/ Criteria	IIRC principles							GRI
	Strategic focus	Connectivity of information	Stakeholder relationships	Materiality	Conciseness	Reliability and completeness	Consistency and comparability	
LOTOS	+/–	+/–	+	+	–	+	+	+
KGHM	–	–	+	+	+	+	+	+
TAURON	–	–	+	+	+	–	+	+
PKN Orlen	+	+/–	+	+	–	+	+	+
Grupa Azoty	–	–	–	+	+	+	+	+

"+" indicates the inclusion of principle in integrated report; "+/–" represents partial inclusion of principle in integrated report; "–" represents the lack of the principle introduced in integrated report

Source: own analysis

As reported in Table 16.7 our sample companies reveal some similarities with respect to the practice of integrated reporting. First, of all sample companies reveal a relatively early stage of experience in integrated disclosure revealing a short history of reporting. Second, taking into account firm characteristics sample companies operate in petroleum, energy or mining sectors featured with potentially negative impact on the natural environment. The engagement in sustainability and integrated disclosure may be viewed as tools for legitimacy of these industries (Caron and Turcotte 2009; Ballou et al. 2006; Thorne et al. 2014). Third, as shown in Table 16.7 majority of sample companies follow 6 of listed 7 IIRC principles, all of them adopt GRI framework, while 4 of 5 companies employed verification by external auditors. It is important to mention that all sample companies follow materiality, while majority of them adopt the stakeholder dialogue in their integrated reporting. Reports also scored high on consistency and comparability. Strategic focus and the connectivity of information appeared to be the most problematic aspect of integrated reporting practice.

16.5 Conclusion

Integrated reporting is understood as a new paradigm to disclose company activity and reveal its performance in a complete, holistic way. The most important qualitative improvements relate to the possibility to use the six capitals framework and strategically links social and environmental activity to market and financial performance. It enhances corporate disclosure and supports the implementation of CSR/sustainability principles. In this way, integrated reporting is also expected to result in the shift in corporate planning and performance measurement leading to the strategic thinking.

While, the concept of integrated reporting is viewed as an important improvement, its implication faces significant shortcomings and challenges. Using the case studies of early adopters our analysis adds to the discussion on the scope and the effectiveness of integrated reporting. We reveal several observations. First, integrated reporting appears still to be significantly limited in terms of small numbers of companies which decide to report in integrated framework. The only five firms which disclose in accordance to integrated reporting framework represent nearly 5% of the overall population of listed companies. Second, of five analyzed cases four companies reveal very short term experience of 2–3 years, while only one company has been publishing integrated report over 7 years. Third, the early adopters of integrated reporting are companies characterized with state control and operation in sectors with highly negative impact on natural environment. Four, the quality of the reports raise certain questions on the adoption of IIRC standards, specifically strategic focus, connectivity of information, conciseness as well as consistency and comparability. In our opinion relating social and environmental performance with strategy and financial results appears to the most problematic issue of integrated reporting. Companies tend to rather combine CSR/environmental performance with traditional financial reporting that to integrate all aspects of firm activity.

In general, our results are in line with previous research. Existing studies emphasize benefits of integrated reporting including improved reputation and profits, effective capital allocation, stakeholder engagement, enhanced loyalty of customers and employees and risk management (Dragu and Tiron-Tudor 2014; Eccles and Krzus 2010; Frias-Aceituno et al. 2014; Higgins et al. 2014; IIRC 2013) which sample companies appear to be aware of. Companies explain adopted business model, strategy, governance and operational performance (Lodhia 2015; Higgins et al. 2014), yet the early practice of integrated reporting stimulates merely incremental changes to processes and structures (Higgins et al. 2014; Stubbs and Higgins 2014). Finally, the early development stage of integrated practice of sample companies exemplify shortcomings raised in the existing literature which indicates instrumental approach by companies (Brown and Dillard 2014; Flower 2015) or insufficient coverage of the decision-making and accountability expectations of various interest (Brown and Dillard 2014).

References

Adams, C. A. (2015). The international integrated reporting council: A call to action. *Critical Perspectives on Accounting, 27*, 23–28.

Adams, S., & Simnett, R. (2011). Integrated reporting: An opportunity for Australia's not-for-profit sector. *Australian Accounting Review, 21*(3), 292–301.

Ballou, B., Heitger, D., & Landes, C. (2006). The future of corporate sustainability reporting: A rapidly growing assurance opportunity. *Journal of Accountancy, 22*(3), 65–74.

Ballou, B., Casey, R., Grenier, J. H., & Heitger, D. L. (2012). Exploring the strategic integration of sustainability initiatives: Opportunities for accounting research. *Accounting Horizons, 26*(2), 265–288.

Beattie, V., & Smith, S. J. (2013). Value creation and business models: Refocusing the intellectual capital debate. *The British Accounting Review, 45*(4), 243–254.

Brown, J., & Dillard, J. (2014). Integrated reporting: On the need for broadening out and opening up. *Accounting, Auditing & Accountability Journal, 27*(7), 1120–1156.

Caron, M., & Turcotte, M. (2009). Path dependence and path creation: Framing the non-financial information market for a sustainable trajectory. *Accounting Auditing and Accountability Journal, 22*(2), 272–297.

Clayton, A. F., Rogerson, J. M., & Rampedi, I. (2015). Integrated reporting vs. sustainability reporting for corporate responsibility in South Africa. *Bulletin of Geography Socio-economic Series, 29*, 7–17.

Demartini, P., & Paolini, P. (2013). *Managing sustainability through the IC practice lens*. Conference paper, International Forum on Knowledge Asset Dynamics, 12-14.07.2013, Zagreb, Croatia.

Dragu, I.-M., & Tiron-Tudor, A. (2013). The integrated reporting initiative from an institutional perspective: Emergent factors. *Procedia - Social and Behavioral Sciences, 92*, 275–279.

Dragu, I.-M., & Tiron-Tudor, A. (2014). Research agenda on integrated reporting: New emergent theory and practice. *Procedia Economics and Finance, 15*, 221–227.

Eccles, R., & Krzus, M. (2010). *One report: Integrated reporting for a sustainable strategy*. Hoboken, NJ: Wiley.

Eccles, R. G., Ioannou, I., & Serafeim, G. (2011). *The impact of corporate sustainability on organizational processes and performance* (Working paper). Harvard University.

Flower, J. (2015). The international integrated reporting council: A story of failure. *Critical Perspectives on Accounting, 27*, 1–17.

Freeman, R. E. (1999). Divergent stakeholder theory. *The Academy of Management Review, 24*(2), 233–236.

Frias-Aceituno, J. V., Rodriguez-Ariza, L., & Garcia-Sanchez, I. M. (2014). Explanatory factors of integrated sustainability and financial reporting. *Business Strategy and the Environment, 23*, 56–72.

Geels, F. (2011). The multi-level perspective on sustainability transitions: Responses to seven criticisms. *Environmental Innovation and Societal Transitions, 1*, 24–40.

Grupa Azoty. (2013). Raport zintegrowany 2013 [*Integrated report 2013*]. Retrieved from http://grupaazoty.com/doc/6f144b08/raport_grupa_azoty_za_2013_pl.pdf

Grupa Azoty. (2014). *Grupa Azoty Integrated Report 2014*. Accessed May 5, 2017, from http://bit.ly/2qSb4kG

Grupa Azoty. (2015). *Grupa Azoty Integrated Report 2015*. Accessed May 5, 2017, from http://bit.ly/2qRYbrz

Higgins, C., Stubbs, W., & Love, T. (2014). Walking the talk(s): Organisational narratives of integrated reporting. *Accounting Auditing & Accountability Journal, 27*(7), 1090–1119.

IIRC. (2013). *The international <IR> framework*. London: International Integrated Reporting Council.

IIRC. (2014). *Assurance on <IR>: An exploration of issues*. London: International Integrated Reporting Council.

IIRC, Black Sun. (2015). *The integrated reporting journey: The inside story*. London: IIRC, Black Sun.

KGHM. (2014). *Integrated Report – KGHM Polska Miedź.SA.2014*. Accessed May 5, 2017, from http://bit.ly/2pmPqE0

KGHM. (2015). *KGHM – Integrated Report – KGHM Polska Miedź.SA.2015*. Accessed May 5, 2017, from http://bit.ly/2pmPqE0

Lodhia, S. (2015). Exploring the transition to integrated reporting through a practice lens: An Australian customer owned bank perspective. *Journal of Business Ethics, 129*(3), 585–598.

LOTOS. (2014). Raport zintegrowany 2014 [*Integrated report 2014*]. Retrieved from http://2014.raportroczny.lotos.pl/en

LOTOS. (2015a). *LOTOS Integrated Annual Report 2014*. Accessed May 5, 2017, from http://raportroczny.lotos.pl/en/downloads

LOTOS. (2015b). *LOTOS Integrated Report 2015*. Accessed May 5, 2017, from http://raportroczny.lotos.pl/en/downloads

ORLEN. (2015). *ORLEN Group 2015 Integrated Report*. Accessed May 5, 2017, from http://raportzintegrowany2015.orlen.pl/en/

Perrini, F., Pogutz, S., & Tencani, A. (2006). *Developing corporate social responsibility: A European perspective*. Cheltenham: Edward Elgar Publishing.

Porter, M. E., & Kramer, M. R. (2011, January-February). Creating shared value. *Harvard Business Review, 89*, 62–77.

Stubbs, W., & Higgins, C. (2014). Integrated reporting and internal mechanisms of change. *Accounting Auditing & Accountability Journal, 27*(7), 1068–1089.

TAURON. (2015). *2015 Integrated Reporting*. Accessed May 9, 2017, from http://raport2015.tauron.pl/

Thorne, L., Mahoney, L., & Manetti, G. (2014). Motivations for issuing standalone CSR reports: A survey of Canadian firms. *Accounting Auditing & Accountability Journal, 27*(4), 686–714.

Villiers, C., Rinaldi, L., & Unerman, J. (2014). Integrated reporting: Insights, gaps and an agenda for future research. *Accounting, Auditing & Accountability Journal, 27*(7), 1042–1067.

Zicari, A. (2014). Can one report be reached? The challenge of integrating different perspectives on corporate performance. In R. Tench, W. Sun, & B. Jones (Eds.), *Communicating corporate social responsibility: Perspectives and practice* [Critical Studies on Corporate Responsibility, Governance and Sustainability, Vol. 6] (pp. 201–216). Bingley: Emerald Group Publishing.

Chapter 17
King Codes on Corporate Governance and ESG Performance: Evidence from FTSE/JSE All-Share Index

Federica Doni, Antonio Corvino, and Silvio Bianchi Martini

17.1 Introduction

Over the last decade, the issue of corporate governance has become of great interest in many countries in the world for a number of reasons, including the growing importance of institutional investors in venture capital for large companies and the globalization of international financial markets (Kojima 1998; Nestor and Thompson 1999). Several issues related to "good corporate governance" have been highlighted by the failure of many well-known companies, and a number of corporate crimes have prompted the establishment of special committees involved in the drafting of reports on corporate governance of large companies in various countries, particularly in the UK (i.e. Cadbury Report 1992; Greenbury Recommendation 1995; Hampel Report 1998; Combined Code 2003; FRC 2012, 2014, 2016, 2017). In the same vein, other countries, including France, Canada, the Netherlands, and Italy, have established committees composed of leading authorities from both the industrial and financial sectors, to set codes of conduct or self-discipline (best practice) on the issues of corporate governance.

The aim of these documents is to build a structure of government that limits the risk of abuse of power by top management or controlling shareholders, which creates a disadvantage for investors. International agreement on the corporate governance

F. Doni (✉)
Department of Business and Law, University of Milano-Bicocca, Milan, Italy
e-mail: federica.doni@unimib.it

A. Corvino
Department of Economics, University of Foggia, Foggia, Italy
e-mail: antonio.corvino@unifg.it

S. Bianchi Martini
Department of Economics and Management, University of Pisa, Pisa, Italy
e-mail: silvio.bianchi@unipi.it

S. O. Idowu, M. Del Baldo (eds.), *Integrated Reporting*, CSR, Sustainability, Ethics & Governance, https://doi.org/10.1007/978-3-030-01719-4_17

principles issued by the G20 and the OECD (2015) is expected to be reached in order to construct an effective corporate governance framework, enhancing the trust and confidence of communities in companies and the financial markets. In December 2017, a public consultation was published by the Financial Reporting Council (comments required by 28 February 2018) for a new-style UK Corporate Governance Code, together with revised Guidance on Board Effectiveness and some questions on the future of the UK Stewardship Code. This document suggests the need for changes in the current UK Corporate Governance Code, especially on the issues of leadership, effectiveness and relations with shareholders, highlighting shareholder engagement as a key aspect of good corporate governance.

From this perspective, and as a result of the increasing significance of companies' commitment to Corporate Social Responsibility (CSR) (Clarkson 1995; West 2009; IFAC 2012; Diale 2012), corporate governance "is not an end in itself" (G20, OECD 2015) but has taken on a broader connotation. The main purpose is not limited to protecting the relations between investors (principal) and management (agent) but also to ensure relationships between the company and stakeholders through the preparation of codes of ethics and codes of conduct. These documents require companies to comply with legal, social, human, institutional and ethical values as an expression of behaviour that is correct and transparent, but also consistent and effective.

These factors are becoming essential, not only in industrialized countries, but also in emerging economies such as the BRICS countries (Ntim et al. 2012). In particular, South Africa stands out as a pioneer, having begun the development of its code of conduct in the 1990s (King I 1994), followed in later years by three further documents King II (2002), King III (2009) and King IV (2016). In particular, the novelty of the *King Code of Governance Principles for South Africa 2009* (IoDSA 2009) was to establish disclosure requirements for good corporate governance, setting a new listing requirement for the companies listed on the Johannesburg Stock Exchange (JSE), i.e. the drawing-up of Integrated Reporting (<IR>) as an innovative form of reporting model (Eccles and Krzus 2010; Eccles et al. 2015a; Adams 2015; De Villiers et al. 2016; KPMG 2017).

The claim for <IR> supported by some international organizations is determining a fundamental shift from traditional reporting practices to an integrated and holistic system of reporting both financial and non-financial information. This initiative could provide a new tool to improve the quantity and quality of reporting, and to tackle the problems linked to the disclosure of nonfinancial information.

Moreover, <IR> could represent an effective tool for stimulating management to adopt an integrated approach in the business context by developing a substantial change towards "integrated thinking" (SAICA 2015). <IR> is able to affect different areas of a company, such as its business model, the organizational aspects of the business activities, the selection of the material information, the connectivity of the information and the integration of financial and non-financial data, communication of multiple capitals, and so on (IIRC 2013; IODSA 2016).

It is important to note that the success of this movement is strongly influenced by the involvement of the CEO and the board in support of this radical change of reporting system. This crucial aspect has been increasingly emphasized by the International Integrated Reporting Council (IIRC), which recently intensified relationships with the organization that is working hard at the international level in corporate governance: the International Corporate Governance Network. On December 2016, the IIRC hosted in partnership with ICGN an international conference that addressed "how to properly integrate the consideration of long term value drivers in pursuing the success of companies—ultimately contributing to a more sustainable capital market system" (http://integratedreporting.org/iirclondondec2016/). The main purpose is to achieve "integrated thinking" across strategy, performance, forward-looking prospects and, in particular, in corporate governance. At the end of February 2018, ICGN, in collaboration with IIRC, will organize a global conference in Japan (https://www.icgn.org/events/icgn-iirc-tokyo-conference-2018). This initiative aims to accelerate governance and stewardship reforms and to evaluate how to mitigate impediments to company and investor engagement efforts.

In the same perspective, the IIRC highlights this crucial statement: "Corporate reporting, and the thinking that has to accompany it, are boardroom issues. This is where strategy, performance and the development and communication of long-term value are best understood, aligned and led" (http://integratedreporting.org/resource/creating-value-value-to-the-board/). One of the most relevant topics linked to the core concept of integrated thinking and reporting is corporate governance as described by Richard Howitt, the CEO of IIRC, in his speech at IIRC-ICGN "Dialogue for longer-term value creation" (Conference, London, on the seventh of December 2016): "But we say Integrated Reporting can be accepted as a principle of twenty-first century corporate governance" (https://integratedreporting.org/news/richard-howitt-ceo-iirc-addresses-the-iirc-icgn-conference/). This perspective is shared by ICGN, which is setting and promoting standards of corporate governance and investor stewardship to advance efficient markets and sustainable economies. The principles of corporate governance (ICGN 2013) include a focus on disclosure and integrated reporting, underlining the importance of the comprehensive disclosure of information for investors and other stakeholders relating to financial statements, strategic and operational performance, corporate governance and material environmental and social factors. Moreover, these principles emphasize a robust audit practice, which is critical for necessary quality standards.

In particular, boards should evidence a commitment to communicating the opportunities and risks linked to Environmental Social and Governance, ESG-related issues that are material to the company's strategy and performance. The emphasis on ESG issues reveals a clear effort on the part of the IIRC to bring about a radical change towards a more sustainable board, and a growing attention to the involvement of the board, particularly the CEO and CFO (IIRC 2014; IIRC and IFAC 2017; IFA 2017), in the process of integrated thinking/reporting. The relationship between <IR> and corporate governance arises from the need to carry out a process of integrated thinking/reporting as corporate reporting is "an essential and inseparable part of corporate governance". The outcome of a corporate governance process is corporate reporting that shows purposes, values and business activities reflected on the behaviours of the board and management team (IIRC 2016).

Recently, this new perspective has been adopted by South Africa, which is the first country in the world to consider corporate reporting, that is, <IR>, "as a mainstream component of corporate governance". The achievement of an integrated and inclusive corporate governance system can provide benefits for both businesses and investors, as <IR> is expected to connect corporate governance and investment stewardship.

Given the importance of the link between corporate governance and <IR>, it is worth assessing how this trend is developing and evolving by identifying the main drivers of this important challenge. This topic attracts attention on South Africa where this relationship seems to achieve a high significance.

To assess this topic, this chapter seeks to analyse in-depth the consequences of the adoption of King III (IODSA 2009) which came into effect in 2010, and the main changes from the recent release of the Corporate Governance Code *Draft King IV™ on Corporate Governance for South Africa 2016* (http://www.iodsa.co.za). The main objective of King III and, afterwards, King IV, is to promote good corporate governance as a driver for ethical and effective leadership at board level. This aim can be achieved through an ethical culture, sustainable performance, adequate control by the governing body and protecting trust in the organization, its reputation and legitimacy (IODSA 2013). Following this initial analysis, we aim to assess the potential correlation between the environmental/social performance and corporate governance practices after the adoption of the King III, which represented a radical shift towards the enhancement of sustainability-related issues in business activities and the corporate governance system.

With a few exceptions, the topic of the interrelationships between <IR>, corporate governance and ESG performance in South African listed companies is still underexplored (Hindley and Buys 2012; Carels et al. 2013; Bianchi Martini et al. 2017; Adams 2017; McNally et al. 2017) and it needs to be investigated with additional empirical studies.

In doing this, the present chapter proceeds as follows: the second section explains the process of developing corporate governance codes in South Africa, and the third section summarizes a number of previous studies on the relationships between corporate governance, reporting and sustainability issues in support of the formulation of our research hypotheses. The fourth section details the research design, while the fifth describes our main findings. Finally, the sixth section portrays the results and offers some concluding remarks.

17.2 Background: The Development of Corporate Governance Code from King III to King IV

Since the end of the twentieth century, there has been a growing interest in corporate governance, the set of principles and rules governing the coordination and control of powers and roles in the business context, even if entrepreneurs and managers have always been interested in the improvement of corporate governance practices.

In particular, the emerging countries are showing a strong commitment to setting or revising corporate governance codes, forcing companies to adopt <IR>. For example, the Securities and Exchange Board of India (SEBI) released the report of the committee on corporate governance (October, 2017) on revising corporate governance principles in Indian companies by suggesting changes in disclosure, transparency, board composition and performance. Moreover, Malaysian companies are being called on to adopt Integrated Reporting as part of the Malaysian Corporate Governance Code, launched in April 2017 by the Securities Commission Malaysia. In the same way, the Institute of Directors of Zambia supports sound corporate governance principles and ethics, as they are key in ensuring proper management, control and accountability for the affairs of private and public enterprises in Zambia. The Institute of Directors of Zambia has affirmed (February, 2015) that sustainable reporting and the development of the concept of integrated reporting are critical, not only to ensure increased levels of transparency and accountability but also to change corporate behaviour.

Among emerging countries, South Africa moved to integrated reporting and integrated thinking in 2010. However, the development of the code of corporate governance in South Africa dates back to 1994, when the King Committee on Corporate Governance published King I, named after the Committee's chairman [Professor] Mervyn E. King. King II was passed in 2002 and in September 2009, the third edition of the King Code of Governance Principles for South Africa 2009, effective from 1 March 2010, appeared. Since March 2014, the Institute of Directors South Africa (IoDSA) has been a member of the Integrated Reporting Committee Council.

The principles of "good governance" signalled by King III—and most recently by King IV (November 2016)—can be connected with an innovative model of corporate reporting, known as Integrated Reporting (<IR>), which became a mandatory listing requirement in South Africa in 2010 (Hindley and Buys 2012; Carels et al. 2013; Rensburg and Botha 2014; Setia et al. 2015; Doni et al. 2016; Raemaekers et al. 2016; Bianchi Martini et al. 2017; Doni and Fortuna 2018). At a time of deep worldwide financial crisis, South Africa sent a signal of renewed trust in the code of self-discipline. This code of conduct is basically voluntary and represents one of the most advanced forms of self-discipline. It affected corporate law in a very specific way, if we consider that the two earlier versions gave way to passing Companies Act No. 71 of 2008. As a pioneer country, South Africa therefore tried early on to promote principles of corporate governance strongly inspired by the Anglo-Saxon model. The basic principle of King III lies in the "comply or explain" approach which originated in the Combined Code of the United Kingdom, which in turn has its roots in the 1992 Cadbury Report. This principle represents the essence of the code's flexibility: the code is not made up of a strict set of rules, but of principles and provisions, quite unlike the governance on a statutory basis model, in which the opposite "comply" or "comply or else" rule is followed. This kind of arrangement is adopted in the US, where governance is partly regulated by the Sarbanes-Oxley Act and non-compliance is subject to sanctions. The "comply or explain" approach has been implemented in South Africa with even greater flexibility, given that the term "apply" replaced the term "comply" in King III, thereby emphasizing the way that principles and recommendations can be applied, rather than focusing on compliance assessment (King Code of Governance 2009: 7). The only obligation placed on directors by law is "to act in the best interests of the company". If the reasons behind the practice adopted are made explicit, and the specifics thereof are duly illustrated, compliance with the principles of King III is assured. However, we shall point out that all the principles included in the code are equally important and contribute collectively to establishing a holistic approach to governance: any "substantial application" of the code does not therefore imply achieving compliance.

The King Committee published the King IV Report on Corporate Governance for South Africa 2016 (IODSA 2016) on 1 November 2016. King IV is effective in respect of financial years commencing on or after 1 April 2017. King IV replaces King III in its entirety. King IV closes the circle of integrated reporting which calls on organizations to prepare an integrated report each year to reflect the understanding that strategy, risk, performance and sustainability are closely connected. King IV is principle- and outcomes-based rather than rules-based and it confirms the strong link with <IR>, (IIRC welcomes the release of King IV as South Africa sets a new global standard for corporate governance) supported by the policies adopted by international corporate governance practices, especially by the International Corporate Governance Network (ICGN). King IV provides a definition of corporate governance as the exercise of ethical and effective leadership by the governing body towards the achievement of the following governance outcomes: 1) ethical culture; 2) good performance; 3) effective control; 4) legitimacy. These outcomes reveal the firm commitment of the IoDSA to a radical shift from financial capitalism to inclusive capitalism, promoting a more sustainable value-creation process.

It is essential to note that the King approach is completely consistent with the listing requirements adopted by the Johannesburg Stock Exchange (JSE), which in 2004 launched the SRI index (JSE 2004), as a system for identifying those companies that incorporate the principles of the triple bottom line and good corporate governance into their business operations. Recently, this approach has been largely adopted by several stock exchanges, especially in the context of the Sustainable Stock Exchanges (SSE) initiative, carried out by the UN Conference on Trade and Development (UNCTAD), the UN Global Compact, the UN Environment Programme Finance Initiative (UNEP FI), and the Principles for Responsible Investment (PRI) network. SSE is a peer-to-peer learning platform for exploring how exchanges, in collaboration with investors, regulators and companies, can enhance corporate transparency—and ultimately performance—on ESG issues and encourage sustainable investment (http://www.sseinitiative.org/).

That said, the requirement for listed companies to adopt <IR> can represent an important turning point, not only in terms of a new corporate reporting model, but also as a "driver" for potential improvements in corporate governance practices. Moreover, <IR> can provide a solution to the lack of adequate information supplied by the Annual Report. The traditional financial reporting model is not able to capture the economic consequences of business innovations (Healy and Palepu 2001) and is becoming more complex and less relevant to shareholders (FRC 2011). A growing number of companies provide non-financial information, but very few organizations are able to integrate financial and non-financial information in an effective way (Investment Responsible Research Center Institute 2013). <IR> can lead to "a concise communication about how an organization's strategy, governance, performance and prospects, in the context of its external environment, lead to the creation of value in the short, medium and long term" (IIRC 2013).

17.3 Literature Review and Research Hypotheses Development

Several studies analyzed different corporate governance issues and practices, nevertheless an interesting field of study focuses on the relationship between corporate governance and the disclosure of information. In this perspective, to enhance corporate governance mechanisms, it is important that the communication of information is as transparent as possible in order to reduce the possible information asymmetries between ownership and management, in accordance with the agency theory (Haniffa and Hudaib 2006; Baek et al. 2009; Garcia-Lara et al. 2009; Beekes et al. 2016; Nagata and Nguyen 2017). In particular, an empirical analysis demonstrated the positive impact of corporate governance factors on transparency and disclosure of forward-looking information (Agyei-Mensah 2017).

Although a number of studies have focused on the impact of corporate governance on the disclosure of intangibles and Intellectual Capital (Cerbioni and Parbonetti 2007; Hidalgo et al. 2011), recently the main interest has tended towards a specific kind of non-financial information: sustainability and Corporate Social Responsibility (CSR) disclosure. Some scholars have demonstrated the positive impact of a high level of corporate governance on legitimacy management as well as the quality of CSR disclosure (Liu and Zhang 2017). More specifically, the empirical evidence highlights a positive association between the CSR approach (Amran et al. 2014) or the stakeholder orientation of the board and sustainability reporting i.e. environmental and social disclosure or performance (Mallin et al. 2013), although some firm-specific factors, such as firm size and industry, exert a strong influence on this relationship (Chan et al. 2014; Kaymak and Bektas 2017; Helfaya and Moussa 2017) and on its effect on financial markets (Rodriguez-Fernandez 2016; Liu and Zhang 2017).

The ESG factors are becoming increasingly significant in emerging economies (Ntim et al. 2012). The novelty of the King Code of Governance Principles for South Africa 2009 and the mandatory drafting of <IR> for JSE-listed companies has been analyzed by some studies on the content of integrated reports and its managerial implications (Hindley and Buys 2012; Carels et al. 2013; Rensburg and Botha 2014; Setia et al. 2015; Raemaekers et al. 2016; Bianchi Martini et al. 2017; Doni and Fortuna 2018). Although the adoption of <IR> is compulsory, it should be highlighted that the process of integrated reporting/thinking requires a holistic approach to the disclosure process. The effective integration of financial and non-financial information should be the end result of an active commitment to <IR> on the part of the board and senior managers, promoting a radical shift towards a more inclusive organizational structure (Feng et al. 2017) and greater cultural control (Dumay and Dai 2017). A comprehensive adoption of integrated thinking across all organizational functions seems to be essential (Dumay and Dai 2017) as well as useful for attracting longer-term investors by improving management ability to implement a strategy of stakeholder engagement and investment (Knauer and Serafeim 2014).

From a practical point of view, the strong link between integrated thinking/reporting and corporate governance has been established by the collaboration between the IIRC and ICGN (London, December 2016, http://integratedreporting.org/iirclondondec2016/). In this view, the ICGN revised Global Governance Principles include the recommendation that boards should produce integrated reports. In addition, this aspect has been assessed in the academic context, as recent theoretical and empirical studies have demonstrated growing evidence of the benefits of adopting <IR> to corporate governance systems and organizational processes (Frias-Aceituno et al. 2013; Churet and Eccles 2014; Dumay and Xi Dai 2014; Eccles et al. 2015b; Nazari et al. 2015; Haji and Hossain 2016; Eccles and Youmans 2016; Dumay et al. 2016). In particular, the perception of a sample from South African institutional investment industry on the first sets of integrated reports being prepared by JSE-listed companies shows a shift of interest towards sustainability issues and <IR>, but that there are also some difficulties and obstacles in delivering high-quality reporting (Atkins and Maroun 2015).

Given these practical difficulties, it is possible to highlight that the successful implementation of <IR> and integrated thinking is strongly affected by the involvement of the board in sustainable practices and in the development of sustainability issues in corporate reporting. To do this, it is essential to enhance sustainability performance, not in isolation from the main organizational performance (Cheng et al. 2014) or as mere operation of "impression management" (Hooghiemstra 2000; Merkl-Davies and Brennan 2007; Melloni 2015) or an acitviity of "greenwashing" (Lyon and Maxwell 2011), but as an integral part of the value-driving initiatives and management processes of organizations (Eccles and Krzus 2010; Eccles et al. 2015b; Frias-Aceituno et al. 2014).

Given these premises, this research is focused on an analysis of the integrated reports drawn up by a sample of JSE-listed companies, to evaluate whether the adoption of King III and <IR> does in fact enhance ESG disclosure and improve corporate governance practices.

Our analysis can be divided into two steps. First, we consider the impact of King III on ESG performance, and then we investigate the influence of King III on certain features and functions of corporate governance, identifying specific issues.

Our first group of research hypotheses is supported by a large body of literature from which it is possible to argue the crucial role played by the Board of Directors in the good practice of corporate social responsibility by adopting policies of stakeholder engagement and holistic disclosure transparency processes (Frias-Aceituno et al. 2013). Moreover, King III established principles encouraging boards to appreciate the link between strategy, risk, performance and sustainability, and to emphasize the influence of the perception of stakeholders on a firm's reputation (Adams 2017).

Given these premises, we first formulate three hypotheses, on governance, on environmental performance and on social performance. We expect that the principles of King III encourage board directors and senior managers to incorporate ESG factors into reporting processes and corporate strategy development. The hypotheses are formulated as follows:

1a) the adoption of King III exerts an influence on governance performance;
1b) the adoption of King III exerts an influence on environmental performance;
1c) the adoption of King III exerts an influence on social performance.

As a second step, we aim to investigate the impact of King III on good and effective corporate governance practices by examining board functions, responsibilities, compensation and vision. For example, we expect the principles of King II to positively influence board commitment as evidenced in the establishment of essential committees with clear allocation of tasks and responsibilities. In addition, we evaluate the influence of King III on board structure through a well-balanced membership, evidenced by the adequate presence of independent and diverse directors, ensuring a completely free exchange of ideas and an independent decision-making process. We also consider good corporate governance practices in relation to competitive and proportionate management compensation. This reflects a company's capacity to attract and retain executives and board members with the necessary skills by linking their compensation to individual or company-wide financial or extra-financial targets. Finally, we expect to find a positive effect from the adoption of King III on the development of corporate strategies, emphasizing the capacity of the board to convincingly demonstrate the integrated approach in all dimensions—financial, social and environmental—in its day-to-day decision-making processes (Adams 2017).

Therefore, we formulate the following hypotheses:

2a) the adoption of King III exerts an influence on the ratio Board of Directors/ Board Functions;
2b) the adoption of King III exerts an influence on the ratio Board of Directors/ Board Structure;
2c) the adoption of King III exerts an influence on the ratio Board of Directors/ Compensation;
2d) the adoption of King III exerts an influence on the ratio Integration/Vision and Strategy.

Because of potential correlations between corporate governance performance and social/environmental—i.e. sustainability factors—confirmed by large literature (i.e. Cowen et al. 1987; Buckholtz et al. 2008; Michelon and Parbonetti 2010; Mallin and Michelon 2011; Mallin et al. 2013), we aim to investigate the influence of corporate governance performance on the two pillars of CSR, that is, on environmental and social performance.

Therefore, we formulate the following hypotheses:

3a) corporate governance performance is positively associated with environmental performance;
3b) corporate governance performance is positively associated with social performance.

Finally, we aim to investigate the potential correlation between corporate governance features and practices and the two pillars of the sustainability concept in order to evaluate and compare the board attitude towards environmental or social issues.

We therefore formulate the following hypotheses:

4a) the ratio Board of Directors/Board Functions is positively associated with environmental performance;
4b) the ratio Board of Directors/Board Functions is positively associated with social performance;
5a) the ratio Board of Directors/Board Structure is positively associated with environmental performance;
5b) the ratio Board of Directors/Board Structure is positively associated with social performance;
6a) the ratio Board of Directors/Compensation is positively associated with environmental performance;
6b) the ratio Board of Directors/Compensation is positively associated with social performance;
7a) the ratio Integration/Vision and Strategy is positively associated with environmental performance;
7b) the ratio Integration/Vision and Strategy is positively associated with social performance.

17.4 Data Collection and Sample Analysis

The dataset is made up of South African companies listed on the FTSE/JSE All-Share Index. This index was selected because it accounts for a large percentage of the market capital value (99%: FTSE Russell, 29 September 2017) and highlights the performance of the main industry segments of the South African market. In terms of net market capitalization, among the top ten constituents (equal to 54.19% of the total), the industries positioned from the first to the third rank are the media (17.47%), personal goods (8.81%) and mining (6.98%). Overall, these three

Table 17.1 Crafting process of the sample

Steps	Description	Observations
1	Total South African listed companies included in the FTSE/JSE All-Share Index	163
2	Data availability with reference to ESG scopes	130
3	Data collected with reference to ESG scopes	65
Representativeness of the sample related to step 1		39.88%
Representativeness of the sample related to step 2		50.00%

Source: Authors' elaboration

segments yield 163 companies. The most representative industries, with a number of constituents above 15, are basic resources (e.g. mining), industrial goods and services, retail, real estate and financial services.

Table 17.1 shows the crafting process used in the empirical analysis. We adopted the ASSET4 database managed by Thomson Reuters. Of 163 listed companies, ESG data was available just for 130. Given that our empirical study is still at an early stage, we collected data just for 65 companies. The representativeness of the sample is however passing as the amount of observations is 39.88% of the statistical population. In this regard, it should be noted that the data collection activities require some steps, in order to build a longitudinal dataset that covers the 2012–2016 period. The latter period is mainly conditioned by the full availability of ESG data and allows us to investigate the effects of the introduction of King III (in 2009) up to 2016, the year in which King IV was released.

In the research design, drawing upon the literature review and the research hypotheses mentioned in the previous sections, the key dependent variables are the environmental and social scores. The independent variables concerning the corporate governance model are its score and the following ratios: Board of Directors/Board Functions, Board of Directors/Board Structure, Board of Directors/Compensation Policy and Integration/Vision and Strategy.

Table 17.2 depicts an accurate description for each of the above variables, in order to show the salient points of our empirical study.

From the methodological standpoint, our findings are derived from the calculation of bivariate analyses. Specifically, two non parametric tests—the Wilcoxson and the Sign—as well as the Spearman correlations were computed. The choice of these non-parametric tests is importantly influenced by the sample size and by the need to analyse a paired sample, as a consequence of the aim to explore the first seven research hypotheses. Furthermore, in this study, the Sign test can be considered a control test, given that it is widely considered more powerful than the Wilcoxcon test (Bryman and Cramer 2001) and consequently can corroborate the empirical evidence.

Finally, the SPSS (Statistical Package for Social Science, version 20.0) software was used for carrying out the econometric estimations.

Table 17.2 Description of Variables

Variables	Code	Description
Dependent variables		
Environmental Score	*Environmental_Score*	The environmental pillar measures a company's impact on living and non-living natural systems, including the air, land and water, as well as complete ecosystems. It reflects how well a company uses best management practices to avoid environmental risks and capitalize on environmental opportunities in order to generate long-term shareholder value
Social Score	*Social_Score*	The social pillar measures a company's capacity to generate trust and loyalty with its workforce, customers and society, through its use of best management practices. It is a reflection of the company's reputation and the health of its license to operate, which are key factors in determining its ability to generate long-term shareholder value
Independent variables		
Corporate Governance Score	*CG_Score*	The corporate governance pillar measures a company's systems and processes, which ensure that its board members and executives act in the best interests of its long term shareholders. It reflects a company's capacity, through its use of best management practices, to direct and control its rights and responsibilities through the creation of incentives, as well as checks and balances in order to generate long-term shareholder value
Board of Directors/ Board Functions	*BoD_BFunctions*	The board of directors/board functions category measures a company's management commitment and effectiveness towards following best practice corporate governance principles related to board activities and functions. It reflects a company's capacity to have an effective board by setting up the essential board committees with allocated tasks and responsibilities
Board of Directors/ Board Structure	*BoD_BStructure*	The board of directors/board structure category measures a company's management commitment and effectiveness towards following best practice corporate governance principles related to a well-balanced membership of the board. It reflects a company's capacity to ensure a critical exchange of ideas and an independent decision-making process through an experienced, diverse and independent board

(continued)

Table 17.2 (continued)

Variables	Code	Description
Board of Directors/Compensation Policy	*BoD_Compensation*	The board of directors/compensation policy category measures a company's management commitment and effectiveness towards following best practice corporate governance principles related to competitive and proportionate management compensation. It reflects a company's capacity to attract and retain executives and board members with the necessary skills by linking their compensation to individual or company-wide financial or extra-financial targets
Integration/ Vision and Strategy	*Integration_Vision_Strategy*	The integration/vision and strategy category measures a company's management commitment and effectiveness towards the creation of an overarching vision and strategy integrating financial and extra-financial aspects. It reflects a company's capacity to convincingly show and communicate that it integrates the economic (financial), social and environmental dimensions into its day-to-day decision-making processes

Source: ASSET4 (Thomson Reuters 2017)

17.5 Results

Figure 17.1 sets out the descriptive statistics for the environmental score from 2012 to 2016. On average, an interesting jump ahead between 2014 and 2015 emerges.

By contrast, the social score displays a slight decrease in the mean between 2015 and 2016. In addition, the maximum is almost constant (Fig. 17.2).

Figure 17.3 shows that the mean as well as the minimum record a positive peak in 2015, the first year of the release process of King IV.

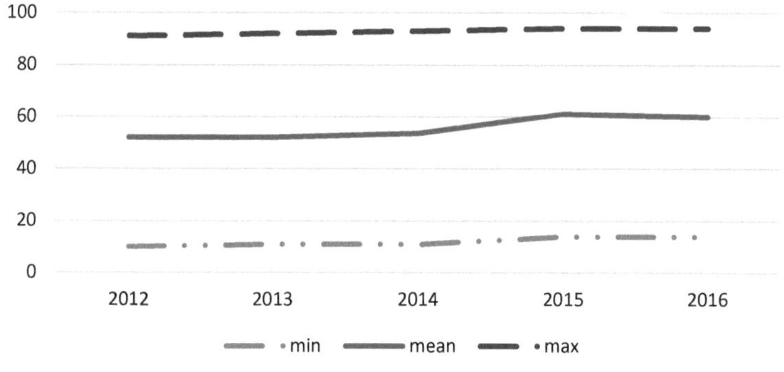

Fig. 17.1 Environmental score (2012–2016)

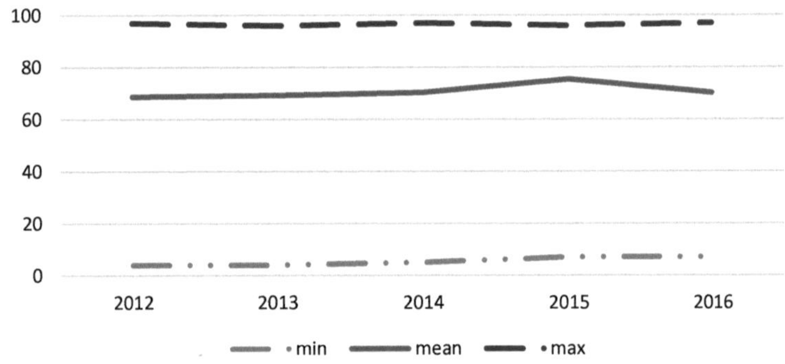

Fig. 17.2 Social score (2012–2016)

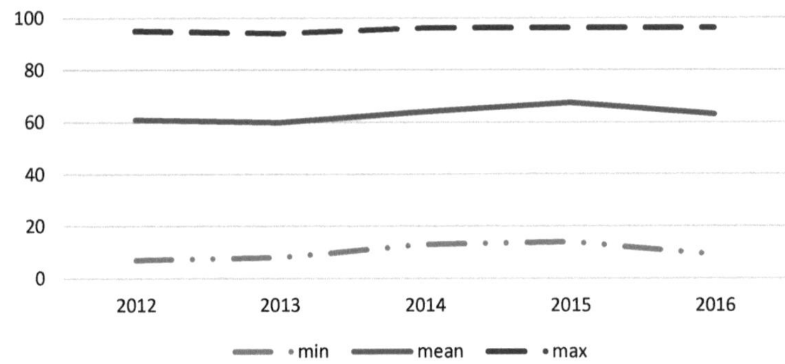

Fig. 17.3 Corporate governance score (2012–2016)

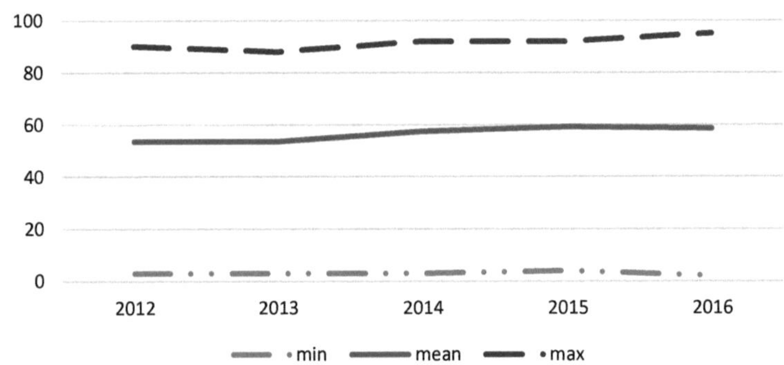

Fig. 17.4 Board of directors/board functions (2012–2016)

The ratio Board of Directors/Board Functions features for a growing trend of the mean between 2012 and 2016. The positive peak takes place in year 2013–2014 (Fig. 17.4).

Figure 17.5 shows that the positive peak of the mean occurs in 2015. Furthermore, in terms of minimum, there is a light performance from 2012 to 2015.

The ratio Board of Directors/Compensation Policy records a negative performance of the mean in 2016. Moreover, the minimum presents an undulating trend where the lowest points are in both 2014 and 2016. By contrast, the positive peak takes place in 2015 (Fig. 17.6).

Figure 17.7 highlights that the trends of minimum and maximum values are similar and growing. As explained earlier, the positive peak of the mean happens in 2015.

The findings listed in Table 17.3 support the first two out of seven research hypotheses. In other words, the Wilcoxon and Sign tests (the latter as control method for testing the reliability of findings) yield evidence that, from the comparison between 2012 and 2016 (i.e. the first year where ESG data is available after the introduction of King III and the year corresponding to the launch of King IV), there

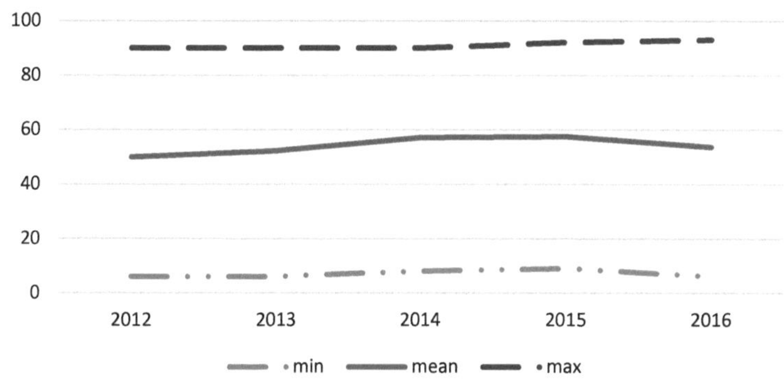

Fig. 17.5 Board of directors/board structure (2012–2016)

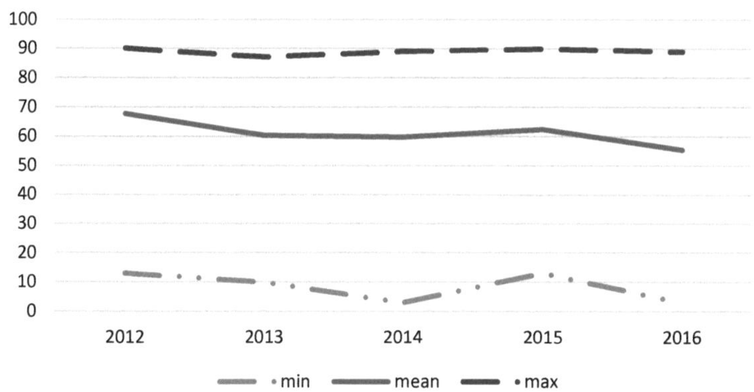

Fig. 17.6 Board of directors/compensation policy (2012–2016)

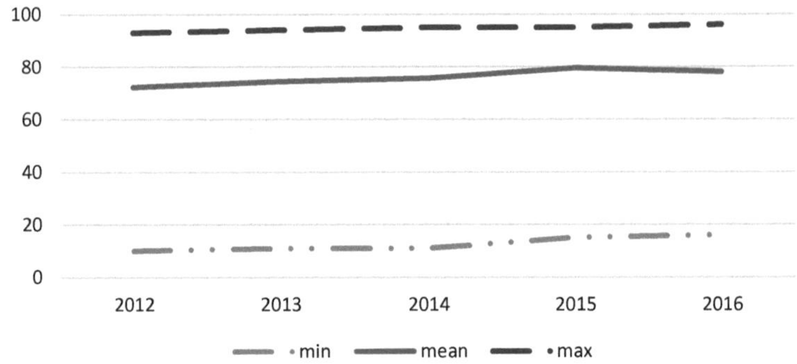

Fig. 17.7 Integration/vision and strategy (2012–2016)

is a statistically significant difference for each single variable included in our empirical analysis. Therefore, King III exerts an influence on both environmental and social performance and on corporate governance practices.

Table 17.4 depicts the results stemming from the calculation of Spearman's correlations where the dependent variable is the environmental score. In particular, hypotheses 3a, 4a and 7a are confirmed, while hypotheses 5a and 6a are rejected, as the *p*-value is not statistically significant. Therefore, the increase in the corporate governance score recorded in 2015 tends to boost the environmental score of the following year. Similarly, this circumstance occurs with reference to the correlations between the dependent variable and the following ratios: Board of Directors/ Board Functions and Integration/Vision and Strategy. Furthermore, such correlations feature for a significant *p*-value (broadly below the critical threshold of 0.001).

As before, Table 17.5 sets out the findings as regards the adoption of Spearman correlations in which the dependent variable is the social score. Clear evidence emerges for hypotheses 3b, 4b and 7b. However, with reference to hypothesis 6b), there is only slight confirmation. Accordingly, we can say that corporate governance performance in 2015 is positively associated with social performance in 2016 (both measured by a specific score computed by ASSET4). The same results can be observed for the correlations between social performance and each of the following ratios: Board of Directors/Board Functions and Integration/Vision and Strategy. As above, the *p*-value of the latter correlation is below the critical boundary of 0.001. In contrast to Table 17.4, there is a positive correlation between the dependent variable and the ratio Board of Directors/Compensation Policy, even if it is worthwhile pointing out that the *p*-value is only just below the threshold of 0.10 (i.e. 0.092).

Finally, it should be noted that the ratio Board of Directors/Board Structure (hypothesis 5b) is never statistically significant, either for environmental or social performance.

Table 17.3 Non parametric tests: Wilcoxson and sign, comparison 2012–2016

				Wilcoxson Test		Sign Test	
Variables	HPs	Expected Results	Findings	Z	p-value	Z	p-value
CG_Score	1a	Presence of a difference	Presence of a difference	−2.511	0.012*	−3.022	0.003**
Environmental_Score	1b	Presence of a difference	Presence of a difference	−4.463	0.000***	−3.946	0.000***
Social_Score	1c	Presence of a difference	Presence of a difference	−2.621	0.009**	−3.081	0.002**
BoD_BFunctions	*2a*	Presence of a difference	Presence of a difference	−2.618	0.009**	−2.313	0.021*
BoD_BStructure	2b	Presence of a difference	Presence of a difference	−2.127	0.033*	−2.586	0.010*
BoD_Compensation	2c	Presence of a difference	Presence of a difference	−3.042	0.002**	−1.923	0.054^
Integration_Vision_Strategy	2d	Presence of a difference	Presence of a difference	−4.858	0.000***	−5.307	0.000***

Significance level: $^\wedge p < 0.1$; $*p < 0.05$; $**p < 0.01$; $***p < 0.001$

Table 17.4 Spearman' correlations, dependent variable: environmental score

Variables	HPs	Expected Sign	Findings	Rho	p-value	N
CG_Score 2015—Environmental_Score_2016	3a	+	Confirmed	0.399	0.003**	55
BoD_BFunctions 2015—Environmental_Score_2016	4a	+	Confirmed	0.512	0.000***	55
BoD_Structure 2015—Environmental_Score_2016	5a	+	Rejected	−0.058	0.674	55
BoD_Compensation 2015—Environmental_Score_2016	6a	+	Rejected	0.202	0.139	55
Integration_Vision_Strategy—Environmental_Score_2016	7a	+	Confirmed	0.700	0.000***	55

Significance level: ^$p < 0.1$; *$p < 0.05$; **$p < 0.01$; ***$p < 0.001$

Table 17.5 Spearman correlations, dependent variable: social score

Variables	HPs	Expected Sign	Findings	Rho	p-value	N
CG_Score 2015—Social_Score_2016	3b	+	Confirmed	0.417	0.002**	55
BoD_BFunctions 2015—Social_Score_2016	4b	+	Confirmed	0.417	0.002**	55
BoD_Structure 2015—Social_Score_2016	5b	+	Rejected	0.015	0.915	55
BoD_Compensation 2015—Social_Score_2016	6b	+	Confirmed	0.230	0.092^	55
Integration_Vision_Strategy—Social_Score_2016	7b	+	Confirmed	0.460	0.000***	55

Significance level: ^$p < 0.1$; *$p < 0.05$; **$p < 0.01$; ***$p < 0.001$

17.6 Discussion and Conclusion

Our study highlights the association between corporate governance mechanisms, ESG factors and <IR> following the adoption of King III and King IV in South Africa. The research was focused on a period that started in 2012 and ended in 2016. While King III was introduced in 2009, the ASSET4 database holds full ESG data only from 2012 onwards. 2016 is the year in which King IV was launched. The comparison between the first and the last years of the timeframe allowed us to explore ESG performance and a number of other corporate governance practices implemented by South African companies listed on the FTSE/JSE All-Share Index and obliged, therefore, to adopt the King Code on Corporate Governance.

Indeed, the findings of the descriptive statistics highlight positive peaks in 2015 with reference to overall corporate governance performance, the appropriate balance of membership on the board (i.e. the ratio Board of Directors/Board Structure), competitive and appropriate management compensation (i.e. the ratio Board od

Directors/Compensation Policy) and effectiveness in formulating a strategic vision in which financial and non-financial elements are closely integrated (i.e. the ratio Integration/Vision and Strategy).

A comparison of the first and last years in which the adoption of King III was compulsory, reveals the presence of a statistical difference with respect to ESG performance and various corporate governance practices pertinent to the balance of membership on the board, board activities and functions, competitive and appropriate management compensation and the integration of financial and non-financial elements in the strategic vision. In other words, our findings corroborate the mainstream research that finds that the mandatory approach can foster a constructive updating or re-engineering of the corporate governance model. In this perspective, our findings are not consistent with McNally et al. (2017), who demonstrate that <IR> is not considered a natural part of internal business processes and reporting protocols, but in fact limits the development of management control systems and accounting infrastructures.

From this point of view, such results furthermore reveal a growing commitment in our sample to many of the specific objectives of King IV, i.e. "reinforce corporate governance as a holistic and interrelated set of arrangements to be understood and implemented in an integrated manner" and "present corporate governance as concerned not only structure and process, but also with an ethical consciousness and conduct" (IODSA 2016, King IV: 22).

The empirical evidence also shows intriguing associations between the implementation of certain governance principles issued in the code and ESG performance. Indeed, the "evolution" from King III to King IV, measured by the comparison between the last year of King III and the first year of King IV adoption, indicates that certain corporate governance practices (such as board functions and structure, compensation policy and the integration process of strategy/vision) can represent a crucial pillar for environmental and social performance (Peters and Romi 2014). In other words, it is insightful to note that there is a substantial alignment of results ascribable to the foregoing dimensions of firm performance, i.e. environmental and social performance. Indeed, the presence of an active board that allocates tasks and responsibilities to different committees and the intention to create an all-encompassing vision with the aim to integrate financial and non-financial perspectives—an integrated thinking approach—are discernible in both environmental and social performance (Oliver et al. 2016; Adams 2017; Dumay and Dai 2017).

In contrast to some prior studies (Bondy et al. 2008), our empirical evidence can provide useful insights about the role of corporate governance codes that should not be considered as tools for governing traditional business issues such as ensuring compliance with laws and regulations, but for the social and environmental commitment to a better integration between the financial and non-financial dimensions.

In terms of managerial implications, the implementation of a corporate governance model intended to generate long-term shareholder value, board activities as well as the ability to integrate the economic, social and environmental dimensions into the corporate strategic design can be considered key "requirements" in a value-creation process that is focused on the achievement of environmental and social goals.

While this study provides useful pointers as regards the improvement of corporate governance mechanisms and practices in an environmental and sustainability perspective, it is centered on a dataset made up of listed companies in a single emerging country which obliged companies to draw up a specific model of reporting, i.e. <IR>. Moreover, the sample size is limited by the partial availability of ESG data, despite the adoption of a dependable secondary database, ASSET4.

Finally, further and stimulating research routes could include the exploration of the comparison between small and medium-sized entities (SMEs), with listed companies and, therefore, the question of the possible relevance of firm size.

Acknowledgment Although the chapter is the result of a conjoint analysis, Sects. 17.1 and 17.2 have been written by Silvio Bianchi Martini, Sects. 17.3 and 17.4 by Federica Doni and Sects. 17.5 and 17.6 by Antonio Corvino.

References

Adams, C. A. (2015). The international integrated reporting council: A call to action. *Critical Perspectives on Accounting, 27*, 23–28.

Adams, C. A. (2017). Conceptualising the contemporary corporate value creation process. *Accounting, Auditing & Accountability Journal, 30*(4), 906–931.

Agyei-Mensah, B. K. (2017). The relationship between corporate governance, corruption and forward-looking information disclosure: A comparative study. *Corporate Governance: The International Journal of Business in Society, 17*(2), 284–304.

Amran, A., Lee, S. P., & Devi, S. S. (2014). The influence of governance structure and strategic corporate social responsibility toward sustainability reporting quality. *Business Strategy and the Environment, 23*(4), 217–235.

Atkins, J. F., & Maroun, W. (2015). Integrated reporting in South Africa in 2012: Perspectives from South African institutional investors. *Meditari Accountancy Research, 23*(2), 197–221.

Audit Committees, Combined Code Guidance. (2003, January). A report and proposed guidance by an FRC- appointed group chaired by Sir Robert Smith. *The financial reporting council limited.* Accessed January 30, 2018, from https://www.riskavert.com/wp-content/uploads/2011/10/Smith-Report.pdf

Baek, H. Y., Johnson, D. R., & Kim, J. W. (2009). Managerial ownership, corporate governance, and voluntary disclosure. *Journal of Business and Economic Studies, 15*(2), 44–71.

Beekes, W., Brown, P., Zhan, W., & Zhang, Q. (2016). Corporate governance, companies' disclosure practices and market transparency: A cross country study. *Journal of Business Finance & Accounting, 43*(3–4), 263–297.

Bianchi Martini, S., Corvino, A., & Doni, F., (2017). Integrated Reporting, corporate governance practices, social sustainability policies and environmental disclosure. The case of South Africa, European Academy of Management. *EURAM 2017 making knowledge work.* University of Strathclyde Business School, 21st–24th of June 2017, 1–27.

Bondy, K., Matten, D., & Moon, J. (2008). Multinational corporation codes of conduct: Governance tools for corporate social responsibility? *Corporate Governance an International Review, 16*(4), 294–311.

Bryman, A., & Cramer, D. (2001). *Quantitative data analysis with SPSS release 10 for Windows: A guide for social scientists*. Hove: Routledge.

Buckholtz, A. K., Brown, J. A., & Shabana, K. M. (2008). Corporate governance and corporate social responsibility. In A. Crane, A. McWilliams, D. Matten, J. Moon, & D. S. Siegel (Eds.), *The Oxford handbook of CSR* (pp. 327–345). Oxford: Oxford University Press.

Cadbury Report. (1992, December). *Report of the committee on the financial aspects of corporate governance*. London: Brugens Science Press. Accessed January 30, 2018, from http://www.ecgi.org/codes/documents/cadbury.pdf

Carels, C., Maroun, W., & Padia, N. (2013). Integrated reporting in the South African mining sector. *Corporate Ownership & Control, 11*(1), 947–961.

Cerbioni, F., & Parbonetti, A. (2007). Exploring the effects of corporate governance on intellectual capital disclosure: An analysis of European biotechnology companies. *European Management Review, 16*(4), 791–826.

Chan, M. C., Watson, J., & Woodliff, D. (2014). Corporate governance quality and CSR disclosures. *Journal of Business Ethics, 125*(1), 59–73.

Cheng, M., Green, W., Conradie, P., Konishi, N., & Romi, A. (2014). The international integrated reporting framework: Key issues and future research opportunities. *Journal of International Financial and Management Accounting, 25*, 90–119.

Churet, C., & Eccles, R. G. (2014). Integrated reporting, quality of management, and financial performance. *Journal of Applied Corporate Finance, 26*(1), 56–64.

Clarkson, M. B. E. (1995). A stakeholder framework for analyzing and evaluating corporate social performance. *Academy of Management Review, 20*(1), 92–117.

Cowen, S. S., Ferreri, L. B., & Parker, L. D. (1987). The impact of corporate characteristics on social responsibility disclosure: A typology and frequency-based analysis. *Accounting, Organizations and Society, 12*(2), 111–122.

De Villiers, C., Venter, E. R., & Hsiao, P. K. (2016). Integrated reporting: Background, measurement issues, approaches and an agenda for future research. *Accounting and Finance, 57*, 937–959.

Diale, A. J. (2012). The relevance or irrelevance of corporate social responsibility (CSR) to enhance government-business relations in South Africa: A theoretical exposition. *Global Conference on Business and Finance Proceedings, 7*(1), 759–764.

Doni, F., & Fortuna, F. (2018). Corporate governance code in South Africa after the adoption of integrated reporting, evidence from the mining industry. *International Business Management, 12*(1), 68–81.

Doni, F., Gasperini, A., & Pavone, P. (2016). Early adopters of integrated reporting: The case of the mining industry in South Africa. *African Journal of Business Management, 10*(9), 187–208.

Dumay, J., & Dai, T. (2017). Integrated thinking as a cultural control? *Meditari Accountancy Research, 25*(4), 574–604.

Dumay, J., & Xi Dai, T. M. (2014). Integrated thinking as an organisational cultural control?. In *Critical perspectives on accounting conference*, 7th-9th July, Toronto, Canada.

Dumay, J., Bernardi, C., Guthrie, J., & Demartini, P. (2016). Integrated reporting: A structured literature review. *Accounting Forum, 40*(3), 166–185.

Eccles, R. G., & Krzus, M. P. (2010). *One report: Integrated reporting for a sustainable strategy*. New York: Wiley.

Eccles, R. G., & Youmans, T. (2016). Materiality in corporate governance: The statement of significant audiences and materiality. *Journal of Applied Corporate Finance, 28*, 39–46.

Eccles, R. G., Krzus, M. P., & Ribot, S. (2015a). *The integrated reporting movement, meaning, momentum, motives and materiality*. New York: Wiley.

Eccles, R. G., Krzus, M. P., & Ribot, S. (2015b). Models of best practices in integrated reporting, 2015. *Journal of Applied Corporate Finance, 27*(2), 103–115.

Feng, T., Cummings, L., & Tweedie, D. (2017). Exploring integrated thinking in integrated reporting: An exploratory study in Australia. *Journal of Intellectual Capital, 18*(2), 330–353.

FRC, Financial Reporting Council. (2011, December). *Developments in corporate governance 2011: The impact and implementation of the UK corporate governance and stewardship codes*. Accessed January 30, 2018, from https://www.frc.org.uk

FRC, Financial Reporting Council. (2012, September). *UK stewardship code*. Accessed January 30, 2018, from https://www.frc.org.uk/getattachment/d67933f9-ca38-4233-b603-3d24b2f62c5f/UK-Stewardship-Code-(September-2012).pdf

FRC, Financial Reporting Council. (2014, September). *The UK corporate governance code*. Accessed January 30, 2018, from https://www.frc.org.uk/getattachment/59a5171d-4163-4fb2-9e9d-daefcd7153b5/UK-Corporate-Governance-Code-2014.pdf

FRC, Financial Reporting Council. (2016, April). *The UK corporate governance code*. Accessed January 30, 2018, from https://www.frc.org.uk/getattachment/ca7e94c4-b9a9-49e2-a824-ad76a322873c/UK-Corporate-Governance-Code-April-2016.pdf

FRC, Financial Reporting Council. (2017, April). *Annual review of corporate reporting 2016–2017*. Accessed January 30, 2018, from https://www.frc.org.uk/getattachment/311af48c-bdfa-4484-8e7d-6de689fd8f4b/Annual-Review-of-Corporate-Reporting-2016-17.PDF

Frias-Aceituno, J. V., Rodriguez-Ariza, L., & Garcia-Sanchez, I. M. (2013). The role of the board in the dissemination of integrated corporate social reporting. *Corporate Social Responsibility Environmental Management, 20*, 219–233.

Frias-Aceituno, J. V., Rodriguez-Ariza, L., & Garcia Sanchez, I. M. (2014). Explanatory factors of integrated sustainability and financial reporting. *Business Strategy and the Environment, 23*, 56–72.

Garcia-Lara, J. M., Garcia-Osma, B., & Penalva, F. (2009). Accounting conservatism and corporate governance. *Review of Accounting Studies, 14*(1), 161–201.

Greenbury Recommendations. (1995). Accessed January 30, 2018, from http://www.ecgi.org/codes/documents/greenbury_recommendations.pdf

Haji, A. A., & Hossain, D. M. (2016). Exploring the implications of integral reporting on organisational reporting practice: Evidence from highly regarded integrated reporters. *Qualitative Research in Accounting & Management, 13*(4), 415–423.

Hampel Committee. (1998). *Final report*. Accessed January 30, 2018, from http://www.ecgi.org/codes/documents/hampel_index.htm

Haniffa, A. J., & Hudaib, M. (2006). The impact of culture and governance on corporate social reporting. *Journal of Accounting and Public Policy, 24*(5), 391–430.

Healy, P. M., & Palepu, K. G. (2001). Information asymmetry, corporate disclosure, and the capital markets: A review of the empirical disclosure literature. *Journal of Accounting and Economics, 31*(1–3), 405–440.

Helfaya, A., & Moussa, T. (2017). Do board's corporate social responsibility strategy and orientation influence environmental sustainability disclosure? UK evidence. *Business Strategy and the Environment, 26*(8), 1061–1077.

Hidalgo, R. L., Garcìa-Meca, E., & Martìnez, I. (2011). Corporate governance and intellectual capital disclosure. *Journal of Business Ethics, 100*(3), 483–495.

Hindley, T., & Buys, P. W. (2012). Integrated reporting compliance with the global reporting initiative framework: An analysis of the South African mining industry. *International Business & Economics Research Journal, 11*(11), 1249–1260.

Hooghiemstra, R. (2000). Corporate communication and impression management – New perspectives why companies engage in corporate social reporting. *Journal of Business Ethics, 27*(1–2), 55–68.

ICGN, International Corporate Governance Network. (2013). *Guidance on institutional investors responsibilities*. Accessed January 30, 2018, from https://www.icgn.org/sites/default/files/ICGN%20Institutional%20Investor%20Responsibilities_0.pdf

IODSA, Institute of Directors in Southern Africa. (2016). *King IV report on corporate governance for South Africa*. Accessed January 30, 2018 from https://www.iodsa.co.za/page/KingIV

IFA, Institute Français des Administrateurs. (2017). *The board of directors and integrated reporting*. IFA publications. Accessed January 30, 2018, from http://integratedreporting.org/wp-content/uploads/2017/07/IFA_Reporting-Integre%CC%81-2017_EN.pdf

IFAC, International Federation of Accountants. (2012, October). *Integrated governance for sustainable success*. Accessed January 30, 2018, from http://html5.epaperflip.com/?docid=22019bb1-4c53-45c9-bc45-a56000fb4779#page=0

IIRC & IFAC, International Integrated Reporting Council. (2017). *Creating value: CFO Leadership in <IR>*. Accessed January 30, 2018, from www.integratedreporting.org

IIRC, International Integrated Reporting Council. (2013, December). *The international <IR> framework*. Accessed January 30, 2018, from http://integratedreporting.org/wp-content/uploads/2013/12/13-12-08-THE-INTERNATIONAL-IR-FRAMEWORK-2-1.pdf

IIRC, International Integrated Reporting Council. (2014). *Creating value: Value to the board*. Accessed January 30, 2018, from www.integratedreporting.org

IIRC, International Integrated Reporting Council. (2016, October). *Creating value: The cyclical power of integrated thinking and reporting*. Accessed January 30, 2018, from www.integratedreporting.org

Investment Responsible Research Center Institute (IRRCI) – Sustainable Investment Institute (SII). (2013). *Integrated financial and sustainability reporting in the United States*. Accessed January 30, 2018, from https://www.frc.org.uk/getattachment/ca7e94c4-b9a9-49e2-a824-ad76a322873c/Uf

IODSA, Institute of Directors in Southern Africa. (2009). *King code of governance for South Africa*. Accessed January 30, 2018, from http://www.ecgi.org/codes/documents/king3.pdf

IODSA, Institute of Directors in Southern Africa. (2013, March). *Perception and practice of King III in South African companies*. Accessed January 30, 2018, from http://c.ymcdn.com/sites/www.iodsa.co.za/resource/collection/DD8B591E-3D00-48D5-B2E9-663FEDCFF131/Perceptions_and_practice_of_King_III.pdf

Johannesburg Stock Exchange (JSE). (2004). *Socially Responsible Investment (SRI) index*. Accessed January 30, 2018, from https://www.jse.co.za/services/market-data/indices/socially-responsible-investment-index

Kaymak, T., & Bektas, E. (2017). Corporate social responsibility and governance: Information disclosure in multinational corporations. *Corporate Social Responsibility and Environmental Management, 24*(6), 555–569.

Knauer, A., & Serafeim, G. (2014). Attracting long-term investors through integrated thinking and reporting: A clinical study of a biopharmaceutical company. *Journal of Applied Corporate Finance, 26*(2), 57–64.

Kojima, K. (1998). *Corporate governance: An international comparison*. Kobe University Research Institute for Economics and Business Administration.

KPMG. (2017). *The road ahead. The KPMG survey of corporate responsibility reporting 2017*. Accessed January 30, 2018, from kpmg.com/crreporting

Liu, X., & Zhang, C. (2017). Corporate governance, social responsibility information disclosure, and enterprise value in China. *Journal of Cleaner Production, 142*, 1075–1084.

Lyon, T. P., & Maxwell, J. W. (2011). Greenwash: Corporate environmental disclosure under threat of audit. *Journal of Economics & Management Strategy, 20*, 3–41.

Mallin, C., & Michelon, G. (2011). Board reputation attributes and corporate social performance: An empirical investigation of the US best corporate citizens. *Accounting and Business Research, 41*(2), 119–144.

Mallin, C., Michelon, G., & Raggi, D. (2013). Monitoring intensity and stakeholders' orientation: How does governance affect social and environmental disclosure? *Journal of Business Ethics, 114*(1), 29–43.

McNally, M. A., Cerbone, D., & Maroun, W. (2017). Exploring the challenges of preparing an integrated report. *Meditari Accountancy Research, 25*(4), 481–504.

Melloni, G. (2015). Intellectual capital disclosure in integrated reporting: An impression management analysis. *Journal of Intellectual Capital, 16*(3), 661–680.

Merkl-Davies, D. M., & Brennan, N. M. (2007). Discretionary disclosure strategies in corporate narratives: Incremental information or impression management? *Journal of Accounting Literature, 27*, 116–196.

Michelon, G., & Parbonetti, A. (2010). Stakeholder engagement: Corporate governance and sustainability disclosure. *Journal of Management and Governance, 16*(3), 477–509.

Nagata, K., & Nguyen, P. (2017). Ownership structure and disclosure quality: Evidence from management forecasts revisions in Japan. *Journal of Accounting and Public Policy, 36*(6), 451–467.

Nazari, J. A., Herremans, I. M., & Warsame, H. A. (2015). Sustainability reporting: External motivators and internal facilitators. *Corporate Governance: The international journal of business in society, 15*(3), 375–390.

Nestor, S., & Thompson, J. K. (1999). *Corporate governance patterns in OECD economies: Is convergence under way?* Paris: OECD Publications.

Ntim, C. G., Opong, K. K., & Danbolt, J. (2012). The relative value relevance of shareholder versus stakeholder corporate governance disclosure policy reforms in South Africa. *Corporate Governance: An International Review, 20*, 84–105.

OECD. (2015). *G20/OECD principles of corporate governance*. Paris: OECD Publishing. Accessed January 30, 2018, from https://doi.org/10.1787/9789264236882-en

Oliver, J., Vesty, G., & Brooks, A. (2016). Conceptualising integrated thinking in practice. *Managerial Auditing Journal, 31*(2), 228–248.

Peters, G. F., & Romi, A. M. (2014). Does the voluntary adoption of corporate governance mechanisms improve environmental risk disclosures? Evidence from greenhouse gas emission accounting. *Journal of Business Ethics, 125*(4), 637–666.

Raemaekers, K., Maroun, W., & Padia, N. (2016). Risk disclosures by South African listed companies post-King III, South African listed companies post-King III. *South African Journal of Accounting Research, 30*(1), 1–20.

Rensburg, R., & Botha, E. (2014). Is integrated reporting the silver bullet of financial communication? A stakeholder perspective from South Africa. *Public Relations Review, 40*(2), 144–152.

Rodriguez-Fernandez, M. (2016). Social responsibility and financial performance: The role of good corporate governance. *Business Research Quarterly, 19*(2), 137–151.

SAICA. (2015). *Integrated thinking: An exploratory survey*. Accessed January 30, 2018, from https://www.saica.co.za/Portals/0/Technical/Sustainability/SAICAIntegratedThinkingLandscape.pdf

Setia, N., Abhayawansa, S., Joshi, M., & Huynh, A. V. (2015). Integrated reporting in South Africa: Some initial evidence. *Sustainability Accounting, Management and Policy Journal, 6*(3), 397–424.

Thomson Reuters. (2017). *ASSET4 ESG research data*. Accessed January 30, 2018 from https://financial.thomsonreuters.com/en/products/dataanalytics/company-data/esg-research-data.html

West, A. (2009). Corporate governance convergence and moral relativism. *Corporate Governance: An International Review, 17*(1), 107–119.

Chapter 18
Integrated Reporting in India: Research Findings and Insights

Sumona Ghosh

18.1 Introduction

Integrated Reporting (IR) is an emerging phenomenon in the world of corporate reporting that has gained significant momentum in the last 10 years. Integrated reporting brings together information about an organization's strategy, governance, performance, risks and opportunities in such a manner that it reflects the economic, social and environmental context within which it operates. It emerges from the historical development of financial and non-financial reporting initiatives by business. Organizations need to respond to stakeholder needs. However, the needs of stakeholders are dynamic and its changing with time and therefore, the organizations also need to change the way they communicate information to them. Stakeholders were previously primarily interested in financial statements, now there is growing attention for other nonfinancial, organizational aspects such as social and environmental consequences and governance structures as well. Stakeholders are questioning the relevance and reliability of annual financial reports and the ability of an organization to operate sustainably in the future. Therefore a rising need was felt for a more holistic, integrated form of accountability and business reporting model which will merge financial and nonfinancial information in a meaningful and in an integrated manner. This requirement brought forth the advent of one specific concept called 'Integrated Reporting'. With the inclusion of this introduction, the paper has been organized into six sections. Section 18.2 gives a brief insight into the work that has been done in this area. Section 18.3 outlines the objectives. Section 18.4 states the methodology. Section 18.5 delves into empirical results. Section 18.6 presents the conclusion.

S. Ghosh (✉)
St. Xavier's College, Kolkata, India
e-mail: sumonaghosh@sxccal.edu

S. O. Idowu, M. Del Baldo (eds.), *Integrated Reporting*, CSR, Sustainability, Ethics & Governance, https://doi.org/10.1007/978-3-030-01719-4_18

18.2 Literature Review

18.2.1 Need for Integrated Reporting

With a rise in socio-economic and environmental challenges, the traditional shareholder centric corporate reports, which places most of the emphasis on financial parameters and fail to make a clear link between financial, social, environmental and ethical matters (Cleverly et al. 2010) have been questioned. Such questions led the companies globally to produce standalone *sustainability reports*. Over 80% of the Global Fortune 250 now publish sustainability reports. Even in India the company's publishing sustainability reports have significantly increased (Ghosh 2016). Researchers have shown through their studies that mandatory sustainability reporting have encouraged corporations to adopt more environmentally and socially responsible practices, to improve on relevant performance measures, get better access to finance and firms experience a decrease in cost of equity capital after issuing a sustainability report (Ioannou and Serafeim 2012; Ioannou and Serafeim 2015; Yu et al. 2014; Dhaliwal et al. 2011, 2012).

However despite the growing use of sustainability reports there was a significant debate on whether or not they actually help corporations to meaningfully improve their performance on sustainability dimensions. Critics argue that sustainability reports are not of use to those stakeholders who want to hold corporations accountable for their actions. Instead, the incentives are for corporations to produce a "high volume and low quality of information," which stakeholders find difficult to assess in terms of veracity and completeness (Mitchell et al. 2012; Luke 2013; Siebecker 2009; Adams 2004; IODSA 2009; Eccles and Krzus 2010; Eurosif and ACCA 2013). These corporate reporting challenges have highlighted the need for integrated reporting which will effectively merge financial and nonfinancial information in a meaningful manner (IRCSA 2011; Solomon and Maroun 2012; Eccles and Armbrester 2011).

18.2.2 Evolution of Integrated Reporting

IR has a short history, its meaning is still evolving, and only recently has a framework been developed that can provide companies guidance on what constitutes an IR. The first companies to produce a self-declared IR were the Danish enzymes company Novozymes (in 2002), the Brazilian cosmetics fragrances company Natura (in 2003), the Danish pharmaceutical company Novo Nordisk (in 2004), The Crown

Estate (UK), SAP (Germany) and Port of Rotterdam Authority (the Netherlands). The concept of integrated reporting was introduced in South Africa through King III, the code of corporate governance in 2009. In 2010, all South African companies listed on the Johannesburg Stock Exchange were required to issue an IR or explain why they were not doing so. There is not, at present, a universally agreed framework or template for an integrated report, although IIRC members agreed to an International <IR> Framework in December 2013. The fundamental concepts of integrated reporting are represented by three aspects, which are (1) the six capitals that an organization uses and affects, (2) the organization's business model and (3) the creation of value over time (Busco et al. 2013a, b).

Integrated reporting can be described as: "*a process founded on integrated thinking that results in a periodic integrated report by an organization about value creation over time and related communications regarding aspects of value creation.*" (IR 2016). The main aspect of integrated reporting is value creation and impact of an organization. Therefore, organizations are using integrated reporting to communicate how all of their resources are creating value. It helps businesses to think holistically about their strategy and plans, make informed decisions and manage key risks which helps in building investor and stakeholder confidence and improve future performance. Integrated reporting incorporates the work of the Global Reporting Initiative (GRI), The World Business Council for Sustainable Development, The World Resources Institute, the Carbon Disclosure Project and the UN Global Compact.

According to Eccles and Krzus (2010), integrated reporting represents the reconciliation of two main theories, namely: *shareholder theory and stakeholder theory*. The shareholder theory primarily focuses on what the companies need to do to maximize shareholder's value. From the perspective of this theory, the role of integrated reporting would be to approach environmental, social and governance issues that are considered to influence the value of the company. Therefore, when setting the framework for integrated reports, one should also consider the additional value for shareholders. The stakeholder theory creates value for all the stakeholders. This theory incorporates environmental, social and governance issues. By introducing integrated reporting, companies would become more aware of their stakeholders, who might impact the decision-making process.

From the Indian perspective we find that on 7 February 2017, SEBI (Securities and Exchange Board of India) has issued a circular advising top 500 listed companies which are required to prepare BRR (Business Responsibility Report) to adopt IR on a voluntary basis from the financial year 2017–18. The information related to IR may be provided in the following ways:

- As part of annual report with a separate section on IR
- Incorporating in management discussion and analysis, or
- By preparing a separate report (annual report prepared as per IR framework).

In case the company has already provided the relevant information in any other report prepared in accordance with national/international requirement/framework, it may provide appropriate reference to the same in its integrated report so as to avoid duplication of information.

18.2.3 Studies on Integrated Reporting

The first publication about integrated reporting was in June 2005, by Allen White, one of the co-founders of the GRI, where he had observed integrated reporting as the future of corporate reporting. Globally speaking studies on integrated reporting can be divided into three categories. First we have the *conceptual research papers* emphasizing on its history, framework, content, benefits, challenges and its value relevance (Eccles and Krzus 2010; Loska 2011; Adams and Simnett 2011; Jensen and Berg 2012; Owen 2013; Abeysekera 2013; Frías-Aceituno et al. 2013a, b; Ioana and Tiron-Tudor 2013; Busco et al. 2013a, b; De Villiers et al. 2014; Cheng et al. 2014; Frías-Aceituno et al. 2014; Stubbs and Higgins 2014; Brown and Dillard 2014; Van Bommel 2014; Carels et al. 2013; Churet et al. 2014; Flower 2015; Adams 2015; Barker and Kasim 2016; Dumay et al. 2016; Perego et al. 2016). Second the *case study research* investigated the internal and disclosure mechanisms employed by early adopters of integrated reporting (Higgins et al. 2014; Stubbs and Higgins 2014; Janeka et al. 2016). Third *Content analyses of integrated reports* aim to find compliance levels with guidelines (Hindley and Buys 2012; Van Zyl 2013; Maubane et al. 2014; Adams et al. 2016) and levels of integration (Gurvitsh and Sidorova 2012). Fourth we have *research on integrated reporting and firm characteristics* showed that the following factors affected the implementation of integrated reporting: firm size and industry (Frías-Aceituno et al. 2013b; Sierra-García et al. 2015), financial performance (Dragu and Tiron-Tudor 2013a), the intensity of market coordination and ownership concentration (Jensen and Berg 2012), corporate governance mechanisms (Frías-Aceituno et al. 2013b; Velte 2014), and country, political, and cultural factors (Eccles and Serafeim 2011; Jensen and Berg 2012; Dragu and Tiron-Tudor 2013b; Frías-Aceituno et al. 2013a; Garcia-Sánchez et al. 2013) and firm valuation (Lee and Yeo 2016).

From the Indian perspective we find very limited research has gone in to the aspect of integrated reporting. For instance we have Thiagarajan and Baul (2014) providing a comprehensive overview and argument for the implementation of integrated reporting or value reporting, with a specific focus on intellectual capital. Kiran and Goud (2015) studied integrated reporting amongst banks and found that the extent of integration amongst most of the Indian banks is limited and they now should move from limited or partial integration to full integration. Raju (2015) studied the integrated reporting practices followed by the Tata Steel Company Ltd. Athma and Laxmi (2013) in their paper concluded that creating awareness about integrated reporting and its importance as a better communication tool between the companies and the investors would enhance the company's reputation and long-term sustainability of the business. Therefore our study would not only contribute to the existing literature but also enhance the understanding of integrated reporting in the context of selected companies in India.

18.3 Objectives

Even though there has been a growing importance of integrated reporting globally but there has been limited research conducted with respect to IR. Conceptual papers or case study based papers or papers on content analysis of integrated reports are limited in India. With these in the background, the present study will be concerned with the following objectives:

Objective 1: To estimate the percentage of companies who have adopted and disclosed their integrated report as a part of their annual reports on the basis of the IR (Integrated Reporting) framework.
Objective 2: To calculate the percentage of companies that provide information about the different content elements of integrated reporting.
Objective 3: To determine the percentage of companies that provide information about the different forms of capital.
Objective 4: To construct Integrated Reporting Index for different time periods across all companies and to differentiate the companies on that basis.

18.4 Methodology

18.4.1 Data Source and Study Design

An empirical and analytical study was undertaken for the financial years 2010–2016 to give us an overview about integrated reporting for said period. The study was based on secondary sources, i. e., by analyzing the company's annual reports. In India integrated reporting normally has been asked to be made a part of the annual report and hence the company annual reports have been studied and analyzed. Data was then generated from such an analysis using Longitudinal Qualitative Document Analysis to show trends as well as absolute integrated reporting levels of Indian companies for the 6 year period. Qualitative Document Analysis according to Glaser and Strauss (1967) described the meanings, prominence and the theme of messages and emphasized the understanding of the organization as well as how it was presented. Four categories had been identified for this study. Detailed exposition of these categories is provided below.

18.4.1.1 Presence of Integrated Report Category

Through this category we tried to observe the presence of integrated report in the company's annual reports which would help us to analyze the priority given by the companies to explain to providers of financial capital how an organization creates value over time because it befits all stakeholders interested in an organization's ability to create value over time, including employees, customers, suppliers, business partners, local communities, legislators, regulators and policy-maker. We assigned 1 if integrated report was present in the annual reports. Otherwise we assigned 0.

18.4.1.2 Content Element Category

Through this category we tried to explore the presence of the content elements prescribed by the IIRC in the integrated report if the company has published one and made it a part of the annual report or if not then we have tried to explore the presence of the content elements prescribed by the IIRC in the various segments of the annual reports. The various segments include Chairman/CEO message, Director's report, and letter addressed to the shareholders, letter addressed by the CEO/CFO/President and "management discussion and analysis". The content elements identified for this part are:

- *Organizational overview and external environment.*
- *Governance.*
- *Business model.*
- *Risks*
- *Opportunities.*
- *Strategy and resource allocation.*
- *Performance.*
- *Outlook.*

We assigned 1 if the companies disclosed such elements otherwise 0.

18.4.1.3 Disclosure of Capital Category

All organizations depend on various forms of capital for their success. These capitals are stores of value that, in one form or another, become inputs to the organization's business model. They are also increased, decreased or transformed through the activities of the organization. In the Framework, the six capitals recognized are:

- *Financial capital.*
- *Manufactured capital.*
- *Intellectual capital.*
- *Human capital.*
- *Social and relationship capital.*
- *Natural capital.*

Through this category we tried to observe the kind of capital/s that the companies have disclosed. We assigned 1 if the companies disclosed about such capitals otherwise 0.

18.4.2 Extent of Integration Category

We have tried to capture the extent of integration with respect to the information disclosed by the companies about the different content elements as discussed above as well the kinds of capital as required by the IR framework. This information helps in assessing an organization's ability to create value. We have identified four main stages of integration—high integration level, medium integration level, progressive integration and low integration level. Companies with IR score higher than the grand average score for each of the 6 years under consideration was regarded to be companies which were highly integrated in terms of their disclosure as per the IR (Integrated Reporting) framework. Similarly companies whose IR score was lower than the grand average score for each of the 6 years, we regarded such companies to have a low integration level, companies with scores gradually improving over the time period were regarded as companies with progressive integration and remaining companies were regarded to have moderate integration.

18.4.3 Selection of Companies

Top 500 companies were selected from the ET 500 list published by Economic Times for the years 2015–2016 and 2016–2017. Now from the list of 500 most valuable companies for the years 2015–2016 and 2016–2017 the common companies which had succeeded in maintaining its rank and position within the list of "500 most valuable companies" for the 2 years were selected and from this the top 135 companies were considered for the study. Banks were excluded from the list of 135. The final select data set consisted of 102 companies. These companies were ranked on the basis of market capitalization. Market capitalization refers to stock price multiplied by the number of outstanding shares. This parameter was chosen to rank companies since it gives us an indication of not only the present but future prospects of the company as well. The companies were analyzed on the basis of the study design specified above for the time period 2010–2016 i.e. 6 years.

18.4.4 Method

The study employed Longitudinal Qualitative Document Analysis for the period 2010–2016. With respect to the *presence of integrated report category* we calculated the proportion of companies who had disclosed integrated report in their annual reports for each of the years and have tried to find out if there is any change under this category for the study period. For *the content element category* we calculated the proportion of companies who had disclosed the content elements as per the IR framework in their annual reports for each of the years. For the disclosure of *capital category* we calculated the proportion of companies who had disclosed different forms of capital as per the IR framework in their annual reports for each of the years. Besides this, the study constructs an *Integrated Reporting Index (IRI)* in examining annual reports of102 most valuable companies ranked on the basis of market capitalization.

$$\mathrm{IRI} = \left(\sum \mathrm{di/n}\right) \text{ x } 100 = (\mathrm{TS/n}) \text{ x } 100$$

Where:
IRI = Integrated Reporting Index
di: 1 if item i is disclosed; 0 if item i is not disclosed
n = no of items = Maximum Score
TS = Total Score

The main criteria used for measuring the integration level are (ACCA 2012: 13): mission and strategy, management approach, performance tracking, risk management, stakeholder engagement, format public reporting. In our study we have taken the 14 items of disclosure as provided by the IIRC (The International Integrated Reporting Council). Since there are 14 items of disclosure (Appendix) the maximum score for each sampled company will be 14. Therefore the maximum IRI score is "100" and the minimum score is "0". Hence a score of 100 or closer to it suggests high level of integration by the company. A score of "0" or closer to it suggests low level adoption of integration. We also calculated the grand mean score for each of the years. From here we identified the four levels of integration which are as follows:

- Companies with score above the grand average score for each year reflected high integration.
- Companies with score below grand average score for each year reflected low integration.
- Companies with scores gradually improving over the time period considered for the analysis reflected progressive companies in respect of integration.
- The remaining companies reflected moderate integration

18.5 Analysis and Discussions

With respect to the presence of integrated report category we observe from Table 18.1 that percentage of companies including integrated reports in the annual reports increased from nil in 2010–2011 to 4% in 2015–2016. A graphical representation of the above has also been provided in Fig. 18.1 below.

In Table 18.2 we present summary statistics on the percentage of companies that provide information about the different content elements of integrated reporting. From the table we observe that governance and performance were the two most well reported content elements throughout. In contrast, information about business model and strategy and resource allocation was the two least well reported content elements. Most of the companies failed to disclose any information on their business model and value creation since most of them didn't have one. Besides companies also showed reluctance to report on where do they intend to go and how. A graphical representation has also been provided in Fig. 18.2 below.

In Table 18.3 we present summary statistics on the percentage of companies that provide information about the different capitals required to be disclosed in an integrated report. From the table we observe that financial and human capital were the two most well reported capital form disclosed by the companies throughout. Across capitals, one can observe that companies are providing less information about manufactured and intellectual capital. While the relative deficiency of information about manufacturing capital may be due to the fact that in the knowledge economy there is a decline in the importance of manufacturing asset but the

Table 18.1 Presence of Integrated Report in the company's annual reports for the time period 2010–2016

Year	Integrated report
2010–2011	0%
2011–2012	1%
2012–2013	2%
2013–2014	2%
2014–2015	2%
2015–2016	4%

Source: Authors own computation

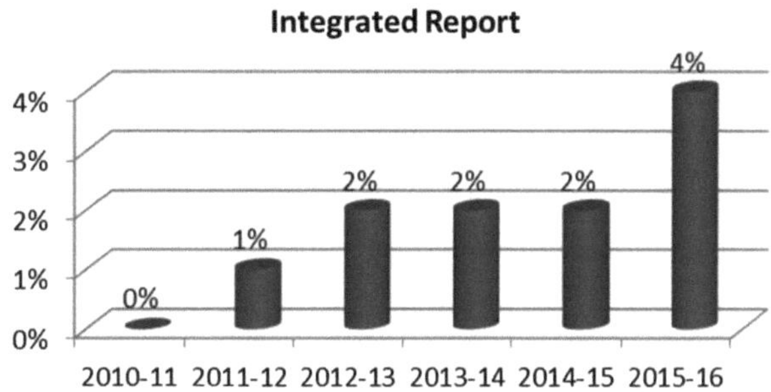

Fig. 18.1 Presence of Integrated Report in the company's annual reports for the time period 2010–2016

Table 18.2 Percentage of companies reporting content element specific information for the time period 2010–2016

	2010–2011	2011–2012	2012–2013	2013–2014	2014–2015	2015–2016
Organizational overview and external environment	70%	70%	72%	75%	75%	79%
Governance	100%	100%	100%	100%	100%	100%
Business model	0%	1%	3%	4%	4%	9%
Risks	88%	89%	91%	91%	92%	94%
Opportunities	57%	57%	58%	54%	56%	61%
Strategy and resource allocation	34%	32%	37%	42%	42%	31%
Performance	100%	100%	100%	100%	100%	100%
Outlook	78%	81%	77%	79%	81%	81%

Source: Authors own computation

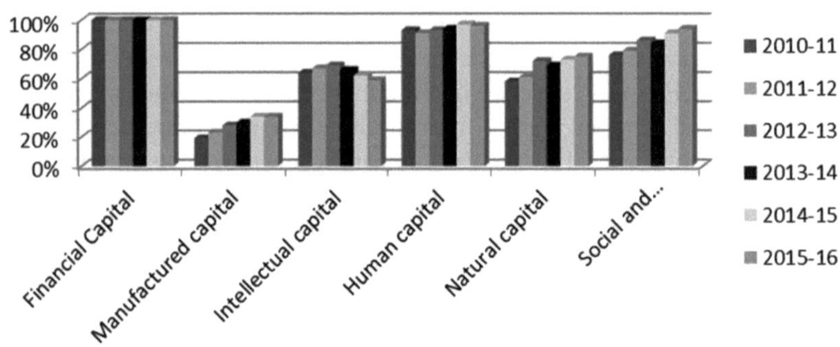

Fig. 18.2 Percentage of companies reporting content element specific information for the time period 2010–2016

Table 18.3 Percentage of companies reporting capital specific information for the time period 2010–2016

	2010–2011	2011–2012	2012–2013	2013–2014	2014–2015	2015–2016
Financial capital	100%	100%	100%	100%	100%	100%
Manufactured capital	19%	23%	28%	30%	34%	34%
Intellectual capital	64%	67%	69%	66%	62%	59%
Human capital	93%	91%	93%	94%	97%	96%
Natural capital	58%	61%	72%	69%	73%	75%
Social and relationship capital	76%	79%	86%	84%	91%	94%

Source: Authors own computation

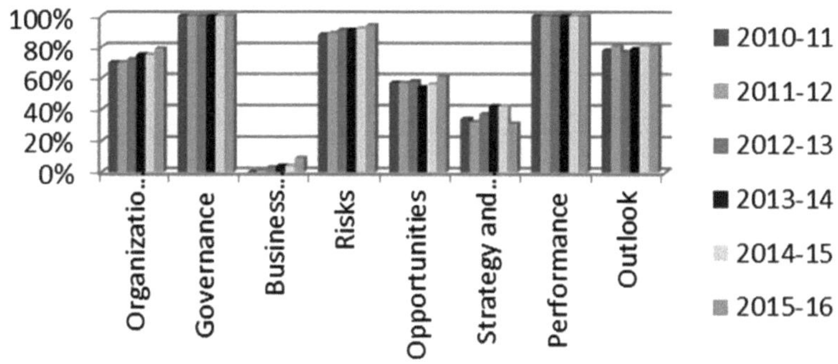

Fig. 18.3 Percentage of companies reporting capital specific information for the time period 2010–2016

deficiency of information on intellectual capital was unexpected. Intellectual capital in the form of innovation capabilities, research and development and outcomes are important value drivers for many businesses now days. A graphical representation has also been provided in Fig. 18.3 below.

Tables 18.4 and 18.5 discloses the individual Integrated Reporting Index (IRI) for the companies for the time period 2010–2016 and the extent of integration of the companies for the time period 2010–2016. The study revealed an average IRI of 70%. (Table 18.4). Table 18.5 showed the extent of integration whereby the companies were differentiated on the basis of their IRI into high integration, low integration, progressive integration and moderate integration. Companies with score above the grand average score for each year reflected high integration. Hence companies with a score above 67% for 2010–2011, 68% for 2011–2012, 70% for 2012–2013, 70% for 2013–2014, 72%for 2014–2015 and 73% for 2015–2016 respectively were companies with high integration. Similarly companies with score below grand average score for each year reflected low integration, companies with scores gradually improving over the time period considered for the analysis reflected progress companies in respect of integration and the remaining companies reflected moderate integration. From Table 18.5 we observe that 39 companies fell into the category of high integration, 21 companies fell into the category of moderate integration, 15 companies fell into the category of progressive integration and 27 companies fell into the category of low integration.

Thus from the above table we can summarize the extent of integration of our select data set in the following manner.

Table 18.4 Individual IRI for the companies for the time period 2010–2016

Company	2010–2011	2011–2012	2012–2013	2013–2014	2014–2015	2015–2016	Average	Integration level
Indian Oil Corporation Ltd	79	79	79	79	79	79	79	HI
Reliance Industries Ltd	86	79	79	79	86	100	85	HI
Tata Motors	79	79	79	79	79	79	79	HI
Bharat Petroleum Corporation Ltd	71	71	79	79	79	79	76	HI
Hindustan Petrolium Corporation	50	50	50	71	71	71	61	LI
Rajesh Exports Ltd	50	36	36	36	36	36	38	LI
Ongc Ltd	79	79	79	79	79	79	79	HI
Tata Steel Ltd	86	86	100	100	100	100	95	HI
Tcs Ltd	79	57	79	71	71	71	71	PI
Bharti Airtel Ltd	64	64	71	71	71	71	69	MI
Lansen And Toubro Ltd	86	86	86	71	71	71	79	MI
Hindalco Industries Ltd	71	79	79	79	79	79	77	HI
Coal India Ltd	79	79	86	86	86	86	83	HI
Ntpc Ltd	86	86	86	86	86	86	86	HI
Mahindra And Mahindra Ltd	79	79	79	79	79	79	79	HI
Vedanto Ltd	64	79	79	100	100	100	87	PI
Infosys Ltd	86	86	86	86	86	86	86	HI
Maruti Suzuki India Ltd	71	71	71	71	71	71	71	MI
Gail India Ltd	64	64	79	79	79	79	74	PI
Wipro Ltd	79	79	79	79	79	100	82	HI
Hdfc	64	64	64	50	57	57	60	LI
Adani Enterprises Ltd	64	71	71	71	100	79	76	PI
HclTechnolagies Ltd	71	71	71	71	71	79	73	PI
Jsw Steel Ltd	71	79	79	79	79	79	77	HI
Itc Ltd	50	57	57	64	64	64	60	LI

Mangalore Refinery And Petro-chemicals Ltd	79	79	79	79	79	79	79	HI
Steel Authority Of India Ltd	86	86	86	86	86	86	86	HI
Tata Power Co Ltd	79	79	79	79	79	79	79	HI
MothersonSumi Systems Ltd	79	79	79	79	79	86	80	HI
Grasium Industries Ltd	79	79	79	79	79	79	79	HI
Idea Cellular Ltd	50	50	64	64	64	64	60	LI
Redington India Ltd	43	43	43	43	79	79	55	PI
Hindustan Unilever Ltd	71	71	86	86	86	86	81	HI
Ruchi Soya Industries Ltd	79	79	79	93	86	86	83	HI
Hero Motocorp Ltd	71	71	71	71	71	71	71	MI
Sun Pharmaceutical Industries Ltd	43	43	57	57	71	79	58	PI
Bharat Heavy Electricals Ltd	93	93	93	93	93	93	93	HI
Power Finance Corporation Ltd	71	71	71	71	71	71	71	MI
PetronetLng Ltd	57	64	64	64	64	64	63	LI
Tech Mahindra Ltd	71	79	79	79	79	79	77	HI
Chennai Petroleum Corporation Ltd	79	79	79	79	79	79	79	HI
Ultra Tech Cement Ltd	71	71	71	71	79	79	74	HI
Adani Power Ltd	50	50	50	50	43	71	52	LI
Rural Electrification Corporation Ltd	79	79	79	71	71	71	75	MI
Aditya Birla Nuvo Ltd	71	79	79	79	79	86	79	HI
Bajaj Auto Ltd	43	43	43	43	43	43	43	LI
Jet Airways India Ltd	43	43	43	43	43	43	43	LI
Reliance Communication Ltd	79	79	79	79	79	79	79	HI
Power Grid Corporation Of India Ltd	71	71	71	71	71	71	71	MI

(continued)

Table 18.4 (continued)

Company	2010–2011	2011–2012	2012–2013	2013–2014	2014–2015	2015–2016	Average	Integration level
Ashok Leyland Ltd	57	57	50	50	64	64	57	LI
Tata Communication Ltd	71	71	71	64	64	71	69	MI
Reliance Infrastructure Ltd	57	57	64	64	64	64	62	LI
Mmtc Ltd	50	36	36	36	64	64	48	LI
Jindal Steel And Power Ltd	79	79	71	100	93	93	86	HI
Tata Chemicals Ltd	79	79	71	71	71	86	76	PI
Jai Prakash Associates Ltd	71	71	71	71	71	71	71	MI
Hindustan Zinc Ltd	71	71	86	86	86	86	81	HI
Asian Paints Ltd	64	64	64	57	64	64	63	LI
Dr Reddy Laboratories Ltd	71	71	79	86	86	86	80	HI
Eid Parry India Ltd	50	50	86	71	71	71	67	MI
Lupin Ltd	71	71	79	79	79	79	76	HI
Videocon Industries Ltd	71	71	71	71	71	71	71	MI
Amtek Auto Ltd	50	50	50	50	50	50	50	LI
Gitanjali Gems Ltd	50	36	50	57	64	64	54	LI
AurobindoPharma Ltd	71	71	79	79	79	79	76	HI
Ptc India Ltd	36	36	36	36	36	36	36	LI
Cipla Ltd	57	57	57	64	79	79	65	PI
Gmr Infrastructure Ltd	71	79	79	71	71	71	74	PI
Mrf Ltd	79	79	79	79	79	79	79	HI
Bajaj Finserv Ltd	50	50	50	50	50	43	49	LI
Upl Ltd	64	79	79	86	79	71	76	MI
Alok Industries Ltd	64	93	93	86	86	64	81	PI
Bharti Infertel Ltd	64	64	57	64	64	64	63	LI
Sundaram Clayton Ltd	64	64	64	64	64	64	64	LI

Eicher Motors Ltd	57	57	57	57	57	71	60	LI
Lic Housing Finance Ltd	64	64	71	64	64	79	68	MI
Cesc Ltd	50	50	50	50	50	50	50	LI
Torrent Power Ltd	71	71	71	71	43	43	62	MI
Acc Ltd	86	86	86	86	86	86	86	HI
Apollo Tyres Ltd	71	71	79	79	79	79	76	HI
Bhushan Steel Ltd	50	57	64	43	50	50	52	LI
Max Financial Services Ltd	86	86	86	86	86	86	86	HI
Coromondal International Ltd	71	64	57	71	79	86	71	PI
Tvs Motor Company Ltd	64	57	57	57	57	57	58	LI
Siemens Ltd	50	64	57	57	50	71	58	LI
Titan Company Ltd	43	50	64	71	71	64	61	MI
Godrej Industries Ltd	79	79	79	79	79	79	79	HI
Oil India Ltd	71	79	79	71	71	71	74	MI
Reliance Power Ltd	50	50	57	57	57	57	55	LI
Bosch India Ltd	79	79	79	79	79	79	79	HI
Suzlon Energy Ltd	86	64	71	79	79	79	76	PI
State Trading Corporation Of India	50	50	50	50	50	50	50	LI
United Spirits Ltd	43	43	57	57	86	79	61	PI
Chambal Fertilisers And Chemicals Ltd	71	71	71	71	71	71	71	MI
Cairn India Ltd	79	79	79	79	79	79	79	HI
Rain Industries Ltd	64	71	71	71	71	71	70	MI
Jsw Energy Ltd	57	86	86	71	79	86	77	PI
Shriram Transport Finance Company Ltd	50	57	79	57	71	71	64	MI
Nhpc Ltd	71	86	79	79	79	79	79	HI

(continued)

Table 18.4 (continued)

Company	2010–2011	2011–2012	2012–2013	2013–2014	2014–2015	2015–2016	Average	Integration level
Zuari Agro Chemicals Ltd	50	50	57	43	50	64	52	LI
Reliance Capital	57	57	57	71	71	71	64	MI
Cadila Health Care Ltd	57	79	64	64	64	64	65	MI
Total	67	68	70	70	72	73	70	

Source: Authors own computation

HI high integration, *MI* moderate integration, *PI* progressive integration, *LI* low integration

Table 18.5 Classification of companies by extent of integration for the time period 2010–2016

High integration	Moderate integration	Progressive integration	Low integration
Indian Oil Corporation Ltd	Bharti Airtel Ltd	TCS Ltd	Hindustan Petroleum Corporation
Reliance Industries Ltd	Larsen And Toubro Ltd	Vedanta Ltd	Rajesh Exports Ltd
Tata Motors	Maruti Suzuki India Ltd	Gail India Ltd	HDFC
Bharat Petroleum Corporation Ltd	Hero Motocorp Ltd	Adani Enterprises Ltd.	ITC Ltd
ONGC Ltd	Rural Electrification Corporation Ltd	HCL Technologies Ltd	PetronetIng Ltd
Tata Steel Ltd	Power Grid Corporation Of India Ltd	Redington India Ltd	Adani Power Ltd
Hindalco Industries Ltd	Tata Communication Ltd	Sun Pharmaceutical Industries Ltd	Bajaj Auto Ltd
Coal India Ltd	Jai Prakash Associates Ltd	Tata Chemicals Ltd	Jet Airways India Ltd
NTPC Ltd	Eid Parry India Ltd	Cipla Ltd	Ashok Leyland Ltd
Mahindra And Mahindra Ltd	Videocon Industries Ltd	GMR Infrastructure Ltd	Reliance Infrastructure Ltd
Infosys Ltd	UPL Ltd	Alok Industries Ltd	MMTC Ltd
Wipro Ltd	LIC Housing Finance Ltd	Coromondal International Ltd	Asian Paints Ltd
JSW Steel Ltd	Torrent Power Ltd	Suzlon Energy Ltd	Amtek Auto Ltd
Mangalore Refinery And Petrochemicals Ltd	Titan Company Ltd	United Spirits Ltd	Gitanjali Gems Ltd
Steel Authority Of India Ltd	Oil India Ltd	JSW Energy Ltd	PTC India Ltd
Tata Power Co Ltd	Chambal Fertilizers And Chemicals Ltd		Bajaj Finserv Ltd
MothersonSumi Systems Ltd	Rain Industries Ltd		Bharti Infertel Ltd
Grasim Industries Ltd	Shriram Transport Finance Company Ltd		Sundaram Clayton Ltd
Hindustan Unilever Ltd	Reliance Capital		Eicher Motors Ltd
Ruchi Soya Industries Ltd	Cadila Health Care Ltd		CESC Ltd
Bharat Heavy Electricals Ltd	Power Finance Corporation Ltd		Bhushan Steel Ltd
Tech Mahindra Ltd			TVS Motor Company Ltd
Chennai Petroleum Corporation Ltd			Siemens Ltd
Ultra Tech Cement Ltd			Reliance Power Ltd
Aditya Birla Nuvo Ltd			State Trading Corporation Of India

(continued)

Table 18.5 (continued)

High integration	Moderate integration	Progressive integration	Low integration
Reliance Communication Ltd			Zuari Agro Chemicals Ltd
Jindal Steel And Power Ltd			Idea Cellular Ltd
Hindustan Zinc Ltd			
Dr Reddy Laboratories Ltd			
Lupin Ltd			
Aurobindo Pharma Ltd			
MRF Ltd			
ACC Ltd			
Apollo Tyres Ltd			
Max Financial Services Ltd			
Godrej Industries Ltd			
Bosch India Ltd			
Cairn India Ltd			
NHPC Ltd			

Source: Authors own computation

18.6 Conclusion

Society demands holistic and integrated reporting of financial and non-financial information. It needs to know the value business is creating over time. As a result, progressive organizations should be moving forward to IR even prior to the establishment of any formal requirements. A recent research conducted by MIT Sloan Management Review, the Boston Consulting Group, and the United Nations Global Compact (UNGC) suggests that business sustainability is moving away from isolated and opportunistic efforts with a main focus on CSR and toward a more integrated, holistic, and strategic approach embracing all dimensions of sustainability performance and engaging diverse stakeholders (Kiron et al. 2015). In our study we observed that integrated reporting is slowly becoming a growing phenomenon amongst the top valued companies since presence of IR increased from 0% in 2010–2011 to 4% in 2015–2016. From our study we can also conclude on a positive note that Indian companies are nearly well equipped to adopt and implement "integrated reporting" as required by the IIRC framework. The study revealed an average IRI of 70%. 38.24% of the companies chosen for our analysis falls under "highly integrated companies". Responsiveness towards imbibing a new approach towards reporting involves inspired leadership and commitment from the top management. The leaders must necessarily institutionalize their own inspirational energy down the line in the organization to sustain this culture and this was observed in such companies. But at the same time 26.47% of the companies chosen for our analysis falls under "low integrated companies". Hence preparation of "Integrated Report" needs its promotion amongst such companies in the Indian context so that it encourages them to step forward and adopt this method of reporting. Besides from our study we do observe that the companies in the Indian context have failed to disclose their business models and strategy as required by IIRC which would affect any stakeholder from understanding how the organizations are creating value over time and how are they going to achieve it. The most important thing about integrated reporting today is that it is an emerging social movement. The meaning of integrated reporting will only be developed and its implementation be successful if this movement is an effective one. This will require energy, enthusiasm, trust, courage, persistence and collaboration amongst every person and organization who believes that integrated reporting can play an important role in ensuring that we have a sustainable society.

Appendix

There are 14 parameters required to be disclosed by IIRC:

- Financial capital
- Manufactured capital
- Intellectual capital
- Human capital
- Social and relationship capital
- Natural capital
- Organizational overview and external environment
- Governance
- Business model
- Risks
- Opportunities
- Strategy and resource allocation
- Performance
- Outlook

References

Abeysekera, I. (2013). A template for integrated reporting. *Journal of Intellectual Capital, 14*(2), 227–245.

ACCA. (2012). *Integrating material sustainability information into corporate reports should be a key and critical outcome of Rio+20*. Retrieved from http://www.Accaglobal.Com/en/discover/news/2012/05/sustainability-rio.html

Adams, C. A. (2004). The ethical, social and environmental reporting-performance portrayal gap. *Accounting, Auditing & Accountability Journal, 17*(5), 731–757.
Adams, C. A. (2015). The international integrated reporting council: A call to action. *Critical Perspectives on Accounting, 27*, 23–28.
Adams, S., & Simnett, R. (2011). Integrated reporting: An opportunity for Australia's not-for-profit sector. *Australian Accounting Review, 21*(3), 292–301.
Adams, C. A., Potter, B., Singh, P. J., & York, J. (2016). Exploring the implications of integrated reporting for social investment (disclosures). *The British Accounting Review, 48*(3), 283–296.
Athma, P., & Laxmi, N. R. (2013). Integrated reporting: The global scenario. *The IUP Journal of Accounting Research & Audit Practices, XII*(3), 7–18. Available at SSRN. Retrieved from https://ssrn.Com/abstract=2357840.
Barker, R., & Kasim, T. (2016). Integrated reporting: Precursor of a paradigm shift in corporate reporting? In *Integrated reporting* (pp. 81–108). Basingstoke: Palgrave Macmillan.
Brown, J., & Dillard, J. (2014). Integrated reporting: On the need for broadening out and opening up. *Accounting, Auditing and Accountability Journal, 27*(7), 1120–1156.
Busco, C., Frigo, M. L., Quattrone, P., & Riccaboni, A. (2013a). *Integrated reporting: Concepts and cases that redefine corporate accountability*. Cham: Springer.
Busco, C., Frigo, M. L., Quattrone, P., & Riccaboni, A. (2013b). Redefining corporate accountability through integrated reporting: What happens when values and value creation meet? *Strategic Finance, 95*(2), 33–42.
Carels, C., Maroun, W., & Padia, N. (2013). Integrated reporting in the South African mining sector. *Corporate ownership Control, 11*, 991–1005.
Cheng, M., Green, W., Conradie, P., Konishi, N., & Romi, A. (2014). The international integrated reporting framework: Key issues and future research opportunities. *Journal of International Financial Management and Accounting, 25*(1), 90–119.
Churet, C., Robeco, S. A. M., & Eccles, R. G. (2014). Integrated reporting, quality of management, and financial performance. *Journal of Applied Corporate Finance, 26*(1), 56–64.
Cleverly, P., Phillips, D., & Tilley, C. (2010). *Tomorrow's corporate reporting. The landscape of integrated reporting reflections and next steps* (pp. 197–202). Massachusetts: The President and Fellows of Harvard College Cambridge.
De Villiers, C., Rinaldi, L., & Unerman, J. (2014). Integrated reporting: Insights, gaps and an agenda for future research. *Accounting, Auditing & Accountability Journal, 27*(7), 1042–1067.
Dhaliwal, D., Li, O. Z., Tsang, A., & Yang, Y. G. (2011). Voluntary nonfinancial disclosure and the cost of equity capital: The initiation of corporate social responsibility reporting. *The Accounting Review, 86*(1), 59–100.
Dhaliwal, D., Radhakrishnan, S., Tsang, A., & Yang, Y. G. (2012). Nonfinancial disclosure and analyst forecast accuracy: International evidence on corporate social responsibility disclosure. *The Accounting Review, 87*(3), 723–759.
Dragu, I. M., & Tiron-Tudor, A. (2013a). GRI compliance and prerequisites of integrated reporting for Asian-Pacific companies. *Annales Universitatis Apulensis Series Oeconomica, 15*(2), 432–442.
Dragu, I. M., & Tiron-Tudor, A. (2013b). The integrated reporting initiative from an institutional perspective: Emergent factors. *Procedia-Social and Behavioral Sciences, 92*, 275–279.
Dumay, J., Bernardi, C., Guthrie, J., & Demartini, P. (2016, September). Integrated reporting: A structured literature review. *Accounting Forum, 40*(3), 166–185.
Eccles, R., & Armbrester, K. (2011). Integrated reporting in the cloud. *IESE Insight, 8*, 13–20.
Eccles, R. G., & Krzus, M. (2010). *One report: Integrated reporting for a sustainable strategy*. Hoboken, NJ: Wiley.
Eccles, R. G., & Serafeim, G. (2011). *Leading and lagging countries in contributing to a sustainable society*. Boston, MA: Harvard Business School, Working Knowledge.
Eurosif and ACCA. (2013). *What do investors expect from non-financial reporting?* Retrieved from www.eurosif.org and www.accaglobal.com

Flower, J. (2015). The international integrated reporting council: A story of failure. *Critical Perspectives on Accounting, 27*, 1–17.
Frías-Aceituno, J. V., Rodríguez-Ariza, L., & García-Sánchez, I. M. (2013a). Is integrated reporting determined by a country's legal system? An exploratory study. *Journal of Cleaner Production, 44*, 45–55.
Frías-Aceituno, J. V., Rodríguez-Ariza, L., & García-Sánchez, I. M. (2013b). The role of the board in the dissemination of integrated corporate social reporting. *Corporate Social Responsibility and Environmental Management, 20*(4), 219–233.
Frías-Aceituno, J. V., Rodríguez-Ariza, L., & García-Sánchez, I. M. (2014). Explanatory factors of integrated sustainability and financial reporting. *Business Strategy and the Environment, 23*, 56–72.
Garcia-Sánchez, I. M., Rodríguez-Ariza, L., & Frias-Aceituno, J. V. (2013). The cultural system and integrated reporting. *International Business Review, 22*(5), 828–838.
Ghosh, S. (2016). Reporting of CSR activities in India: Are we still at a Nascent stage even after the legal mandate? In M. Nayan & S. René (Eds.), *Corporate social responsibility in India: Cases and development after the legal mandate*. Cham: Springer.
Glaser, B. G., & Strauss, A. (1967). *The discovery of grounded theory: Strategies for qualitative research*. Chicago, IL: Aldine Publishing.
Gurvitsh, N., & Sidorova, I. (2012). Survey of sustainability reporting integrated into annual reports of Estonian companies for the years 2007-2010: Based on companies listed on Tallinn Stock Exchange as of october 2011. *Procedia Economics and Finance, 2*, 26–34.
Higgins, C., Stubbs, W., & Love, T. (2014). Walking the talk(s): Organizational narratives of integrated reporting. *Accounting, Auditing and Accountability Journal, 27*(7), 1090–1119.
Hindley, T., & Buys, P. W. (2012). Integrated reporting compliance with the global reporting initiative framework: An analysis of the South African mining industry. *International Business & Economics Research Journal, 11*(11), 1249–1260.
Ioana, D., & Tiron-Tudor, A. (2013). New corporate reporting trends. Analysis on the evolution of integrated reporting. *Annals of the University of Oradea, Economic Science Series, 22*(1), 1221–1228.
Ioannou, I., & Serafeim, G. (2012). What drives corporate social performance? The role of nation-level institutions. *Journal of International Business Studies, 43*(9), 834–864.
Ioannou, I., & Serafeim, G. (2015). The impact of corporate social responsibility on investment recommendations: Analysts' perceptions and shifting institutional logics. *Strategic Management Journal, 36*, 1053–1081.
IoDSA. (2009). *King Code of Corporate Governance for South Africa2009*. Johannesburg: Institute of Directors in Southern Africa.
IR. (2016). *Integrated reporting*. Accessed April 25, 2016, from http://integratedreporting.org/
IRCSA (Integrated Reporting Committee of South Africa). (2011). Framework for integrated reporting and the integrated report. Accessed May 18, 2017, from http://www.sustainabilitysa.org
Janeka, C., Riccerib, F., Sangiorgia, D., & Guthrie, J. (2016). Sustainability and integrated reporting: A case study of a large multinational organisation.
Jensen, J. C., & Berg, N. (2012). Determinants of traditional sustainability reporting versus integrated reporting: An institutionalist approach. *Business Strategy and the Environment, 21* (5), 299–316.
Kiran, V. U., & Goud, M. M. (2015). A study of integrated reporting in Indian banks. *The Management Accountant, 50*, 29–33.
Kiron, D., Kruschwitz, N., Haanaes, K., Reeves, M., FuiszKehrbach, S.-K., & Kell, G. (2015). Joining forces: Collaboration and leadership for sustainability. *MIT Sloan Management Review*, The Boston Consulting Group, and the United Nations Global Compact. Retrieved from http://marketing.mitsmr.Com/PDF/56380-MITSMR-BGC-UNGCSustainability2015.pdf?cid=1
Lee, K. W., & Yeo, G. H. H. (2016). The association between integrated reporting and firm valuation. *Review of Quantitative Finance and Accounting, 47*(4), 1221–1250.

Loska, T. (2011). *Integrated reporting. Towards a framework for a sustainable international corporate reporting*. Munich: Grin, Verlag GMBH.
Luke, T. W. (2013). Corporate social responsibility: An uneasy merger of sustainability and development. *Sustainable Development, 21*(2), 83–91.
Maubane, P., Prinsloo, A., & Van Rooyen, N. (2014). Sustainability reporting patterns of companies listed on the Johannesburg Securities Exchange. *Public Relation Review, 40*(2), 153–160.
Mitchell, M., Curtis, A., & Davidson, P. (2012). Can triple bottom line reporting become a cycle for "double loop" learning and radical change? *Accounting, Auditing & Accountability Journal, 25* (6), 1048–1068.
Owen, G. (2013). Integrated reporting: A review of developments and their implications for the accounting curriculum. *Accounting Education, 22*(4), 340–356.
Perego, P., Kennedy, S., & Whiteman, G. (2016). A lot of icing but little cake? Taking integrated reporting forward. *Journal of Cleaner Production, 136*, 53–64.
Raju, K. K. (2015). Communicating value to the stake holders: A case study of Tata Steel Company Ltd. *The Management Accountant, 50*, 34–35.
Siebecker, M. R. (2009). Trust & Transparency: Promoting efficient corporate disclosure through Fiduciary-based discourse. *Washington University Law Review, 87*(1), 115–174.
Sierra-García, L., Zorio-Grima, A., & García-Benau, M. A. (2015). Stakeholder engagement, corporate social responsibility and integrated reporting: An exploratory study. *Corporate Social Responsibility and Environmental Management, 22*(5), 286–304.
Solomon, J., & Maroun, W. (2012). *Integrated reporting: The influence of King III on social, ethical and environmental reporting*. London: The Association of Chartered Certified Accountants. Retrieved from http://www.accaglobal.com/content/dam/acca/global/PDF-technical/integratedreporting/tech-tp-iirsa.pdf.
Stubbs, W., & Higgins, C. (2014). Integrated reporting and internal mechanisms of change. *Accounting, Auditing and Accountability Journal, 27*(7), 1068–1089.
Thiagarajan, A., & Baul, U. (2014). Holistic intellectual capital conceptual offering for empirical research and business application. *International Journal of Management, 3*(1), 31–50.
Van Bommel, K. (2014). Towards a legitimate compromise? An exploration of integrated reporting in the Netherlands. *Accounting, Auditing & Accountability Journal, 27*(7), 1157–1189.
Van Zyl, A. S. (2013). Sustainability and integrated reporting in the South African corporate sector. *International Business and Economics Research Journal, 2*(8), 903–926.
Velte, P. (2014). Improving corporate governance quality through modern controlling: Integrated reporting in the German two tier system. *Business and Economics Journal, 5*(1), 2–5.
White, L. A. (2005). New wine, new bottles: The rise of non-financial reporting. *A Business Brief for Business for Social Responsibility*. Retrieved from https://www.bsr.org
Yu, K., Du, S., & Bhattacharya, C. B. (2014). Everybody's talking but is anybody listening? Stock market reactions to corporate social responsibility communications. Working paper.

Index

S. O. Idowu, M. Del Baldo (eds.), *Integrated Reporting*, CSR, Sustainability, Ethics & Governance, https://doi.org/10.1007/978-3-030-01719-4

D

E

T

U

MIX
Papier aus verantwortungsvollen Quellen
Paper from responsible sources
FSC® C105338

If you have any concerns about our products,
you can contact us on
ProductSafety@springernature.com

In case Publisher is established outside the EU,
the EU authorized representative is:
Springer Nature Customer Service Center GmbH
Europaplatz 3, 69115 Heidelberg, Germany

Printed by Libri Plureos GmbH
in Hamburg, Germany